Bryophytorum Bibliotheca 10

The Conducting Tissues of Bryophytes

by

CHARLES HÉBANT

with 348 figures on 80 plates

1977 · J. CRAMER
In der A.R. Gantner Verlag Kommanditgesellschaft
FL-9490 VADUZ

Author's address:

Université des Sciences et Techniques du Languedoc
Laboratoire de Paléobotanique
et Évolution des Végétaux
Place E. Bataillon
F - 34060 - Montpellier

Printed in Germany
by Strauss & Cramer GmbH, 6945 Hirschberg II
ISBN 3-7682-1110-X

à Sinske Hattori

en témoignage de reconnaissance
envers son apport irremplaçable
à la recherche bryologique
et à la coopération internationale

à mes Parents, à mes Amis

à la mémoire de Louis Grambast

Gosia Charles Hébant

CONTENTS

AIM OF THIS BOOK

A renewed interest for the conducting tissues of Bryophytes has emerged within recent years. Indeed, this is not surprising, since these plants provide interesting models of "primitive" conducting systems. Correlated structural and physiological investigations on bryophytes are thus likely to contribute, through comparison, to a better understanding of the physiology of conducting systems in land plants (= Archegoniatae *sensu lato*).

An increasing amount of palaeontological evidence tends to demonstrate the existence of striking similarities between certain early tracheophytes, especially the Rhyniophytina, and various contemporary bryophytes. A reasonable hypothesis, which would deserve further investigation, is that extant bryophytes represent remote descendants of those early Archegoniatae from which contemporary tracheophytes were also derived. Thus progress in knowledge of bryophyte anatomy and physiology may throw new light on the biology of early land plants, leading to a better comprehension of the first steps in the adaptation of higher plants to terrestrial life.

This review represents an attempt to summarize the main results of the research performed on the conducting tissues of bryophytes. A special effort has been made to go back to the original publications and to emphasize ancient and little known work. Indeed, early investigators in the field, especially during the second half of the nineteenth century, produced remarkable results, and much of contemporary knowledge originates from their work (more details on historical aspects will be found in Hébant & Schäfer 1974).

From this review, it will also become clear that research has been relatively well developed in certain areas, whereas others remain almost uninvestigated. Similarly, certain species (e.g. *Funaria hygrometrica* and *Polytrichum spp.* among mosses, *Conocephalum conicum* among liverworts) were very commonly selected as material for study and experiment, whereas many others, including representatives of several major groups (e.g. the leafy liverworts), have been very little investigated, or even not at all. Certain of those areas which according to the present reviewer would deserve further investigation have been outlined.

No details on general aspects of structure and life of bryophytes are given in this book since excellent reviews on the subject are available (Watson 1971; Doyle 1970).

REMERCIEMENTS

Cet ouvrage résume une expérience acquise au cours de 14 années de contacts avec l'histophysiologie des plantes terrestres primitives, ainsi qu'avec la bibliographie couvrant ce domaine. Je veux ici exprimer ma gratitude envers tous ceux qui, sous des formes diverses, m'ont apporté leur aide et leur concours pour la réalisation de ce travail.

Louis Grambast m'a accueilli dans son laboratoire dès sa fondation en 1961. Je lui reste très profondément reconnaissant des conditions de travail dont il a su m'entourer, en particulier lors des derniers stades de la préparation de ce livre. Sa sensibilité à certaines expériences humaines ne sera pas oubliée.

Monsieur le Professeur R. Buvat m'a avec générosité ouvert les portes de son Institut de Cytologie. C'est à lui que je dois d'avoir pu entreprendre une étude ultrastructurale approfondie du système conducteur des bryophytes; je le prie de croire en ma réelle gratitude.

Le Professeur W.R. Philipson et tout le personnel du Département de Botanique de l'Université Canterbury de Christchurch (Nouvelle-Zélande) m'ont réservé un accueil inoubliable. Le Dr. B. Fineran, en me laissant le libre accès à son laboratoire de microscopie électronique, a rendu très fructueux mon séjour en Nouvelle-Zélande. A tous les membres de ce Département si dynamique, j'adresse mes très sincères remerciements.

Une aide financière efficace m'a été apportée par divers organismes : C.N.R.S. (Equipe de Recherche Associée no. 114; Laboratoire Associé no. 179); Comité organisateur de l'Advanced Study Institute "Phloem Transport" (voyage au Canada; récolte de *Takakia* vivant); Ministère français des Affaires Etrangères (voyage en Nouvelle-Zélande); Ministère de la Coopération (séjour au Congo-Brazzaville); New Zealand University Grants Committee (post-doctoral research fellowship); Université des Sciences et Techniques du Languedoc; University of Canterbury (Nouvelle-Zélande).

L'aide matérielle de nombreux collègues et amis m'a été bien souvent précieuse, sinon vitale :

Antoine Billard (Marseille-Luminy), et plus tard Sue Bullock (Christchurch) ainsi que Sylvain Halpern (Paris-Sud) m'ont appris avec beaucoup de patience à utiliser seul un microscope électronique.

Françoise Layrisse a réalisé près de 8.000 lames d'histologie de bryophytes du monde entier; cette collection de référence s'est révélé d'une valeur inestimable pour la préparation de l'illustration de cet ouvrage.

Pierrette Lambert et Alexis Saury ont, avec leur compétence et leur gentillesse habituelles, très largement contribué aux travaux photographiques.

C'est au talent, au soin et au dévouement de Jill Courbet que sont dues les nombreuses reproductions de figures anciennes qui illustrent ce livre.

Les personnels de plusieurs bibliothèques françaises et étrangères, et tout particulièrement celles de l'Institut Botanique de Montpellier et de l'U.S.T.L., m'ont beaucoup facilité l'accès à des ouvrages parfois rares et de grande valeur; mention particulière doit être faite de l'aide inestimable apportée par Mlle A.-M. Rosier et ses collègues du service du prêt inter-bibliothèques (Montpellier-Sciences).

J'ai aussi bénéficié, comme tous mes collègues du Laboratoire de Paléobotanique et Evolution des Végétaux, du soutien discret mais si efficace de Nicole Grambast-Fessard, agrégée de l'Université.

Janice Britton (Montpellier) a relu avec une grande patience le manuscrit anglais de ce livre au fur et à mesure de son élaboration progressive. Plusieurs collègues ont bien voulu en revoir des chapitres, et me faire part de leurs critiques constructives : Jean Berthier (Clermont-Ferrand), Pierre Cruiziat (Versailles), Richard Johnson (Aberdeen), G. Miksche (Lund), John R.L. Walker (Chrischurch).

J'ai une grande dette de reconnaissance envers le Dr. E.V. Watson (Reading) qui a eu l'extrême bienveillance de relire entièrement le manuscrit du texte principal et d'en corriger le style. Sa connaissance exceptionnelle du monde des bryophytes m'ont rendu très précieuses ses suggestions et ses critiques.

Monsieur le Professeur J. Bons a très aimablement mis à notre disposition, pour la frappe définitive de cet ouvrage, la machine à composer de l'Institut montpelliérain de l'Ecole Pratique des Hautes Etudes.

De très nombreux autres collègues et amis m'ont souvent apporté leur aide au cours des années (ceux que j'oublie, dont je ne me rappellerai que trop tard, qu'ils ne croient pas être les moins importants) : Joëlle Barthonnet, M. R. Baudoin, Dr. Jean Berthier, Dr. H. Bischler, Prof. A.L.M. Bonnet, Prof. E.-J. Bonnot, M. Etienne Boula-Gamba, Dr. G. Brownlie, Robin Burr, Dr. B.G. Butterfield, M. Claude Chalies, Mrs Anna Cook, Dr. Robert Cook, Marie-Claude Espagnac, M. L. Granel de Solignac, Prof. Francis Hallé, Dr. Sinske Hattori, Dr. Rose Hébant-Mauri, M. John Holmes, Dr. Zennoske Iwatsuki, Ian A. Johnson, Dr. Richard P.C. Johnson, Prof. S. Jovet-Ast, Pierrette Lambert, Suzanne Lowry, Marie de Malet, Dr. Danièle Marty, Dr. Francis Marty, Dr. V. Mazeran, Dr. G. Miksche, Dr. Masami Mizutani, Dr. Francine Prévost, Dr. J.F. Péchère, Danièle Proske, Enid E. Roberts, Dr. G. Roizes, Dr. W.D. Ian Rolfe, Prof. J. Roux, M. Peter A. Schäfer, Prof. W.B. Schofield, Dr. G.A.M. Scott, Prof. Lalit M. Srivastava, Dr. Claude Suire, Prof. Louis Thaler, Prof. G. Valdeyron, Christiane Vanvert, Dr. J.R.L. Walker, Dr. J.S. Womersley, Prof. B.O. van Zanten.

Małgorzata Rucińska a assumé avec une rare compétence les frappes successives du manuscrit. Elle porte la responsabilité de la présentation définitive du texte de ce livre.

Que tous croient en ma profonde gratitude.

PART I

Structures Involved in Conduction

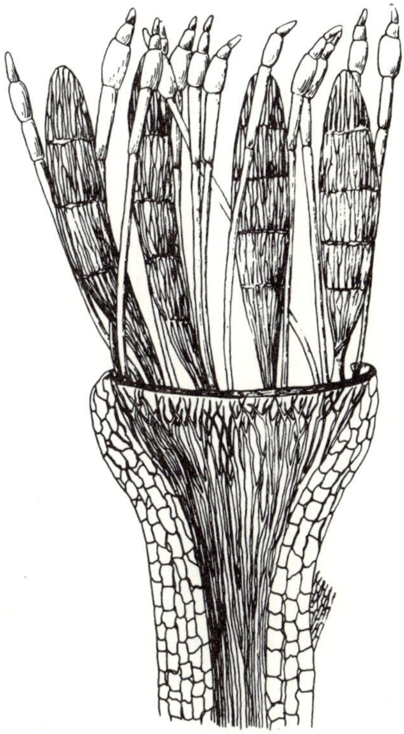

(HEDWIG 1784)

Chapter 1.

PRELIMINARY REMARKS

Only a few problems more or less specific to bryophyte histology are outlined here. A special chapter (p. 6) is devoted to structural variability, which is particularly important in bryophytes.

1.0.1. The concepts of "water-conducting" and "food-conducting cells" in Bryophytes.

Water can usually circulate in even the less specialized tissues, and indeed all living cells are capable of translocating. When we are dealing with such small-sized and relatively little differentiated plants as bryophytes, the question of what to call a "conducting cell" and what to call a "parenchyma cell" is not always clearly answered. By analogy with the tracheary elements of vascular plants, the expression "water-conducting cells" will be restricted in the present review to those elements in bryophytes which are dead and empty at maturity, as the result of a lysosomal breakdown of their protoplast occurring in the last stages of their development (Plate 1). No special terms are employed to designate such cells in hepatics, whereas they are quite numerous for mosses (cf. p. 29), the most commonly in use probably being "hydroids".

In contrast with the water-conducting cells which can easily be recognized on the basis of their lack of a protoplast at maturity, specialized food-conducting cells are much more difficult to identify in bryophytes on a histological basis only. In particular, the recognition of the so-called "leptoids" (=sieve element-like cells) is a critical problem, which is discussed in more detail on page 30. In short, if indisputable "leptoids *sensu stricto*", with typical sieve-element characteristics may be observed – essentially in the gametophytes of various highly developed Polytrichales –, a great many other cell types with structures intermediate between those of leptoids *s.st.* and of conducting parenchyma elements are also encountered.

1.0.2. The concept of "conducting strand" in bryophytes.

The expression "conducting strand", or similar expressions [1], have been variously employed to designate structures which sometimes have very little in common. Such different strands as the highly elaborated one found in the gametophytes of certain polytrichaceous mosses, or the rather poorly developed one encountered in the leafy stem of the Calobryales (Hepaticae), fall under this common denomination. Even the medullary cells, with characteristics of conducting parenchyma, which are found in the gametophytes of various mosses, Andreales, and both leafy and thalloid hepatics, have sometimes been termed "conducting strands". In the present review, this expression will be used only when *at least* a strand of clearly defined "water-conducting cells" – as defined above – is present [2]. Identification of "true" strands on sections of either living or fixed material is easy since their constituent elements are dead and empty when mature. As a consequence of the degeneration of their protoplasts, a slight protrusion of the living elements of the inner cortex within the central strand is commonly observed (e.g. in figs. 1 and 266). This of course is not always evident when herbarium material is used.

At this point, particular mention must also be made of the specialized elements devoid of protoplast at maturity, which are found in such bryophytes as *Sphagnum, Leucobryum,* etc. (cf. Chapter 4) : no confusion with a central strand is possible, since these cells are encountered only in the lamina of the leaves, or in the outer portions of the stems.

(1) Central strand, central cylinder, "faisceau conducteur", "Centralstrang" or "Zentralstrang", etc.

(2) It would seem preferable that the expression "conducting strand" be avoided (as is done in the present review) for the description of cells of living conducting parenchyma, whatever their level of specialization. If used at all, it should always be specified whether a "true" water-conducting strand, or a strand of conducting parenchyma is dealt with. Similarly, expressions like "feebly" or "ill-defined central strand" introduce confusion as they are too imprecise; if present, a "true" water-conducting strand, either "large" or "small" or "reduced", is always clearly defined.

1.0.3. Discrepancies in published works dealing with histology of bryophytes.

Most of the precursory investigations made on the conducting tissues of bryophytes were carried out by rather isolated workers. The latter, especially during the second half of the nineteenth century, developed parallel and distinct terminologies to describe the tissues they were observing. Consequently, a certain state of confusion resulted which has persisted until now. Other sources of complication lie in the following facts :

– **Structural variability** (see next chapter) : this variability is so important that mosses or liverworts which are known on other grounds to be closely related may present widely differing internal structures. Even worse, specimens of a single species growing under quite different environmental conditions may occasionally present such important variations as the presence or the absence of a central strand. Many contradictory reports in early or even recent literature originate from this.

– **Restricted distribution** of many interesting species. As a result, a number of anatomical studies have been made on herbarium material. However, precise identification of certain cell types demands detailed examination of carefully fixed material [1] , and some erroneous interpretations have emerged from the study of herbarium specimens.

– Certain authors have attributed a **physiological significance** to various tissues on the sole basis of histological observations (sometimes of herbarium material !), and in the absence of experimental controls. Some of these inferences proved subsequently to be correct, but some of them did not. For example, the water-conducting cells (=hydroids, or "begleiters", or "stenocystes" [2]) found in the leaf nerves and leaf traces of certain mosses have repeatedly been

(1) The same problem is encountered with the phloem of vascular plants; cf. Esau *et al.* 1953.

(2) For details on terminology, see p. 29.

interpreted as food-conducting cells, the neighboring "deuters" (or "guide cells"; =cells of conducting parenchyma) being correlatively interpreted as *the* water-conducting cells !

More details on these problems will be given as they arise.

Chapter 2

STRUCTURAL VARIABILITY

In bryophytes, structural variability occurs to a really unusual extent. This is particularly evident when conducting tissues are considered. This variability, which had struck many early investigators accounts for the apparent discrepancies which exist in some published papers. I will give a short analysis of the possible causes for these phenomena and will provide some examples to illustrate their extent.

The observed variations depend upon various *interacting factors*, especially : stages of development; environment; genotype. These will be taken into consideration successively. The facts presented here refer mostly to the gametophytic phase of mosses and liverworts. Apart from the differences related to taxonomic position, some of which are considerable, the developmental variability of the sporophyte in bryophytes has apparently been little investigated.

2.0.1. Stages of development.

As in many other land plants, *juvenile forms* in bryophytes show a very simplified organization. Indeed, in mosses and liverworts this goes to such an extent that young plantlets of species which normally have conducting tissues may, at the very beginning of their development, lack any internal differentiation; then, during subsequent stages of development, a more complex organization is progressively established.

Differences are seen in the internal structure of those axes which appear successively in the course of normal development. A long time ago, Lorentz (1867-8) noted that the central strand of hydroids could be absent from stems of smaller diameter of certain mosses which normally have one (Figs. 9-10). Disappearance of the conducting strand has also been observed in certain of the smaller axes in the Rhynie Chert (Fig. 347).

Unger (1861), Davy de Virville (1927-8) and Lorch (1931) similarly observed that the prostrate creeping stems of diverse mosses (e.g. *Mnium undulatum*) possessed smaller central strands than the erect green-leaved portions of the gametophytes (Figs. 6-7). In addition to this, it is not unusual to find species in which the fertile axes of the gametophytes, especially the female ones, are more robust, and therefore possess a better developed conducting strand than the vegetative ones. I have observed this *e.g.* in the liverwort *Haplomitrium gibbsiae* from New Zealand. An extreme example of sexual dimorphism is offered by various species of *Dicranum* (Musci) where male gametophytes are dwarf plants borne by female gametophores (cf. Fleischer 1920).

Structural variability related to developmental stages is particularly spectacular in polytrichaceous mosses, and deserves to be considered in some detail. *P. commune* has been widely studied in this context (cf. Wigglesworth, 1956) and thus may serve as an example (Plates 4 & 5). A very young plantlet, grown in sterile culture from the spores, does not show at first any internal differentiation (Fig. 13). This organization is strongly suggestive of an "advanced" *(neotenic?)* moss such as *Orthotrichum.* Soon, however, a conducting strand of hydroids (but without typical leptoids) differentiates in the young plant of *P. commune.* Its structure, as seen in transection, resembles that of the many mosses of the Arthrodontae which possess a central strand (Fig. 14). The first formed leaf traces are *"false"* leaf traces : they end blindly in the ground-tissue without joining the axial strand. Then, progressively, a more complex structure is established. These observations can be related to those on vascular plants (cf. for example Wardlaw, 1952 and 1968).

If an adult leafy stem of *P. commune,* the apex of which has been severed, is cultivated in a wet environment or under water, development of the previously inhibited lateral branches is observed (Lorch, 1909; Schulz, 1937) (Plate 5). At a short distance from its insertion, the lateral branch shows no internal differentiation (Fig. 20). But very soon, it shows the well-known rhizome-like organization described by various authors (Figs. 18 & 21). In the field, such an organization is self-perpetuating in the subterranean parts of the gametophytes : the branches produced from the basal region of the aerial leafy stems, or from the rhizomes themselves show this organization. The ultimate stages of differentiation reached by those axes which grow upright from the subterranean rhizomes and give rise to the aerial leafy stems, are well known today, notably through the works of Tansley & Chick (1901) and Wigglesworth (1956) on *Polytrichum*, of Goebel (1906) on *Dawsonia*, and of Schulz (1937) on *Dendroligotrichum* and other Polytrichales (see also Figs. 70 to 72 and 78 to 81). In particular these works have shown the important role played by the leaf traces in the progressive acropetal transformations of the anatomical and histological structures. Moreover, in her elegant study of *Polytrichum commune*, Wigglesworth (1956) was able to correlate the anatomical changes to modifications in the apical segmentation patterns (cf. Figs. 70 to 72).

In the "rhizomes" of the Polytrichales, the hydroids of the central strand are intermixed with parenchyma cells which generally have thickened walls (Figs. 85, 87 and 228 to 232). In most cases, these thick-walled elements disappear rapidly from the upright stems giving rise to the aerial leafy portions of the gametophytes, but there are exceptions : the case of those species of *Dawsonia* which belong to the section *Superba* is well known (see Plate 56 and Hébant 1976a). In *Atrichum*, the number of "stereids" of the central strand is considerably reduced at the level of the basal parts of the aerial stems, but later these elements can reappear in abundance, especially in fertile specimens (Figs. 24-25; see also Haberlandt, 1886).

The few examples of structural variations linked to stages of development which have been reviewed are known mostly from a descriptive viewpoint. It would be of great interest to undertake experimental studies designed to throw light on the functional significance of these variations.

2.0.2. Environment.

Bryophytes are very sensitive to their environment. Various field and experimental studies have shown what important structural modifications could be induced by – sometimes relatively minor – environmental changes. Precursory work has been ably reviewed by Douin and Davy de Virville (1924) for hepatics, and by Davy de Virville (1927-8) for mosses.

A few well documented cases of anatomical changes related to environmental factors are known at present :

– **Liverworts.** Disappearance of the conducting strands from the thalli under the influence of strong humidity or liquid culture is known for *Moerckia flotoviana* (cf. de Sloover 1959) and for *Haplomitrium hookeri* (Grubb 1970, p. 315).

– **Mosses.** Davy de Virville (1927-8) studied extensively the action of various factors on several species. He reported many examples of simplification of internal structure under the influence of high humidity, diminished light intensity, burial in soil, etc. (Figs. 6-8 and 11-12). However it must be noted that according to Leach (1930), part of Davy de Virville's results from cultivation in darkness or diminished light could in fact be humidity effects.

These facts concerning structural simplification under the influence of external factors have been overlooked by some recent workers who investigated the anatomy of certain polytrichaceous mosses. The specimens studied by Paolillo and Reighart (1967), and those studied by Eschrich and Steiner (1968a, b), were obviously underdeveloped – perhaps as a result of cultivation in too wet an environment : their descriptions do not match those of previous investigators working with light microscopy on specimens collected in the field (Firtsch 1883; Haberlandt 1886; Tansley and Chick 1901; etc.), and several misleading concepts, especially concerning the structure of the leaf nerves and leaf traces, and of the leptoids, emerged from their studies. But since their results appeared to confirm one another, although issued independently, they have been rather widely accepted ! In a reinvestigation of structural variability in *Polytrichum commune*, Hébant (1973c) found that extremely puzzling modifications could be observed in the structure of the gametophytes of this moss. The table below summarizes some of these variations (see also Figs. 15-16).

Origin of the material studied	Approximate number of hydroids within the central strand	Number of hydroids of each leaf trace joining the central strand	Level of differentiation of leptoids	References
pseudo-alpine grassland	900	7–8	cells with numerous sieve-element characteristics	Hébant (1967b, 1969b, 1970c)
peat-bog	400	3	idem	
specimens cultivated in an artificial peat-bog?	280 (*)	1	"parenchyma-like cells" (Eschrich 1970) with chloroplasts and normal-looking nucleus	Eschrich & Steiner (1967, 1968a–b), Eschrich (1970)
specimens cultivated under water	70	0	parenchymatous cells?	Hébant (1969b, fig. 12)

(*) Total derived from fig. 2 of Eschrich & Steiner (1968a).

Some aspects of structural variability in gametophytes of *Polytrichum commune* from various origins (Reproduced with permission from *J. Hattori Bot. Lab.* 37 [1973], p. 223).

2.0.3. Genotype.

Whatever the environment, a given species will only produce a limited range of types of organization depending upon its genetically controlled potentialities. Indeed, in the field most species usually exhibit a relatively uniform pattern. The variations observed from one taxon to another affect the presence or absence of conducting tissues in one or both generations, and also the relative importance of these tissues if they are indeed present. Thus precise knowledge of taxonomic status is important when anatomical structures are studied. Conversely, the structure of the conducting tissue-system may be used for taxonomic purposes, and a renewed interest for anatomical considerations is evident in many recent systematic revisions (see e.g. Kawai 1965, Schauer 1967, Saito 1975, etc.). At this point, it may be well to recall that the early investigations by Lorentz (1867 sqq.), Morin (1893) and others were already closely related to taxonomic considerations. It must also be noted that detailed analysis has shown that genecological differentiation occurs within Bryophytes (Longton 1974); therefore some of the structural variations observed in the conducting tissue-system of distinct populations of a single species could be at least in part genetically controlled.

Chapter 3.

STRUCTURES INVOLVED IN "EXTERNAL" CONDUCTION

Most bryophytes being of reduced size, processes of external conduction can be particularly efficient in them. An indirect proof of the presence of external films of moisture in these plants lies in the fact that fertilisation by motile sperms, requiring water for transit, occurs in these plants (Bowen 1933); in dioecious forms, this may even intervene between widely separated specimens.

Capillary channels are produced in gametophytes through various adaptive patterns (cf. Oltmanns 1884, Mägdefrau 1935-6, McConaha 1939 & 1941, Bowen 1933 & 1935, Clee 1937 & 1943, etc.) [(1)]:

- Structure and arrangement of leaves, which are often closely imbricated (Fig. 29). In a number of species, the leaf-bases are strongly decurrent, thus forming long and narrow channels. Leaf positions vary with the degree of moisture : dry leaves are often appressed to each other, whereas they expand when wet. Modifications in the structure and efficiency of the capillary channels may result from this (Mägdefrau 1935-6, Bayfield 1973). Various kinds of filaments or lamellae may also be borne by the leaves.
- Rhizoids are frequently present, especially in lower portions of stems (Fig. 29). Specialized filamentous or foliose expansions also exist throughout the stem in a number of species.
- Various branching patterns may facilitate conduction by capillarity. Growth in tufts, with the stems adjoined to various degrees, is also an important factor.
- A number of gametophytes remain prostrate on their substrate (either soil, or rocks, or the bark of trees). Water and the substances it contains can easily circulate by capillarity and is retained in the complex network of rhizoids, scales, soil particles, etc., on their ven-

(1) A number of "arrangements for retention of water", as described by earlier investigators and thoroughly reviewed in Goebel (1905), are interpreted with caution by recent authors (cf. e.g. Watson 1971).

tral side. Such structures facilitating capillarity are particularly efficient in various Marchantiales where "bundles" of rhizoids are seen (cf. e.g. McConaha 1941) (Figs. 26 to 28 and 159).

It is frequently admitted that in many forms, water and the substances dissolved in it can be absorbed at all levels of the gametophytes, either leafy or thalloid. However a cuticle has been shown to be present in a number of bryophytes (Strunk 1914). Recently Schönherr & Ziegler (1975) have demonstrated that hydrophobic cuticular ledges prevent water from entering the air pores of various Marchantiales. Thus preferential pathways and sites of absorption certainly exist, although they have been little investigated. In thalloid liverworts, ventral portions where rhizoids are inserted probably constitute such sites. In mosses, rhizoid-bearing lower portions of stems, as well as leaf-laminas are capable of absorbing water preferentially. In all forms, apices with their protective mucilage (secreted by specialized hairs, see Berthier *et al.* 1974) also appear to attract and retain water (Bowen 1933).

Sporophytes have been little investigated with reference to external conduction. In liverworts, water can circulate between calyptra [1] and the young sporogonium (Clee 1939 & 1943). When close to maturity, the diploid generation of mosses apparently is little adapted to conduction by capillarity, and seems to be rather impermeable. A large amount of the water needed by the capsule for its normal development must therefore come through the lower part of the sporogonium, or "foot", which is deeply inserted within gametophytic tissues. A contrast between the organization of the gametophytes and of the sporophytes in mosses is thus evident in this respect.

(1) *Calyptra* = protective tissue for the young sporophyte. It derives from the archegonium, and is therefore gametophytic in origin.

Chapter 4.

CELLS SPECIALIZED IN WATER STORAGE AND/OR CONDUCTION BUT WHICH CANNOT BE COMPARED IN A PRECISE MANNER TO XYLEM AND PHLOEM IN TRACHEOPHYTES.

Such cells are frequently encountered in bryophytes, where they show a remarkable diversity. They have originated independently within various unrelated groups, and the level of specialization they can show varies greatly from one group to another.

In most cases, these elements are more or less ordinary parenchyma cells, distinct only through the great development of the primary pit fields in their walls. Such cells are found in the gametophytes of various liverworts and Anthocerotes. They are also observed in both generations of various mosses. In addition to this, more highly differentiated cells are present in certain specialized groups, such as the well-known Sphagnales.

▪ **Liverworts**. Most contemporary liverworts lack a true water-conducting strand [(1)] in their gametophytes. None possesses such a strand in its sporophyte – which is always extremely specialized and short-lived.

However internal differentiation exists in hepatics (for reviews, see Herzog 1925 and Winkler 1969). The cells in the center of the stem in leafy forms, and of the nerve in thalloid ones, are usually elongated elements (Figs. 31, 33 and 158) with plasmodesmata in their walls, espacially the transverse or oblique end ones. Pit callose has been demonstrated in the latter by aniline blue fluorescence microscopy in various species (*Madotheca platyphylla, Conocephalum conicum,* etc.). Such cells could be interpreted as "conducting parenchyma" elements.

(1) For comments about the concept of "central strand" in Bryophytes, see p.3.

Various Marchantiales (*Conocephalum, Marchantia, Targionia,* etc.) provide sophisticated examples of internal differentiation reaching, through a remarkable convergent evolution, an organization comparable with that of leaves in phanerogams. Figs. 31 and 153 to 158 summarize some of the structural features of these hepatics. Many authors have reported the presence in these plants of "lignified" thickenings giving a reticulate appearance to certain cells of the ventral parenchyma. However, as early as 1895 Gjokic reported the absence of lignin from *Marchantia* and recent investigators (Kratzl and Eibl 1951, Siegel 1962, etc.) have corroborated this : the cells from the ventral regions of the thalli are not lignified. Light microscope observations both of sections and of macerated material from *Conocephalum conicum* and *Marchantia polymorpha* have shown that the parenchymatous cells of the wings possess large primary pit fields located preferentially on their lateral walls rather than on the upper and lower ones (Figs. 31 and 155 to 157). The "reticulate" appearance of the lateral walls is due to the fact that the regions between primary pit fields are thickened to various degrees. When the latter (the pit fields) are sufficiently numerous and close to each other, they give the impression of being separated by "bars" belonging to a reticular network of "secondary" wall material (Figs. 155-156). Anyway, this distribution of the primary pit fields is remarkable as it probably greatly facilitates transport of water and various substances in a direction parallel to the surface of the thallus rather than in a perpendicular one.

▪ **Anthocerotes.** Little work has been done on the histology of the gametophytes of hornworts, which apparently do not show a high level of internal differentiation. However, Proskauer (1960) was able to describe large primary pit fields, similar in appearance to those known from the thalli of certain Marchantiales, in the gametophytic parenchyma of *Dendroceros crispus* (Fig. 32). According to him, such tissue is of common occurrence in Anthocerotales. Conducting parenchyma also occurs in the sporophytes of Anthocerotes (the latter will be discussed in more detail in the following chapter, cf. p.19).

▪ **Mosses.** Conducting parenchyma is quite common in mosses where its occurrence has been repeatedly reported from stems, leaves (especially leaf nerves) and sporogonia of many species. Haberlandt (1886), among others, has stressed the important role played by the "Leitparenchym" in these plants. Early descriptions of pits and plasmodesmata in parenchyma cells of mosses (e.g. by Kienitz-Gerloff 1902 and Piskernik 1914) (Figs. 34-35) have been confirmed and substantiated by recent electron microscope observations (Bonnot 1967, Paolillo and Reighard 1967, etc.) (Fig. 161) whereas the presence of callose at their level has been demonstrated by fluorescence microscopy (Hébant 1968) (Figs. 163 to 165). From an extensive study of pitting of cells in the gametophores of several species Finocchio (1967, p. 19) concluded : "Mosses with and without central strands were found to have pitted cells,... The development of the end walls, the complementary pits and the articulation of the cells to form short vertical series indicate that these cells may be a type of conducting cell" (cf. also fig. 160) (1). Convergent interpretations were put foward by Noailles (1974) from her ultrastructural investigations of the parenchyma in the leafy stem of *Pleurozium schreberi*: "La vacuolisation très accentuée du parenchyme suggère qu'une partie importante des réserves d'eau se constitue préférentiellement dans ces cellules qui pourraient même avoir un rôle dans la conduction".

At this point it may be noted that existence of pit pairs between adjacent cells is readily identifiable on various figures from early investigators (e.g. Unger 1861, Plate III; "complementary pits" are seen on his drawing of macerated cells from *Hedwigia ciliata*). Necessity for conduction through parenchyma is also evident from description of mosses with "false" leaf traces or without any leaf traces at all (cf. chapter 6). Another vivid example of specialized conducting parenchyma is found in the enlarged apical region of the fertile gametophytes of certain forms where elements with large primary pit fields are encountered.

(1) There is apparently no reason to liken the cells of *conducting parenchyma* described by Finocchio (1963 and 1967) to tracheary elements, as has been done by some recent reviewers on the subject !

In addition to this, mention must be made of the various mosses which possess dead empty cells [1] that are thought to play a role in water storage or conduction. Pores are frequently present in these cells. Several examples of such mosses are given by Lorch (1894) and Goebel (1905) : *Sphagnum* (treated separately below), the Leucobryaceae (Fig. 30), certain Syrrhopodontales such as *Calymperes fasciculatum* or *Syrrhopodon revolutus* (Figs. 37 to 39), etc. With the sole exception of *Sphagnum*, all are known from early light microscope observations only, and electron microscope studies of fresh material are much needed.

■ **Sphagnales**. Early investigators, especially Schimper (1857), have established that the so-called hyaline cells of stems and leaves were dead at maturity, and that true perforations were present in their walls (Figs. 40 to 42 and 166). The latter are strenghthened by various types of "spiral fibres" or thickenings. Recent scanning electron microscope observations of leaves (Mozingo *et al.* 1969, Fabre 1973 and 1974a) and stems (Fabre 1974a) of several species have provided vivid illustrations of the structure of the pores (Figs. 167 to 169). Although these studies dealt mostly with mature elements, younger hyaline cells could also be seen, revealing the presence of a thin membrane over the future pore sites. The factors responsible for the rupture and final disappearance of these membranes were not definitely ascertained by these authors. A brief account of the differentiation of hyaline cells in *Sphagnum palustre* as seen under the transmission electron microscope was also presented by Schnepf (1973) as part of a broader study of leaf development in this plant. According to him, microtubules play a fundamental role in the formation of the "spiral thickenings" by lifting off the plasmalemma from the wall against the turgor pressure to form an extraplasmatic space in which wall material (initially derived from Golgi apparatus) is accumulated. At the emplacement of a future pore, the wall of the

(1) Only elements close to parenchyma cells by their shape and location (in leaves; in outer cortex of stems) are considered here. The tracheid-like hydroids are treated in chapter 6.

young hyalocyte becomes progressively thinner, – although no specialization of the protoplasts is evident at this level. Towards the end of the development, only the cuticle is left, which eventually ruptures. In addition, Schnepf (1973) as well as Fabre (1974b) noted the final disappearance of the protoplast of the hyalocytes.

Cells with large primary pit fields and similar in appearance to those found in the thalli of various Marchantiales exist in the sporogonium of *Sphagnum palustre*, in the region located between the foot and the columella (Figs. 170 and 171).

▪ **Andreaeales**. The occurrence of many pit pairs in the parenchyma cells of the median regions of the gametophytic stems has been depicted by Lorch (1931; cf. Fig. 36). These elements must act as conducting parenchyma cells. Trotter (1948) interpreted them in *Andreaea hartmanii* as constituting a "very ill defined" central strand.

Chapter 5.

ELEMENTS OF DUBIOUS EVOLUTIONARY SIGNIFICANCE, BUT WHICH HAVE BEEN RELATED TO THE CONDUCTING TISSUES OF VASCULAR PLANTS BY CERTAIN AUTHORS

Several authors have compared the *columella* [(1)] in the sporophyte of Anthocerotes to a conducting strand. This columella, as the other aerial green regions of the anthocerotalean sporophyte, originates from a basal intercalary meristem (Fig. 172). It is apparently made of elongated parenchyma cells (Figs. 174-175) which may eventually degenerate (Campbell 1924); – according to Proskauer (1960), in *Dendroceros crispus* this degeneration is a consequence of the natural dehiscence of the sporangium. Goebel (1905, p. 95) wrote : "the columella has ... a nutritive function. It passes below into the basal portion of the sporogonium which is concerned with the taking up of water from the mother-plant. Transpiration from the surface of the sporogonium is considerable, especially as the sporogonium continues to elongate for a long time by intercalary growth; it opens indeed at the apex before the spores are formed in the lower part. The columella is then comparable with the 'central strand' of many of the Musci in serving as a channel for water". Similar views were held by Campbell (1924, 1925), who interpreted as additional evidence for this the remarkable development exhibited by the columella in some exceptionally large (up to 16 cm !) specimens of *Anthoceros fusiformis*. In the latter, he said (1924, p. 477), "The columella is nearly twice the diameter of that of the normal sporophyte, and in longitudinal section it appears as a stout strand of elongated cells, highly suggestive of a simple vascular bundle". According to Campbell, this was particularly evident in the basal part of these giant sporophytes, which was "practically composed of a central stele, enclosed by a massive layer of green tissue, which may perhaps

(1) *Columella* : column of sterile tissue in the central region of the sporangium.

be called a cortex" (loc. cit., p. 478) (cf. Figs. 44-45). Goebel and Campbell assigned without demonstration a truly conducting role to the columella. However, from his experimental investigations in *Phaeoceros (Anthoceros) laevis*, Isaac (1941) was able to establish that the columella is indeed not very efficient as a water-conducting strand.

More recently, Proskauer (1960) expressed reserve as to the comparison of the anthocerote columella with a psilophyte stele. On the other hand, he attributed a conductive role to the elaters, going as far as calling them "tracheidal elaters" (cf. Crow 1946). These elaters, which are also commonly present in liverwort sporophytes [(1)], originate from the same tissue as the spore mother cells and are ultimately found intermixed with the spores (Fig. 176). When mature, these cells usually appear as elongated elements with tapering ends. Various types of thickenings exist in their walls (cf. e.g. Siessegger 1966), which can indeed give them a "tracheid-like" appearance (Fig. 48); however, these thickenings do not seem to be lignified (Proskauer 1960). Recent electron microscope studies on elaters (Lambert 1968, Heckman 1973) have also confirmed that in *Diplophyllum albicans, Lophocolea bidentata* and other species, these elements were truly devoid of protoplasm at maturity [(2)] (Fig. 177). According to J. Denizot (pers. comm.) developing elaters present certain cytological patterns reminiscent of those found in developing tracheids. Proskauer (1960, p. 15) wrote : "I think that in *Dendroceros* the tracheidal elaters connecting horizontally and vertically, in fact constitute the main water conducting tissue, before they ultimately play an active part in dehiscence and spore dispersal [(3)]". Although Proskauer apparently denied any precise significance to this fact, he also

(1) Occurrence of poorly developed elaters, or lack of elaters, in the sporangia of certain hornworts or liverworts, is usually interpreted as a secondary condition. *Anthoceros* itself has only "pseudo-elaters".

(2) Heckman (1973) however noted the persistence of protoplasmic remnants in apparently mature elaters of *Pallavicinia lyellii.*

(3) Earlier investigators had also attributed a "nutritive" role to the elaters (Leclerc du Sablon 1885, Goebel 1895).

noted the occurrence in some *Megaceros* (Anthocerotales) from Hawaii, of spirally thickened cells in the columella (which may even break up and lose its identity as a columella) (Fig. 46). The thickenings here are like those of the elaters at the corresponding level. At this point, it may be recalled that the suggestion has been made (cf. e.g. Chadefaud 1960, p. 957) that distribution patterns of elaters within sporangia of certain liverworts (e.g. *Pellia*) are reminiscent of hypothetical columella which has disappeared. In this respect, mention should also be made of the observation by Cribbs (1918) of a "columella" largely made up of elaters in a teratological sporogonium of *Marchantia polymorpha* (this species normally lacks such a structure) (Fig. 47).

These diverse observations, which link stele, columella, elaters and tracheids, are provocative and lead to speculation. For instance, one could be tempted to interpret the elaters as an "atypical" resurgence of the morphogenetic potentiality to develop tracheids, – potentiality which normally is not expressed in bryophytes –.

In fact, no really firm grounds have existed up to now on which to base a realistic comparison between the structures reviewed above and the conducting tissues of the tracheophytes.

Chapter 6.

CONDUCTING TISSUES OF BRYOPHYTES WHICH ARE MORE OR LESS COMPARABLE WITH XYLEM AND PHLOEM OF TRUE VASCULAR PLANTS

6.1. LIVERWORTS (*HEPATICOPSIDA* Roth.).

Studies on conducting tissues of liverworts are relatively scarce, this probably being related to the restricted distribution of those species which present them. However, representatives of all major groups have been investigated – or reinvestigated – within recent years (cf. Smith 1966, Grubb 1970, Hébant 1972a and 1973d, Burr *et al.* 1974); in particular, a comprehensive study of *Pallavicinia* and *Symphyogyna* (Metzgeriales), including a thorough review of early literature, has been published by J.L. Smith (1966).

6.1.1. Occurrence of conducting strands in hepatics.

No liverwort sporophyte is known to possess a central strand. Indeed this is not surprising, since in most species the "seta" is a rather unusual and specialized organ, which elongates rapidly at the end of sporophyte maturation causing the capsule to break through the calyptra just before dehiscence. "True" water-conducting strands [(1)] do not commonly occur in liverwort gametophytes either. Only a few leafy forms belonging to the genera *Haplomitrium* (Calobryales) and *Takakia* [(2)], and a few thalloid ones – all of them belonging to the order Metzgeriales – are presently known to possess such a strand (Plates 12 to 14, 44, and 49 to 54). However, investigations on hepaticae are so scarce that new findings could be made.

(1) For the concept of water-conducting strand, see p.3.

(2) A controversy exists as to the systematic position of this plant (see Schuster 1966; Hattori *et al.* 1974); as far as the structure of the conducting strand is concerned, it appears akin to the Calobryales.

6.1.2. Development and structure at maturity.

■ **Origin of strands within apical derivatives.**

In most hepatics, growth of the gametophytes is dependent on a single apical cell with two or three lateral cutting faces (1). The central strand, when present, originates from the innermost portion of the segments which are cut off from this apical cell. This has been clearly established for several Metzgeriales such as *Symphyogyna aspera* (McCormick 1914), *Pallavicinia radiculosa* (Campbell and Williams 1914) or *P. lyellii* (Smith 1966) (see also Figs. 53 and 190). According to the descriptions and figures produced by various authors, when first formed the young conducting cells in these plants are rather narrow and elongated elements with a densely staining protoplasm. Their general appearance under the light microscope is reminiscent of that of a strand of procambium in vascular plants (Figs. 53, 54, 190 and 191). They are thus readily distinguishable from the neighbouring parenchyma cells. In *Haplomitrium* and *Takakia*, the central strand similarly originates from the innermost portion of the segments cut off from the apical cell (Figs. 61, 62 and 183). However, the young conducting cells in these leafy hepatics appear less specialized than in *Pallavicinia* and allied forms; under the light microscope they are distinguished from the surrounding ground parenchyma cells mostly because of the precocious degeneration of their protoplasts (Figs. 63 and 185).

Multistrand forms. Some species of *Symphyogyna* from the tropical and southern portions of the New World are known to possess several strands in their gametophytes (Evans 1925, Hässel de Menendez 1961, Smith 1966) (Figs. 52, 56 and 57). In the course of his study of the thalli of *Symphyogyna sp.* from Temuco and Rio Negro, Smith (1966) observed how initiation of the multistrand condition occurs : "The apex of a single-strand thallus may begin to produce

(1) The occurrence of several apical cells has been recorded for some species (Cf. G.M. Smith 1955; – this would perhaps need reinvestigation); none of these hepatics possesses a central strand.

a second strand... in two ways : the original strand may dichotomize..., or a strand may arise *de novo*. In either case both strands are ultimately derived from the single original apical cell (as opposed to the origin from two apical cells in an early stage of bifurcation of the thallus). As the apex changes from the one to the two strand state, a shift in the position of the origin of the original strand occurs. Prior to its forking, it arose within the innermost portion of the derivative segments from both sides of the apical cell. After the dichotomy of the strand, this part of the derivative segments forms the parenchyma cells between the strands, the strands themselves being formed from a slightly more lateral portion of the derivative segments". During the course of development new strands may appear or disappear, resulting in a variable arrangement of the strands within the thalli (Fig. 56).

In *Moerckia flotoviana* two strands are usually present (Fig. 208). Apparently no detailed study of their development within apical derivatives has been made.

- **Branching of gametophytes and its anatomical consequences.**

Metzgeriales. In his study of *Pallavicinia* and *Symphyogyna*, Smith (1966, p. 15) stresses the clear distinction which exists between what he calls terminal and lateral branching : "by terminal branching I denote the bifurcation of the main axis, as opposed to lateral branching, which results in formation of pinnate laterals from primordia initiated at the apex".

When ***terminal branching*** is present in *Pallavicinia* and allied forms, division of the main strand into secondary strands usually occurs well below the level where the two new branches separate. The "dichotomous" (1) branching of the strand thus observed is somewhat reminiscent of that known in the archaic (?) pteridophyte *Rhynia* from the old red sandstones of Scotland (Fig. 342). This pattern is particularly evident in the thalli of certain "dendroid" Metzgeriales, such

(1) "Dichotomous" is used here in a purely descriptive sense. Apical patterns are still incompletely understood (cf. Smith 1966).

as *Hymenophyton flabellatum* (Figs. 199 to 201) or *Symphyogyna hymenophyllum* (Figs. 202-203). When terminal branching occurs in multistrand forms of *Symphyogyna*, an equivalent number of strands normally enters each branch (Fig. 56).

In *Moerckia flotoviana*, according to Cavers (1903, p. 451) : "where forking of the thallus takes place, each strand becomes broader and then divides into two, which pass into the branch on the corresponding side; at this point a cross-section shows four strands lying side by side".

Lateral branching has a superficial origin. Numerous authors (e.g. Farmer 1894, Finlayson 1950, Smith 1966) have noted that in *Pallavicinia* and allied forms, the strands of the lateral branches never unite with the strand of the primary shoot (Figs. 58 and 196). This is not surprising since these branches originate from previously inhibited secondary apical cells which normally initiate their development only when cortical tissues of the bearing shoots are fully mature and apparently incapable of dedifferentiating to further differentiate into conducting cells. Lack of a direct connection between the strands implies the existence of transport phenomena within cortical cells, which therefore must act as conducting parenchyma.

Haplomitrium and Takakia. Grubb (1970) depicts a longitudinal section through an aerial stem and a daughter rhizome of *Haplomitrium mnioides*, where continuity of the central strands is clearly seen. However, examination of numerous specimens of *H. gibbsiae* and *Takakia lepidozioides* has shown that in these two species, a connection of the central strand of the lateral branches with that of the bearing shoot is established in certain cases only (Figs. 179, 186 and 187). This probably is related to developmental patterns (the early development of a branch presumably resulting in a junction of the strands) and would deserve further study.

- **Differentiation and structure at maturity of conducting cells.**

While still young, the future conducting elements are living cells with protoplasm and a prominent nucleus. Those which have been studied under the electron microscope (in *Symphyogyna aspera*, Smith 1966; *Hymenophyton flabellatum* and *Haplomitrium gibbsiae*, unpublished observations) have revealed many typical organelles, such as mitochondria, small plastids, etc., in their cytoplasm. During development, these cells elongate to various extents.

In *Haplomitrium, Takakia* and *Moerckia*, the cells of the central strand remain relatively little specialized, and may even recall by their general appearance, the parenchyma cells of the cortex. Although they frequently become more elongated [(1)] than the neighbouring cortical cells, they usually retain transverse or only slightly oblique end walls [(2)], and they never become "tracheid-like" in appearance (Figs. 60, 63 and 185). The situation is different in *Pallavicinia, Symphyogyna* and *Hymenophyton*, where the conducting cells may become extremely elongated with tapering ends, and thus may superficially resemble the tracheids of vascular plants (Figs. 55, 198 and 204).

Whereas the walls, both lateral and terminal of the water-conducting cells of *Takakia* and *Haplomitrium* remain rather thin (Figs. 181 and 185), those in *Pallavicinia* and allied forms develop important thickenings during the final stages of their differentiation (cf. fig. 205). The presence of "pits" in the latter has long been noted (cf. e.g. Tansley & Chick 1901). Discrete "pits" have also been identified by light microscopy in the conducting cells of *Haplomitrium* (Hébant 1964) and *Takakia* (Winkler 1969). Recent electron microscope studies have established the existence of true perforations in the walls of all these cells. This was first demonstrated in *Symphyogyna circinata* by Smith (1964, 1966) who showed that pores originate by digestion of wall material from around the plasmodesmata. Since then, similar pores have been observed in *Pallavicinia* and *Hymenophyton* (Burr *et al.* 1974, Campbell *et al.* 1975) (Figs. 206 and 207). It was also found that in *Haplomitrium* and *Takakia* (Hébant 1972a,

(1) According to Grubb (1970), the ratio length-breadth of these cells can be 5-10 : 1 in *Haplomitrium gibbsiae*. I observed somewhat shorter elements in the specimens of *Haplomitrium* and *Takakia* that I investigated.

(2) Steeply inclined end walls also occur, but this does not seem to be the rule.

1973d; Burr *et al.* 1974), the small "pits" identified by light microscopy similarly correspond to true perforations in the walls, each of them probably originating from an enlarged plasmodesma (Figs. 182, 188 and 189). However, no such perforations were found in the conducting cells of *Moerckia*; in the latter, the walls present an unusual "swollen" appearance, this probably being a consequence of partial hydrolysis resulting in the removal of interfibrillar substances (Figs. 210 and 211). Nevertheless, traces of plasmodesmata have occasionally been encountered in these walls (unpublished observations).

Cell wall chemistry. According to Smith (1966) who applied specific stain and differential extraction methods, the walls of mature conducting cells of *Symphyogyna circinata* are composed of hemicelluloses, noncellulosic polysaccharides and cellulose. These walls are clearly birefringent. In spite of some early reports of the presence of lignin in the conducting cells of certain hepatics, it is now generally admitted that these elements are not lignified. Finlayson (1950) on *Symphyogyna hymenophyllum*, and Smith (1966) on *Moerckia flotoviana* could not obtain convincing positive phloroglucinol-HCl tests for lignin. I similarly obtained negative results when I tested *Takakia* and *Haplomitrium* with the Wiesner reagent (see also p.64).

Degeneration of protoplasts. During the last stages of their development, the protoplast of the water-conducting cells degenerates; mature elements are thus dead and empty (cf. e.g. figs. 181 and 185). This fact has been repeatedly reported for a number of species [(1)] (cf. e.g. Tansley & Chick 1901, Campbell & Williams 1914, Finlayson 1950, Grubb 1970). It has been confirmed by recent electron microscope studies (Smith 1964, 1966; Hébant 1972a, 1973d; Burr *et al.* 1974) (cf. e.g. fig. 181). A peak of strong acid phosphatase activity is revealed in the central strand at the level where the protoplasts of the conducting cells break down (Hébant 1973b and 1975b, in *Pallavicinia* and *Takakia* respectively; Burr *et al.* 1974 in *Haplo-*

(1) Persistence of a protoplast in the cells of the median region of the stems, that is to say disappearance of a "true" central strand, has been reported by Grubb for certain axes of *Haplomitrium* and *Takakia.*

mitrium and *Hymenophyton* (Figs. 180, 184, 192 and 193). This phosphatase activity probably has a lysosomal significance.

In such liverworts as *Haplomitrium* or *Takakia*, differences in turgor pressure between the dead conducting cells and the living adjacent parenchyma cells lead to the slight protrusion of the latter into the central strand (cf. fig. 1).

6.1.3. Concluding remarks.

Only specialized water-conducting elements have been described herein. Cells of conducting parenchyma are also of common occurrence in liverworts [1] (see chapter 4). Ultrastructural studies of such cells will probably provide new insights into transport phenomena in hepatics.

No liverwort is known with certainty to possess leptoids (=sieve element-like cells). The recent report of the occurrence of leptoids in *Pallavicinia wallisii* by Winkler (1969) needs substantiation. The cells he described as such, from the study of herbarium specimens, are probably ordinary elongated parenchyma cells.

6.2. MOSSES (*BRYOPSIDA* Roth.)

Conducting strands are of common occurrence in mosses. In contrast with liverworts, these plants may possess conducting tissues in either or both generations, haploid and diploid. In addition, not only water-conducting but also specialized food-conducting cells may be present. As in other groups of bryophytes, but perhaps to an even greater extent, structural variability is observed in anatomical patterns of mosses. Before a description of conducting tissues and their vari-

(1) Certain of these groups of cells have even been termed "conducting strands" by some authors (e.g. Herzog 1925). About the concept of central strand in bryophytes, see p.3.

ations is made, some precise definitions of the terms adopted here will be given (see also preliminary remarks, p. 2).

6.2.1. Definitions.

It seems reasonable that among the somewhat complex terminologies elaborated during the second half of the nineteenth century, only those terms which are both really useful and commonly accepted to-day be preferentially retained. Some specialized terminologies, used for instance in the description of the leaf nerves, can be of great interest to taxonomists but are less important to the histophysiologist.

□ **Hydroids :**

Water-conducting cells. They constitute an axial strand in the leafy stem and/or seta of many mosses (Schimper 1848, Unger 1861, Lorentz 1867-8, etc.). Though they may often be observed in the midrib of the leaf, these hydroids do not usually come into contact with the central strand. In various forms which have well developed leaf traces, the hydroids end blindly in the cortex. These are "false" leaf traces (Lorentz, 1867).

The term "hydroid" was created by Potonié (1883, p. 243). Although originally designed to describe the tracheary elements of vascular plants, it has been widely adopted in modern usage to designate the water-conducting cells of mosses (*not* of liverworts; the latter having water-conducting elements with original ultrastructural features, the retention of such a distinction seems reasonable). The generalized adoption of this term probably originates from the fact that Tansley & Chick (1901) used it in a now classical paper on the conducting tissue-system of bryophytes.

The term *hydrom* (Potonié 1883, p. 243) is commonly used to describe the hydroids of the central strand collectively. The hydroids of the leaf traces, when they exist, could also be considered as part of the hydrom of the leafy stem.

Other terms have been proposed to designate the hydroids, some of which have occasionally reappeared in recent publications :

- *"hadrom"* cells (cf. Haberlandt 1879, p. 5-6, and 1883),
- cells of the *"parenchyme médullaire"* (Bastit 1891, p. 266-267),
- *"leptoxylem"* cells (Vaisey, 1888, p. 275-276),
- *"sténocystes"* (Morin, 1893, p. 22),
- *"Zentralzellen"* or *"cellulae centrales"* (Lorentz 1864 sqq.),
- *"Begleiter"* or *"comites"* (Lorentz, 1867-8, p. 378),
- *"tracheids"* (Haberlandt 1886, p. 485, and 1914, p. 311; Strasburger 1891, p. 465).

"Stenocystes" is in common usage among French authors, whereas terms like "Zentralzellen" or "Begleiter" are employed mostly by systematists of various countries for the description of the principal elements of leaf nerves.

□ **Leptoids** :

Food-conducting cells. The *"leptom"* (Haberlandt 1879, p. 5-6) is the ensemble of the leptoids and associated parenchyma cells. The term "leptoid" is derived from "leptom" and apparently appears for the first time in Tansley & Chick (1901, p. 18) : "this little group of sieve-tube-like cells (*leptoids* as we may call them)...".

In contrast with the hydroids, which are always clearly identifiable through many developmental and structural characteristics (cf. p.53) the identification of the leptoids raises so many questions that a real semantic problem exists here. Difficulties in particular arise from the fact that there is no clear-cut border line between the leptoids themselves and the various cells of conducting parenchyma which coexist with them in the leptom *sensu lato*. Indeed, if true sieve element-like leptoids exist, various "intermediate" cell types in which leptoid characteristics are more or less well defined are also – and more commonly – encountered. In a first attempt towards clarification of these problems, I suggest that we call "leptoids *sensu stricto*" (abbreviated form : leptoids *s.str.*) or "sieve element-like leptoids" only those cells which are readily comparable with sieve elements of vascular plants through a whole ensemble of characteristics. I suggest retention of the term "leptoids" for those cells which apparently deserve to be distinguished from cells of conducting parenchyma, and which by their position and general appearance seem akin to leptoids *s.str.* – although they do not really show typical sieve element ultrastructures. The other cell types should be classified as "conducting parenchyma" elements. In any case, the existence of cells with intermediate structure frequently renders difficult a clear identification of the leptoids. It must also be pointed out that some recent investigators have described as "leptoids" cells that the present reviewer would not always interpret as such [(1)].

The term *"leptophloem"* of Vaisey (1888, p. 275-6) designates the leptom (without parenchyma cells) of the seta. The so-called *"zone péricyclique"* of Bastit (1891) seems to correspond, at least in part, to the leptom at the periphery of the central strand in the gametophyte of *Polytrichum*. The terms *"Deuter"* or *"duces"* (Lorentz 1867-8, p. 374), *"socii"* (Lorentz 1867-8) and *"eurycystes"* (Morin 1893, p. 21) are employed to designate diverse elements of conducting parenchyma, specialized to various degrees, which exist in the leaf nerves or leaf traces of many mosses [(2)].

□ **Stereids :**

Supporting cells. These are elongated elements with thickened walls. They may, or may not, retain a living protoplast at maturity. The *"stereome"* is the collective term for stereids.

□ **Notes :**

– Terms like "hydrom", "hadrom" or "leptom" had, in the nineteenth century, a wide acceptance and were at first employed to designate the conducting tissues (= conducting elements + associated cells) of the tracheophytes.

– In mosses, the expression "central strand" usually refers to water-conducting strands of hydroids only. However, in those forms which possess a leptom, this expression may alternatively be used to designate the ensemble hydrom + leptom, *or* the hydroids alone. In the latter case, and to avoid confusion, it is advisable to specify "central strand of hydroids".

---

(1) Cf. e.g. fig. 7 in Stevenson (1974), which shows what I would interpret as a parenchyma cell.

(2) The German word "Deuter", "guide cell(s)", will be anglicized and used throughout the text as *deuter* (without a capital; plural : *deuters*).

6.2.2. Conducting tissues of gametophytes and sporophytes of mosses : occurrence, general organization, variations.

▪ Gametophytes

A conducting tissue-system exists in the gametophytes of a great many species. A typical moss stem comprises, from the outside towards the inside, an epidermis with a thin cuticle but no stomata, an outer cortex which frequently consists of supporting cells with thickened walls ("stereids"), an inner cortex of thin-walled conducting-parenchyma cells, and, in a number of species, a central strand of hydroids. Important variations are seen, which may concern the relative development of the central strand of hydroids (if it exists), the presence or absence of leaf traces, the possible occurrence of a leptom, etc. Whereas such mosses as the Polytrichales show an unusual level of internal complexity, others are completely devoid of conducting tissues.

▫ Distribution patterns of conducting-tissues – especially hydroids – within the different groups of contemporary mosses.

From a survey of earlier literature (especially : Brotherus 1924-5) correlated with a cursory examination of numerous herbarium specimens, Hébant (1966b) concluded that "well-developed" strands are of common occurrence in certain groups (e.g. the Polytrichales or the Eubryales [1]), whereas possession of a small strand, or complete lack of it, is the more frequent condition in others (e.g. the Hypnobryales). Similarly, Frey (1971) stated that many pleurocarpous mosses have reduced strands, or no strand at all, whereas a number of acrocarpous ones possess an important conducting tissue-system. Thus a certain correlation appears to exist between internal structure and taxonomic position. However, any attempt towards systematization would rapidly prove to be unsatisfactory; many examples could be given of forms with a poorly developed strand or none at all in those groups which normally present well-developed conducting

(1) The classification adopted here is a simplified modification of that put foward by Dixon (1932).

tissues, – and *vice versa.* In the same way, the notional contrast between acrocarps and pleurocarps re-emphasized by Frey (1971) suffers from many drawbacks. For instance, *Orthotrichum spp.* are acrocarps without conducting tissues, whereas the pleurocarpous *Spiridens spp.* have a very conspicuous central strand (Figs. 217 and 223). The extent of these variations has been exemplified by several recent systematic revisions in which anatomical data were incorporated (e.g. Kawai 1965, Schauer 1967, Saito 1975). Indeed, the situation is a complex one since forms with and without conducting tissues may apparently coexist even within one single genus (e.g. *Grimmia*, Kawai 1965).

Attempts have also been made to reconcile these variations in anatomical patterns with the ecology of the different forms (cf. Physiological section, p. 88). However, it is not unusual to find mosses whose internal structure cannot be explained by a simple reference to their ecology. As Morin stressed a long time ago (1893, p. 130) : "des mousses de stations, de latitudes et de terrains différents, ont souvent la même structure, tandis que des mousses de même milieu offrent de grandes différences anatomiques".

Thus no simple and definite correlations have been established up to now between the variations in anatomical structures and the taxonomy or the ecology of the forms concerned. Until a detailed statistical analysis of the distribution patterns of the conducting tissues in the different groups has been made together with thorough autecological studies, no really comprehensive survey of these tissues in mosses can be presented. Only a preliminary and very superficial account is given here [1].

(1) *Note of caution* : due to the unusual level of structural variability in these plants (see chapter 2), discrepancies between this account and other reports or new observations will probably be found. Changes in classification are also to be expected.

Distribution of conducting strands in leafy stems of contemporary mosses (Class Musci sensu stricto = Bryopsida Roth.)

I. Nematodonteae

– *Polytrichales* (incl. *Dawsoniaceae*). A highly developed water-conducting system was found in all investigated forms; a typical leptom also exists in a number of species (1) (cf. e.g. Figs. 24, 76, 101, 111, 225, 237-238, 245, 283, and many others throughout the plates).

– *Buxbaumiales*. The gametophytes of *Buxbaumia* are so rudimentary that no typical leafy stems with conducting tissues are produced; there is apprently no strand in the small-sized gametophytes of *Diphyscium foliosum.*

– *Tetraphidales*. A well-developed strand exists in *Tetraphis (= Georgia) pellucida.*

– *Schistostegales*. Noguchi (1974) did not represent a strand in the stem of a Japanese specimen of *Schistostega osmundacea (= S. pennata)* whereas I found one in a strain from Europe.

II. Arthrodonteae Haplolepideae

– *Fissidentales*. A typical central strand exists in many (all?) *Fissidens* (Fig. 216); its development varies from weak to strong. There is no strand in *Octodiceras julianum.*

– *Grimmiales*. A notable variability exists within the order; this has been exploited by taxonomists (e.g. Morin 1891, Kawai 1966). Forms with and without conducting tissues coexist in the genus *Grimmia.*

– *Dicranales*. A typical, but usually not very large, central strand is seen in a number of species (Fig. 220), although some may lack it (e.g. several species of *Seligeria*, Schauer 1967; the *Leucobryaceae).*

– *Pottiales*. A variety of structures exists, ranging from large to small strands, or even to complete lack of conducting tissues (e.g. in *Cinclidotus aquaticus* (cf. e.g. Saito 1975).

(1) No other moss gametophytes are presently known to possess a leptom. Some "reduced" Polytrichales exist (G.L. Smith 1971, p. 70); it is possible that eventually certain of these may prove to be devoid of a central strand.

III. Arthrodonteae Diplolepideae

– *Funariales*. A well developed central strand has been depicted for various Funariaceae and Splachnaceae (Figs. 91, 241-242 and 341). Some "reduced" forms (e.g. *Ephemerum*) however, lack a strand.

– *Eubryales*. A great many species are known to possess a well defined – sometimes large – central strand. Some of the most conspicuous strands are found in *Mnium, Bryum, Timmia*, etc. (Figs. 6, 64 to 66, 68, 92 and 217).

– *Isobryales* (incl. *Orthotrichales*). A majority of these mosses have very reduced strands (e.g. *Climacium dendroides*) or no strand at all (*Hedwigia ciliata, Neckera crispa, Orthotrichum spp., Fontinalis*, etc.) (Figs. 128A, 222 and 223).

– *Hookeriales*. Although a well-developed strand exists in a few forms (e.g. *Cyathophorum bulbosum*, Fig. 218), in others this strand is smaller (e.g. *Pterygophyllum lucens*, Fig. 219) or absent (e.g. *Lepidopilum polytrichoides*).

– *Hypnobryales*. A majority of these mosses possess a very discrete strand only (e.g. *Thuidium tamariscinum*) or no strand at all (e.g. *Amblystegium fluviatile*) (Figs. 9-10, 127A and 221).

□ **Morphogenetic aspects : apical segmentation and origin of conducting tissues; branching and its anatomical consequences; etc.**

The ontogeny of the apex of the leafy stem in mosses has been reinvestigated by a number of authors during recent years (Wigglesworth 1956, Bonnot 1967b and 1968b, Frey 1971, Berthier 1972, Hallet 1972, Hébant 1973c). A thorough and critical review of previous work, as well as an accurate description of apical segmentation in a wide range of species representative of various systematic groups, can be found in Berthier (1972). From these studies, a relatively clear picture of the origin of branching, and of the development of conducting tissues within apical derivatives, can now be presented.

Origin of the axial conducting system

Hydroids. The hydroids of the central strand originate from the most central part of the segments which are cut off from the apical cell (Figs. 22-23, 69, 143, 216 to 219, 224 and 234-235). The relative importance of the young conducting tissues, as compared with the young cortical tissues, varies greatly according to the different species. In fact, this is correlated with early segmentation patterns, and Berthier (1972) points out that a rapid cleavage of the inner part of the segment is a characteristic of "primitive" mosses with a well developed conducting strand.

In the aerial leafy stem of most mosses with conducting tissues, all the cells of the most internal part of each segment cut off from the apical cell differentiate into hydroids. In polytrichaceous mosses, however, and in certain cases only, not all these cells differentiate into hydroids; those which do not become conducting cells yield elements with characteristics of parenchyma (Figs. 22-23 and 224). In mature stems, the latter usually exhibit thickened walls, and are therefore called "stereids". The relative number of hydroids and of stereids within the central strand may vary to a large extent in relation to the species considered, the stage of development, etc. The resulting structure (i.e. the hydroids intermixed with stereids) has been observed in the so-called "rhizome" of all the Polytrichales which have been studied up to now from this point of view : *Atrichum sp., Pogonatum sp., Polytrichadelphus sp., Polytrichum spp.* (Tansley & Chick, 1901; etc.), *Dawsonia sp.* (Goebel, 1906), *Dendroligotrichum sp.* (Schulz, 1937) (Figs. 85, 87 and 228 to 232). It is also encountered in the aerial leafy stems of those species of *Dawsonia* which belong to the section *Superba* (van Zanten 1973, Hébant 1976a) (Figs. 225 to 227). Stereids also reappear amongst hydroids of the central strand in upper regions of upright female axes of certain forms which otherwise lack them (e.g. *Atrichum undulatum*, Fig. 25).

Leptoids. Authentic gametophytic leptoids are known only in Polytrichales. They originate within apical derivatives from just below the level of the first periclinal wall which divides each segment ("a" wall according to Leitgeb 1868) (Fig. 69).

Branching and its anatomical consequences

Lateral branches originate from secondary apical cells which are regularly produced within the segments cut off from the main apical cell (Figs. 73-74) (for review and details, see Berthier 1972). In many mosses, early development of lateral buds is inhibited by apical dominance (cf. Figs. 75 and 236). As in *Pallavicinia* and allied forms (cf. p. 25), lateral branches usually initiate their development only when cortical tissues of the bearing shoots are fully mature and incapable of dedifferentiating and subsequently differentiating afresh into conducting cells. As a result, the strands of the lateral branches if present, never unite with the strand of the main stem (cf. e.g. Figs. 241-242).

In certain cases however, early development of lateral branches is not inhibited by apical dominance. This is the normal condition in the vegetative gametophytes of certain species (e.g. *Dendroligotrichum dendroides*, Figs. 235 and 243). This also occurs in various other forms when the main stem becomes fertile (e.g. in *Polytrichum alpinum, Polytrichadelphus magellanicus, Aulacomnium palustre*, etc.) (Fig. 234). In all these cases, secondary strands are differentiated which unite with the main strand (Figs. 75, 76, 82 to 84, 237-238 and 244-245). These phenomena were first recognized by Schoenau (1912), although some precursory observations on the anatomy of *Dendroligotrichum* had also been made by Bastit (1891). Schulz (1937) subsequently provided interesting details of phyllotaxy and periodicity of branching in the same species. More recently, Berthier *et al.* (1971) and Hébant & Berthier (1972) studied in detail the anatomical consequences of branching in some selected examples of polytrichaceous mosses. The latter authors concluded that within the apical region of gametophytes, young developing lateral branches could induce differentiation of conducting strands in the subjacent ground tissue. "Intermediate" patterns of organization, in which development of the secondary strand is more or less irregular were also occasionally encountered in *Polytrichum alpinum* by these investigators. Frey (1974b) showed that in *Hypnodendron dendroides* another "intermediate" type of organization is constantly present : in growing axes, a parenchymatous tissue is found between the lateral branch complex and the central strand of the main stem; but a secondary connection is soon established through the degeneration of the protoplast of certain "parenchyma" cells situated between the strands of the secondary axes and that of the main stem (Figs. 239-

240). Thus, the developmental pattern exhibited by this moss establishes the link between that found in such forms as *Dendroligotrichum*, in which well-developed secondary strands are differentiated, and that found in the many other mosses where a connection between strands of lateral branches and the main stem is never established (Frey 1974a).

In certain Polytrichales, where both "true" leaf traces (cf. p.41) and secondary strands irrigating lateral branches coexist in the cortical tissues of the main stem, an original pattern is established; the trace of the leaf belonging to the same segment as the secondary branch joins the strand of the latter before it merges with the conducting strand of the main axis (Fig. 77) (for details see Berthier *et al*. 1971, Hébant & Berthier 1972).

Possible influence of archegonia with developing embryos on differentiation of conducting strands.

In the course of a fine investigation on the development of the embryo and the "embryotheca" in mosses, Roth (1969) noted the presence, in *Dicranum undulatum*, of a conducting strand irrigating each sporophyte borne by the gametophore (op. cit., p. 30). He also observed that no such strands were present below sterile archegonia. Roth concluded that presence of a developing embryo within the archegonium could induce the differentiation of a conducting strand. Observations which are in agreement with that interpretation were subsequently made on a strain of *Polytrichum alpinum* by Hébant & Berthier (1972).

□ Internal structure of "rhizomes", main upright axes and secondary branches

In various mosses which possess both creeping axes and aerial leafy stems, the conducting strands in the "rhizomes" are less developed than in the erect portions of the corresponding gametophytes (Unger 1861, Davy de Virville 1927-8, Lorch 1931). *Mnium undulatum* (Figs. 6 and 7) and *Hypopterygium* are classical examples of this. Similarly, the strands of secondary branches borne by the main upright stems are usually reduced in comparison with those of the shoots that bear them, or may even be absent. When central strands are present in these secondary branches, they do not necessarily unite with the conducting strand of the main stem (see above).

Polytrichales provide such a notable example of mosses with a high level of internal complexity in their gametophytes that they deserve to be treated separately. In a number of species which have been investigated, the subterranean axes were found to possess an organization similar in appearance to that of a phanerogamous root, which is rather unusual for a bryophyte. This was first observed by Hy (1880) in *Polytrichum commune*, and has since been found in several other forms (*Dawsonia spp.*, Goebel 1906, Hébant 1976a; *Dendroligotrichum dendroides*, Schulz 1937; *Polytrichadelphus magellanicus*; etc.) (Figs. 85 and 228 to 231). Plate 22 illustrates the general organization and the main cell types found in the "rhizome" of *Polytrichum juniperinum.* One notes that hydroids of the central strand are intermixed with stereids, which retain a living protoplast. Three furrows, in which leptoids are seen together with associated parenchyma cells, exist at the periphery of this strand. The large cells of the internal cortex are strongly suggestive of an endodermis in appearance, and a number of authors have interpreted them as such (Tansley & Chick 1901, Goebel 1906, Waenker von Dankenschweil 1915, Scheirer 1974, etc.). However, in the opinion of the present author equating the internal cortex to an endodermis awaits substantiation. No truly convincing structural evidence proves the validity of this, and no physiological experimentation has apparently been done so far. In an early investigation, Strunk (1914) had also denied the presence of an endodermis in bryophytes.

In Polytrichales, those axes which grow upright and give the aerial leafy stems show progressive transformations, namely :

– disappearance of stereids from the central strand (with the exception of the species of *Dawsonia* belonging to the section *Superba*),

– increasing importance of the leaf traces. In the lower portions of the stems, only small traces irrigating scale-like leaves are seen. In the green-leaved regions, more complex and overlapping leaf traces are present. These changes are correlated with modifications in apical segmentation patterns (especially the transition from a "parallel" to an "encroaching" type of segmentation; cf. Wigglesworth 1956). Some of these transformations are depicted in figures 70 to 72 for *Polytrichum* (see also plate 5), and in figures 78 to 81 for *Dawsonia*.

Green-leaved portions of the best developed gametophytes possess a conspicuous central strand of hydroids at the periphery of which a leptom is present. The hydroids of the median portion of the strand usually exhibit very thick lateral walls, whereas the hydroids of the outer portion, – which are in relation with the leaf traces –, have much thinner walls (Figs. 274 and 275). This has been related to the fact that hydroids of the outer region of the strand differentiate more rapidly, their protoplast degenerating at an earlier stage than that of the hydroids of the central region (Hébant 1966a). In the latter, dictyosome-mediated wall elaborations seem particularly active. As stated above, in many species of *Dawsonia* the hydroids of the central strand are intermixed with stereids; these hydroids remain rather thin-walled, even in the median portion of the strand, as the stereids here contribute to the strengthening of the plant (Figs. 225 to 227). In all polytrichaceous mosses, the central strand is surrounded by living parenchyma cells (cells of the *"hydrom-sheath"*, Tansley & Chick 1901) (Figs. 111 and 274). The leptom at the periphery of the central strand shows a complex organization. In the most highly developed polytrichaceous gametophytes, and only in these, leptoids *sensu stricto*, with clear sieve element-like characteristics are present. In addition to the leptoids *s.str.*, ordinary parenchyma cells, as well as a great many other cell types with "intermediate" structures exist. Part of these establish the connection with the food-conducting cells of the leaf traces.

□ Leaf traces

In many mosses the nerves of the leaves exhibit a complex internal organization. Lorentz (1864 *sqq.*), Morin (1893) and many others have described their structure and the important variations they may show (cf. Figs. 11-12, 95 to 99 and 108). Various taxonomic conclusions have emerged from these studies. According to the species, leaf traces may or may not exist within the ground tissue of the stem, in prolongation of the leaf nerves. If leaf traces are present, they usually show a simpler organization than that of the corresponding nerves.

The first observations concerning leaf traces were generally inaccurate (e.g. Schacht 1856, p. 314). Unger (1861) depicted with precision the pattern of the leaf traces in *Dawsonia superba* (fig. 90). Lorentz (1867 *sqq.*) conducted the first detailed study of leaf traces in mosses. He introduced the now classical distinction between "false" leaf traces *(falsche Blattspuren)*, which do not come into contact with the central strand, and "true" leaf traces *(echte Blattspuren)*, which join it (Figs. 91-92). In Lorentz's mind, this distinction referred primarily to the hydroids of the leaf traces. Lorentz was also the first to describe (on *Atrichum undulatum*) the progressive simplification in structure which affects the leaf traces of polytrichaceous mosses in the course of their penetration within the cortical tissues of the gametophytic stems. He states (Lorentz 1867-8, p. 432) that "die Centralgruppen an Zahl abnehmen"; – according to his terminology, each "Centralgruppe" of the leaf trace comprises one hydroid and the associated "parenchyma" cells ("deuters" and "socii" see pp. 31 and 43) –. Simplification of the leaf traces through decrease in the number of hydroids was also recorded by Eschrich & Steiner (1968a) for underdeveloped (cultivated) specimens of *Polytrichum commune.* In the material they studied, one hydroid only of each leaf trace could eventually join the central conducting system. Hébant (1969b) reinvestigated several strains of the same species and found that, in accordance with the descriptions of several earlier authors (e.g. Tansley & Chick 1901), well developed leaf traces occurred in the most differentiated forms (e.g. specimens from pseudoalpine grassland, Fig. 246). In the latter, 7-8 hydroids of each trace were found to be connected with the central strand, whereas 3 such hydroids only were found in a somewhat less differentiated strain (specimens from a peat bog). In specimens cultivated under water no leaf trace joins the central strand. Figs. 101-102 and 104 to 107 illustrate some of the variations pertaining to hydroids of leaf traces in *Polytrichum* (see also table on page 10).

Lorentz described "true" leaf traces in *Splachnum luteum, S. sphaericum* and *Voitia nivalis* (Fig. 91). In the study of complete series of transverse sections of *vegetative* stems of *Splachnum luteum, S. ampullaceum, S. vasculosum* and *Voitia nivalis*, the present reviewer never observed the junction of a leaf trace with the central strand. These discrepancies are perhaps another consequence of the exceptional structural variability which exists in mosses, even at the intraspecific level. Indeed, Haberlandt (1886) noted that in *Funaria hygrometrica*, a species normally with "false" leaf traces, a junction of the latter with the central strand could sometimes be observed [(1)].

Thus, up to now, "true" leaf traces have been found with some constancy only in the most developed Polytrichales. *Funaria hygrometrica* (see above) and those polytrichaceous mosses in which only a few hydroids (sometimes just one) of each trace reach the central strand can be considered as interesting intermediates between mosses with "true" and mosses with "false" leaf traces. From the available information, and considering only the hydroids of the leaf traces, the following types of organization can be distinguished in mosses (after Hébant 1969b) (Fig. 112) :

a) Mosses in which the leaf traces link up with the central cylinder in a fairly definite way (the most developed Polytrichales) (Fig. 112A).

b) Mosses exhibiting only a loose or very loose connection between the leaf traces and the central strand (various Polytrichales) (Fig. 112B).

c) Mosses in which the leaf traces end blindly in the ground tissue without joining the axial strand (various Funariales and Eubryales; *Polytrichum commune* cultivated under water) (Fig. 112C).

d) Mosses in which the leaf possesses hydroids in the midrib, but where the latter barely penetrate the ground tissue of the stem (various Eubryales) (Fig. 112D).

e) Mosses in which the stem still exhibits a strand of hydroids, but where the latter are totally absent from the leaves (numerous Dicranales) (Fig. 112E).

f) Mosses having no hydroids in either stem or leaves (numerous Orthotrichales) (Fig. 112F).

Food-conducting cells of leaf traces

Cells of specialized conducting parenchyma are present in the leaf

(1) Funariaceae are akin to Splachnaceae. Both *Splachnum* and *Voitia* belong to the latter family.

nerves of a great many mosses. Such cells are usually termed "eurycystes" or "deuters" by taxonomists [1] (cf. p. 31). They are usually wide and elongated elements which retain a living protoplast. Numerous plasmodesmata are present in their transverse or somewhat oblique end walls. These cells have sometimes been erroneously interpreted as the water-conducting cells of the leaf nerves and leaf traces (for review, see Hébant 1972b, p. 736). Symplastic transport of organic compounds synthesized within green leaves probably occurs preferentially through them. In most mosses, once a leaf has merged with the stem, these elements rapidly become indistinguishable from the ground tissue of the cortex. Transport of organic substances towards "sinks" such as growing apices or subterranean axes must therefore take place within cortical parenchyma of stems.

With the exception of polytrichaceous mosses, no detailed ultrastructural studies have apparently been made of these cells of conducting parenchyma. The most highly developed Polytrichales are notable by the fact that certain of the deuters [2] in their leaf nerves and leaf traces may show cytological features reminiscent of those of the leptoids *s.str.* [3] (Figs. 247 to 253). In *Polytrichum commune*

(1) In a majority of the polytrichaceous mosses I could investigate (*Dawsonia spp., Polytrichadelphus magellanicus, Polytrichum spp., Psilopilum australe,...*), the so-called "socii" of the leaf nerves and leaf traces appeared as rather ordinary parenchyma cells accumulating reserves (lipid droplets in the cytoplasm; starch in the well developed plastids). By their cytological characteristics such socii seem akin to the elements of the hydrom sheath, with which in fact they show a spatial relationship.

(2) The term "deuter" is retained here to designate these cells as they correspond to a rather well defined category of elements within moss leaves and leaf traces. For *Polytrichum*, Bastit (1891) spoke of "cellules en tibia", and Tansley & Chick (1901) of "leptoids" of the leaf traces. Goebel (1906) similarly used the term "leptoid" for *Dawsonia*. However these elements are not always leptoid-like in character. Therefore, retention of the more general term "deuter" seems justified.

(3) Identification of a clearly differentiated layer of food-conducting cells within leaves of *Polytrichum* was first made by Firtsch (1883), followed by many other investigators (cf. review in Hébant 1972b). Such a layer was not recognized by Paolillo & Reighart (1967) in a recent ultrastructural study, apparently because they investigated underdeveloped specimens.

such elements occur only in the median region of the leaf nerve; the deuters on both sides of the latter show intermediate characteristics and progressively establish the transition to elements of ordinary conducting parenchyma (Fig. 108). In this moss, the deuters remain clearly identifiable during the penetration of the leaf traces within the ground tissues of the stem (Fig. 109). They eventually come into relation with the leptoids of the main axial conducting system (Figs. 101-102); cells of specialized conducting parenchyma, the most differentiated of which may also show leptoid characteristics, establish the connection [1]. Typical zones of transition, and gradients of differentiation are seen at these levels (Figs. 110-111). Together the leptoids and the cells of conducting parenchyma constitute, at the periphery of the axial strand of hydroids, a continuous network through the apertures of which run the hydroids of the leaf traces (Fig. 103).

The above description refers mostly to *P. commune.* A similar organization has been observed in the green-leaved portions of *Polytrichadelphus magellanicus.* A rather comparable, although slightly different pattern has also been found in some "giant" species of *Dawsonia* (Hébant 1976a). A cursory examination of several other polytrichaceous mosses (*Atrichum spp., Dendroligotrichum dendroides,* etc.) has shown that somewhat simpler types of organization may also exist; in the latter mosses, the general level of differentiation, especially within cells of conducting parenchyma, appears to be less remarkable.

- **Sporophytes**

In nearly all moss species, the life span of the sporophytes is shorter than that of the gametophytes which bear them. Whereas the latter may frequently continue their growth year after year, the sporophytes are always of limited duration and degenerate immediately after the spores are shed. In addition to this, a number of species exist in which sporogonia are produced only rarely; in some they are

(1) This was clearly depicted by Tansley & Chick (1901); Eschrich & Steiner (1968a) proposed the expression "leptoid bridge" *(Leptoidenbrücke)* for this zone.

even totally unknown. These facts do not facilitate histological studies on the diploid generation of mosses. Indeed, if our knowledge of the gametophytes is still unsatisfactory, that of the sporophytes is even more incomplete, and published works dealing with the conducting tissue-system in the setae remain scarce. For this reason, no systematic account of the distribution of conducting tissues in the sporogonia of the various groups will be given here.

A well developed moss sporophyte comprises a "foot" deeply inserted within gametophytic tissues, an axis or "seta" of variable length (usually from a few millimiters to a few centimeters) and a "capsule" or "sporangium" where the spores are produced. It has been suggested (e.g. by Haberlandt 1886) that the development of a seta is an adaptative mechanism facilitating dissemination of spores by the wind. In certain mosses, however, the seta is very short or almost non-existent.

The precursory investigations of Schimper (1848), Unger (1861) and Lorentz (1867 *sqq.*) established the presence of a strand in the median region of the seta of several species (cf. e.g. Figs. 118 and 119). Haberlandt (1886) also paid some attention to the structure of this strand in *Funaria, Meesia* and certain Polytrichales. The first detailed study of the conducting tissue-system in the sporophytes of mosses was undertaken by Vaisey (1887a-c *sqq.*); his untimely death at the age of 26 unfortunately put an abrupt end to an excellent investigation. A renewed interest in the conducting tissues of the sporophytes has arisen within recent years (cf. e.g. Favali & Gianni 1973, Favali & Bassi 1974, Bassi & Favali 1974, Hébant 1975a, Schulz & Wiencke 1976) (1).

□ Structure of foot

The foot is an extremely important organ, not only in mosses but also in all other bryophytes, even those devoid of specialized food-

(1) The recent work of Kawai, part of which deals with the seta (Kawai 1969), being primarily that of a taxonomist is of limited value to the histologist. Although this author has published a number of remarkably elegant drawings, the various tissues are not always properly identified in them, and some misleading statements concerning their structure or role are found in certain of his papers.

and water-conducting tissues. A special chapter is devoted to its organization (p. 69).

□ Conducting tissues in setae

A central strand is very commonly encountered in setae, even in those of species which lack conducting tissues in their gametophytes (Oltmanns 1884, Ruhland 1924) (Plates 32-33, 35-36 and 63). Indeed, it remains to be determined whether well developed setae without a strand are of frequent occurrence or even exist at all.

When "mature" setae are studied under the electron microscope, various cell types which would be expected to retain a living protoplast, e.g. the leptoids or the elements of the internal cortex, frequently appear to have degenerated. It is not known with certainty whether this is the result of poor fixation or if precocious degeneration of these cells occurs naturally. The fact is that collapse of the median region of the seta, sometimes resulting in the formation of a lacuna, has been recorded in aged sporophytes of several mosses (cf. Lorch 1931).

The most highly developed conducting strands are found in the setae of various Polytrichales (Figs. 113, 114 and 259 to 262). An accurate description of these strands, as they appear under the light microscope, was made by Vaisey (1888) on *Atrichum* and *Polytrichum*. More recently, Favali & Bassi (1974) on *Polytrichum*, and Hébant (1975a) on *Dawsonia* and *Dendroligotrichum*, extended these observations to the ultrastructural level. In all these mosses, the strand of the seta constitutes a solid core anchored in the middle of a lacunate inner cortex. The description below is taken from Vaisey (op. cit., p. 266-7) and refers to the seta of *Atrichum undulatum* (cf. Fig. 114):

"The central strand is surrounded by a single layer of thin-walled cells (...) similar to the cortical cells, except that they are on the average longer than the innermost of those cells (...). These further resemble the cortical cells by containing starch-forming plastids, and so differ from the tissues of the central strand. [...].

"The central strand itself consits of certain tissues, which in the seta are characterized by the absence of intercellular spaces; it contains two kinds of tissue. One of these forms a solid axial cylinder (...), called Leptoxylem [1]; the other a hollow cylinder concentric with the former (...) called Leptophloëm [2].

"The Leptoxylem consists of very thin-walled, much elongated prosenchymatous cells, whose cell-walls are, as far as can be made out, all that remain in the mature seta (...). These cells do not contain any living protoplasm, but only here and there remnants of disorganized protoplasm. This conclusion has been come to after very careful examination and comparison of sections, both transverse and longitudinal, in which nothing more than a few granules, the last remains of the protoplasm of the meristematic cell, were found.

"The Leptophloëm is a more extensive tissue-system than the leptoxylem, and it consists of thick-walled, considerably elongated cells (...), forming in transverse section a zone two, rarely three or more, cells deep. The innermost cells of this layer are thicker than the more external. In longitudinal section (...) the cells of this tissue show their peculiar form, being slighly dilated at one, and sometimes at both ends. In the latter case these cells may be described as being somewhat dumbbell-shaped. Their transverse walls are not usually horizontally placed, and in longitudinal optical section have often more or less the appearance of an elongated S".

An organization somewhat similar to that described here for *Atrichum* exists in *Polytrichum* (Vaisey 1888, Favali & Bassi 1974) (Fig. 113), *Dawsonia* and *Dendroligotrichum* (Hébant 1975a) (Figs. 259 to 262). In all these mosses, the leptom cylinder ("leptophloem" of Vaisey) consists of what the present author has tentatively interpreted as poorly differentiated leptoids (Hébant 1975a) (see also p. 62). No ordinary parenchyma cells are intermixed with them [3].

(1) *Leptoxylem* = hydrom.

(2) *Leptophloem* = leptom.

(3) Favali & Bassi (1974) claimed that in *Polytrichum commune* the leptoids were intermixed with parenchyma cells. This is true in the gametophyte, but *not* in the seta : careful examination, both by light and electron microscopy, of the leptom in sporophytes of *Polytrichum spp.*, as well as of *Dawsonia* and *Dendroligotrichum*, shows that there is no reason to establish such a distinction between two cell types in this tissue (cf. Hébant 1975a).

The relative importance of the hydrom and of the leptom varies according to the species. A notable development of the strand is observed in the setae of several species of *Dawsonia* (Figs. 259-260).

A simpler organization, but somewhat reminiscent of that just described above for Polytrichales is known in the seta of certain Funariales (e.g. *Splachnum luteum*, Vaisey 1890b; *Funaria hygrometrica*) and Eubryales (e.g. *Meesia longiseta*). In the latter species, in addition to a strand of hydroids and external to it, a cylinder of "leptoids" is also seen. However, the "leptoids" are here generally one layer thick only, and intercellular spaces are no longer present in the internal cortex (Figs. 115 to 117 and 126B).

An even simpler organization is found in a majority of contemporary mosses, in the seta of which a strand of hydroids only, without leptoids, is identifiable (cf. e.g. Lorch 1931) (Figs. 118, 119, 127B and 128B).

In the present review, "leptoids" have been identified in the setae of a number of forms, including some species of the genera *Funaria, Meesia* and *Splachnum.* In this respect, the early viewpoint of Vaisey (1888, 1890b) has been adopted. However, not all investigators have accepted this interpretation. A number of authors (e.g. Haberlandt 1886, in *Funaria* and *Meesia,* and Schulz & Wiencke 1976, in *Funaria*) have interpreted these cells as constituting a protective sheath *("Schutzscheide")* for the hydroids. Schulz & Wiencke have correlatively interpreted as "leptoids" the cells of conducting parenchyma of the inner cortex (op. cit., fig. 8, p. 52). According to the present reviewer, the following facts are in favour of Vaisey's concept :

a) Identification of these cells as leptoids in the seta of such mosses as *Polytrichum* or *Dawsonia* presents little difficulty. By both their position and general structure, they seem to be the counterpart of the gametophytic leptoids. No sheath apparently exists in the seta of these mosses between the leptoids and the central strand of hydroids.

b) In the seta of such mosses as *Funaria, Splachnum* or *Meesia,* these cells occupy the same position, and have the same general appearance as those in *Polytrichum* or *Dawsonia.* The major difference is that the "leptom" cylinder is here only one layer of cells thick (Figs. 115 to 117 and 126B). But less developed Polytrichales such as *Atrichum* (Fig. 114) establish the transition between the *Polytrichum* type and the *Funaria* type. One is thus tempted to interpret these cells in *Funaria, Meesia* and comparable forms as *residual leptoids*[1]. Haberlandt (1886) himself noted that (op. cit., p. 382) (cf. also Fig. 116) : "Auf Längsschnitten sieht man, dass die Scheidenzellen von langgestreckter, aber nicht prosenchymatischer Form sind; die dünnen Querwände sind höchstens ein wenig schief gestellt. *Bemerkenswert ist die an Siebröhrenglieder erinnernde Anschwellung der Zellenden*" (underlined by me).

c) In the capsules of certain species which have been investigated *(Polytrichum spp., Dawsonia spp., Dendroligotrichum dendroides, Funaria hygrometrica)* the "leptoids" of the seta are in direct continuity with cells which are strongly suggestive of elements of specialized conducting parenchyma; the latter establish the connection with the other types of parenchymatous elements in the sporangium (Figs. 256-258).

d) Evidence for the conductive role of the leptoids as identified here in the seta of *Polytrichum commune* has been produced by Eschrich (1975 – see physiological section p. 95).

Remark : Claims have recently been made for the occurrence of leptoids intermixed with parenchyma cells of the inner cortex in the setae of *Tortula muralis* (Favali & Gianni 1973) and *Mnium orthorhynchum* (Bassi & Favali 1974). However, no convincing evidence has been put foward for these assertions; the fact that some elements of conducting parenchyma may show somewhat inclined end-walls is not sufficient in itself for these cells to be identified as leptoids.

□ Capsule

The conducting strand of the seta usually terminates just below

(1) Whether or not such "residual" leptoids play an effective role in the translocation of organic substances remains to be determined. In this respect, it may be noted that the "residual" strands of hydroids found in certain mosses have apparently little functional value (Zacherl 1956; see also chapters 8 and 11).

the columella, shortly after entering the capsule. There are probably few fields in bryophyte histology which have been so little investigated as the structure of the conducting tissues at this level. No published electron microscope study is available.

In *Funaria hygrometrica* and *Polytrichum commune,* the termination of the strand of hydroids within the capsule is dilated in the form of a small ampulla consisting of short and wide elements (Figs. 256 to 258). This probably acts as a reservoir redistributing water and substances dissolved in it to various parts of the capsule. A similar ampulla was not found in other mosses such as species of *Dawsonia, Dendroligotrichum* or *Fissidens.*

As mentioned above, in various forms in which the seta possesses leptoids, the latter are found in direct continuity with cells of what appears to be a specialized conducting parenchyma, which makes the transition towards other cell types within the capsule (Figs. 256 to 258).

Lack of specialized conducting tissues within the capsule implies that conduction at this level is performed by parenchyma cells (cf. Haberlandt 1886). Indeed, the small size of the sporangium allows this type of conduction. The columella in the median region of the capsule has also been allegedly reported to be involved in the water economy of the capsule.

▪ Distribution patterns of conducting tissues within both gametophytes and sporophytes

The analytical studies summarized above clearly show that a tremendous variability exists in organization patterns of conducting tissues in gametophytes and sporophytes of mosses. If both setae and leafy stems are taken into consideration, a number of distinct combinations are to be found. Some parallelism in the structures may exist between the two generations. This is not surprising since the same genetic systems, either in haploid or diploid complement, control their differentiation. However, differences are also frequently found, which could perhaps be interpreted as the result of specialization. In a general manner, the setae appear as more "conservative" than the corresponding gametophytes, viz. they present a conducting strand in a number of species where the latter has disappeared from the leafy stem (cf. Ruhland 1924). Here again, a number of fundamental types of organization can be distinguished :

a) Mosses in which both phases present hydroids and leptoids *(Polytrichum spp., Atrichum spp.)* (Fig. 125).

b) Mosses in which the gametophyte possesses a well-developed strand of hydroids but no leptoids, the seta still having both hydroids and "leptoids" *(Funaria spp., Splachnum spp.)* (Fig. 126).

c) Mosses in which both the leafy stem and seta possess a strand of hydroids but no leptoids *(Dicranum spp., Brachythecium spp.)* (Fig. 127).

d) Mosses in which conducting tissues are absent from the gametophyte but present in the seta *(Neckera crispa, Buxbaumia spp.)* (Fig. 128).

e) Mosses no longer presenting any conducting tissues in either of their two phases *(Orthotrichum spp.)*.

f) Mosses in which the leafy stem exhibits a conducting strand but in which the seta has regressed or disappeared (various *Grimmia* – Lorch, 1931).

Several of these types of organization frequently coexist – in varying proportions – within the different taxa recognized by the systematists, – e.g. within one family.

6.2.3. Development and ultrastructure of conducting tissues.

Although the structure of mature conducting tissues in mosses, especially of hydroids, has been studied by a number of light microscopists (see above), developmental and ultrastructural investigations remain particularly scarce. In addition, a majority of these studies have been restricted to Polytrichales (attention has been focused on these mosses because of their remarkable level of internal differentiation).

■ Hydroids and associated elements

□ Development of hydroids

Differentiation of hydroids has been investigated in some detail in the gametophytes of *Polytrichum commune* (Hébant 1967b and 1974). A complex and ordered sequence of events is observed during their development : within apical derivatives (cf. pp. 35-36), the young hydroids of the differentiating central strand increase in size isodiametrically, and then in length. They rapidly acquire steeply inclined terminal walls. Towards the end of their elongation phase, these elements, especially those of the median portion of the strand, elaborate thick lateral walls. Then an abrupt lysosomal degeneration of their protoplast occurs, whereas their thin end walls are partially hydrolysed. Mature conducting cells are thus dead and empty.

Among the most noteworthy cytological events which accompany the differentiation of the hydroids (for details see Hébant 1967b and 1974), one may note :

– The remarkable development of their nuclei (Plates 17, 38, and Fig. 267). Hallet (1972) has shown that these nuclei become endopolyploid (up to 8C) in *P. formosum.* Endopolyploidy is commonly observed in differentiating tracheary elements of vascular plants (List 1963).

– The extensive development of their vacuolar system (Plate 17, and fig. 270).

– The active synthesis of wall contituents. Golgi-derived vesicles containing polysaccharides are involved (Fig. 268). A direct participation of the plasmalemma is also probable. Lomasomes and other structures are encountered (Fig. 269).

– The final breakdown of their protoplast. A peak of strong lysosomal acid phosphatase activity is revealed at this level (1) (Plate 68). Hydrolysis phenomena also affect the thin end walls of contact between hydroids. Pectic and various wall substances are removed, leaving only a delicate and highly permeable network (Plate 66).

The above refers mostly to the gametophytes of polytrichaceous mosses. In the leafy stems of many other mosses, the hydroids of the central strand do not develop thick lateral walls. However, a comparable sequence of events leading to the final autolysis of the hydroid

(1) Electrophoretic studies have further shown that a particular band of activity, probably corresponding to the lysosomal phosphatase, is highly developed at this stage in *Polytrichum* (Kahlem & Hébant 1973) (Plate 40).

protoplast is observed in these plants (Plate 37). In particular, a peak of strong lysosomal acid phosphatase activity has been revealed in the differentiating central strand of various species selected as representative of a wide range of systematic groups (Hébant 1973b) (cf. Figs. 279 and 280). These facts also apply to the hydroids of the leaf traces (Figs. 277 and 279).

Differentiation of the hydroids in the seta of *Funaria* was recently investigated by Schulz & Wiencke (1976). The latter authors concluded that in spite of some peculiarities, the general pattern was comparable with that in the gametophytes of polytrichaceous mosses.

□ Structure of hydroids at maturity

Excellent descriptions and accurate illustrations of hydroids at maturity have been produced by several investigators in the second half of the nineteenth century (e.g. by Haberlandt 1886, and Vaisey 1888). Some of these authors even attempted to establish a comparison with the tracheids of vascular plants. Haberlandt (1886, p. 485) writes : "Der einfache Centralstrang besteht ausschliesslich aus wasserleitenden Zellen, die man als Tracheiden einfachster Art bezeichnen darf". Recent electron microscope investigations have confirmed earlier reports based on light microscope studies and have also produced new information, especially on the structure of the walls (Hébant 1970c and 1974, Scheirer 1973, Hébant & Johnson 1976).

The general characteristics of hydroids at maturity can be summarized as follows (cf. plates 64, 66 and 67) :

a) They are elongated elements, with more or less inclined terminal walls (Figs. 68, 130, 263).

b) Their end walls are usually thin, their lateral walls often so. In various mosses they present, as a result of partial hydrolysis occurring in the final stages of their development, a characteristic aspect comparable with that of the non-lignified primary walls of the tracheary elements in vascular plants (O'Brien 1970) (Figs. 227, 254 and 271 to 275). Large gaps appear between microfibrils of such hydrolysed walls (Fig. 273); their consequent high permeability contributes to making hydroids preferential pathways for water and solutes within stems and setae. Early investigators (e.g. Gjokić 1895 in *Polytrichum commune*) had demonstrated the cellulosic nature of these walls – although they could not recognize with the light microscope that they were hydrolysed ! –. Thick lateral walls are present in the hydroids of the median portion of the strand of the gametophytes of various Polytrichales (Figs. 100, 265 and 275). Whether or not these walls must be termed "secondary" has not been clearly established (cf. Hébant 1974, Scheirer 1975). They are usally not hydrolysed, being apparently protected by what has been shown to be non-lignin polyphenols (Erikson & Miksche 1974ab). *No lignified secondary thickenings ("ornamentations") in the form of rings, helices or reticulum have ever been observed in the hydroids of a moss.*

c) Mature hydroids normally lack a living protoplast (Figs. 227, 249, 254, 255, 262, 266 and 273 to 275). However, the persistence of various residual bodies and "lipid" droplets is commonly observed in them (Fig. 272). The nature and significance of the "lipid" droplets, and to what extent they are related to the residual bodies are not known. One possible cause for the persistence of these inclusions lies perhaps in the fact that protoplasm autolysis is not as thorough in hydroids as in tracheary elements of most vascular plants. Another possible cause is the lack of a strong transpiration stream capable of eliminating them, this being partly related to the lack of a true root system in moss gametophytes (cf. e.g. Necker 1775, Vaisey 1887b).

From the structural and developmental evidence summarized above, it is clear that in spite of one major difference – the lack of lignified secondary "ornamentations" –, hydroids have much in common with the tracheids of primitive vascular plants.

□ Associated elements

The central strands of hydroids of the leafy stems and setae are surrounded by living tissues. In addition, a leptom may or may not exist at the periphery of the hydrom (see above). In any case, the internal cortex of the gametophytic axes and setae acts as a conducting parenchyma. In the leafy stems of various highly developed polytrichaceous mosses, the cells which surround the hydrom have been interpreted as a *"hydrom-sheath"* (cf. Tansley & Chick 1901, Guttenberg 1943) (Fig. 111). Under the electron microscope, these elements present a well-structured protoplast (Fig. 274). Starch grains are frequently seen in their plastids, as well as lipid droplets in their cytoplasm. In *Polytrichum commune*, various enzyme activities (phosphatases, peroxidases, respiratory enzymes) have been revealed in these cells (Figs. 333 to 335). Interestingly, in the course of development, these activities become intense approximatively at the time when the protoplast of the hydroids degenerates (Fig. 143).

In polytrichaceous gametophytes, stereids are intermixed in certain circumstances with the hydroids of the central strand (cf. pp. 8 and 36). These stereids are living cells; a nucleus and various organelles (especially plastids, in which starch grains are frequently seen) are identifiable in them (Fig. 227). Their lateral walls are usually thickened to various degrees and a strengthening role has been allegedly attributed to them. The fact is that when such stereids are present, the hydroids of the median portion of the strand do not show thick lateral walls – as is frequently the case in these mosses when stereids are absent (compare e.g. Figs. 227 and 275). In *Dawsonia superba* var. *pulchra*, an acid phosphatase activity has been revealed in the stereids of mature portions of the strand (Hébant 1973b).

The hydroids of leaf nerves and leaf traces of mosses are usually associated with cells of conducting parenchyma (= "deuters" or "eurycystes"). Within leaf nerves, stereids may also be present, but they frequently disappear when the leaf traces penetrate the ground tissues of the stems. In Polytrichales, "deuters" and "socii" accompany the hydroids (cf. pp. 31 and 43) (Figs. 108 and 247 to 249). In the most developed forms, the socii apparently show some resemblance to the parenchyma cells of the "hydrom-sheath", and are therefore quite distinct from the deuters. In *Atrichum undulatum*, however, which is a less developed polytrichaceous moss, deuters and socii of leaf nerves are more alike, and seem both to correspond to conducting parenchyma cells.

■ Leptoids and associated elements

□ Development of leptoids *sensu stricto*

Only a few developmental studies are available, which deal primarily with gametophytic leptoids. Here again, species of *Polytrichum* were the most extensively investigated material (Eschrich & Steiner 1968ab (1), Hébant 1967b, 1970a, 1973e and 1976b). Some published work however concerns *Atrichum undulatum* (Stevenson 1974), as well as *Dendroligotrichum dendroides* and other forms (Hébant 1976ab).

Within apical derivatives, the young leptoids are easily identifiable by their precocious elongation and the great development of their nucleus and nucleolus (Figs. 69, 137 to 141 and 284). In *Polytrichum commune*, the protoplasts of the very young food-conducting cells present ultrastructures which do not differ markedly from those of the neighbouring parenchyma cells. One may note however the smaller size of leptoid plastids; although grana may be present in them at this stage, they are never particularly well-developed, as may be the case in the neighbouring parenchyma cells. This, together with the relative importance of their nucleus and the precocious appearance of inclusions similar to the refractive spherules of pteridophyte phloem, facilitates the identification of leptoids in *P. commune* under the electron microscope.

A number of cytological events characterize the differentiation of the leptoids *s.str.*, which may be summarized as follows :

○ The cells first elongate actively. In the meantime, their end walls, at first transverse, become inclined to various extents (Figs. 137 to 141). Towards the end of the elongation phase, thickening of the lateral walls occurs; dictyosomes are involved in this phenomenon (Hébant 1970a). The lateral walls of the leptoids frequently become thicker than those of the neighbouring parenchyma cells. The end walls of the young leptoids are perforated by a great many plasmo-

(1) Unfortunately Eschrich & Steiner did not recognize many sieve-element-like characteristics of the leptoids, one possible cause for this being that the specimens available to them were underdeveloped (see pp. 9-10).

desmata. During development an enlargement of these plasmodesmata occurs which is initiated in their median region : "median cavities" may thus be established (Fig. 292). Further enlargement may lead to the formation of small "pores" (Fig. 293), but this occurs only in the most highly developed leptoids, and not necessarily in all the leptoids of a single specimen. The diameter of these "pores" is never very large (less than 0.2 μm in the specimens studied).

○ The nucleus of the leptoids enlarges considerably during the early stages (in *P. formosum*, it has been shown to become endopolyploid; Hallet 1972) and remains conspicuous during the entire elongation phase (Figs. 137 to 141 and 284); but subsequently it may present symptoms of degeneration. – Such a late degeneration is observed in the most differentiated leptoids only; sometimes the whole nucleus diminishes in size, thus giving rise to a small dense "pycnotic" body (Figs. 142 and 301). In other cases the nucleus remains rather large or even swells, whereas the nuclear sap becomes progressively lighter (*chromatolysis*) (Fig. 303). However clumps of dense chromatin may persist, preferentially appressed to the nuclear envelope. Various types of degenerated nuclei intermediate between the two types described above may be found (e.g. pycnotic nuclei with a clear median zone). Degenerating nuclei of variable morphology may coexist within one single gametophyte, although one type frequently predominates, depending on the species, its ecology, etc.

Since, as in vascular plants (Esau & Gill 1973), degeneration occurs in recognizably graded steps, and nuclei showing the structures described above are found in apparently little disturbed leptoids, there is an increased probability that the features observed here are not merely artifacts.

○ During the maturation of leptoids, a partial (controlled) autolysis of their protoplast occurs. That a correlation exists between this phenomenon and the degeneration of the nucleus constitutes a reasonable hypothesis. Autophagic vacuoles were also seen in the vicinity of highly active dictyosomes in differentiating leptoids of *P. commune* (Fig. 308). These phenomena are certainly complex since persistence of various organelles, although in a modified form, is observed at the same time as autolysis of hyaloplasm and disappearance of ribosomes, dictyosomes, etc. Plastids are among the persisting organelles. In the course of differentiation of leptoids *s.str.* a regression of their thylakoids is observed, so that no grana are normally seen in plastids of mature leptoids. Mitochondria are also present at all stages of the development of the food-conducting cells.

○ The endoplasmic reticulum of the very young leptoids is similar to that of neighbouring parenchyma cells. In particular, it shows comparable relationships with plasmodesmata in both types of cells. During more advanced stages of differentiation this ER becomes extensively developed, and stacks soon begin to appear (Fig. 308). Various forms of aggregated ER, such as stacked saccules, convoluted tubules or complicated networks are progressively established. They persist in mature leptoids.

Ribosomes are commonly seen attached to the ER of young leptoids but they progressively disappear and in fully differentiated elements the endoplasmic reticulum is normally found to be smooth (cf. e.g. Figs. 294 and 306).

○ Inclusions similar to the "refractive spherules" of pteridophyte phloem have been occasionally observed in the leptoids of several species of Polytrichales, but the development of these organelles was more particularly studied in a strain of *Polytrichum commune* where they are abundant. In the young leptoids of this moss, the "spherules" seem to originate in direct association with the ER (Fig. 304). Very early during development tubules or sheets of ER become closely appressed to them. Ribosomes are frequently identifiable on the outer membrane of these sheets (i.e. not on the membrane closest to the "spherules"). In more aged leptoids, sheets of smooth ER may persist appressed to the "spherules" (Fig. 305). Although dictyosomes were frequently seen in the vicinity of the "spherules" in differentiating leptoids, no definite evidence of direct participation of these organelles in their formation was obtained.

The developmental features summarized above, which concern leptoids *s.str.* only, are strongly reminiscent of those known for the sieve elements of vascular plants (summarized e.g. by Esau 1969, and Srivastava 1975).

The differentiation of the many cells with an "intermediate" organization which are found in the leafy stems of Polytrichales (cf. pp. 30 and 44) involves patterns which are less specialized than those described for leptoids *s.str.* Phenomena like regression of the plastids, or a certain development of the ER, may however be observed in them. This concerns in particular certain cells located on the adaxial side of the leaf traces and which, in the most highly developed Polytrichales, establish the junction between the "deuters" of the leaf traces and the typical leptoids of the central system (cf. p. 42 *sqq.*).

□ Structure of leptoids *s.str.* at maturity

On the "mature" condition of leptoids : Leptoids will here be termed "mature" when their protoplasm has reached, at the end of their development, an organization which is both highly specialized and relatively stable. However it is known that leptoids, like sieve elements of vascular plants (Kollmann 1965), can translocate before this stage is reached. This was experimentally shown by Eschrich & Steiner (1967) who observed a heavy labelling of the leptom region in still immature portions of gametophytes of *Polytrichum commune*.

Maturation of leptoids in *setae* has been very little investigated. In *gametophytes*, "mature" leptoids are normally found in those regions of leafy stems which appear brown or reddish-brown; "immature" leptoids are found in younger, greenish portions of stems. The distance from the apex to the regions where "mature" leptoids are found depends upon many factors (species, stage of development, environment, etc.) and may vary within wide limits, from several millimeters to a few centimeters. In all cases, the hydroids complete their differentiation before the leptoids (Fig. 143). A delayed maturation of the phloem, as compared with that of the xylem, is also known from certain vascular plants (Esau 1969). In this respect, I have observed that in a strain of *Lycopodium clavatum* collected during the period of active growth, most of the sieve elements of the metaphloem were still immature 5 cm below the apex.

Gametophytic leptoids : When in their most highly differentiated form, the leptoids *s.str.* found in the gametophytes of certain Polytrichales share an impressive ensemble of characteristics with the sieve elements of true vascular plants. Their main features are summarized below [(1)] :

○ They are elongated elements, which may show somewhat enlarged extremities (giving them a "tibia-like" appearance) (Figs. 142, 285 and 286).

○ Their end walls are inclined to various extents, sometimes considerably so [(2)]. A great many plasmodesmata, which frequently show a "median nodule", exist in these walls (Figs. 288, 289 and 292). As mentioned earlier, these plasmodesmata may enlarge to such an extent as to approach the structure of small "pores", but the diameter of such pores is never very large (less than 0.2 μm in the specimens studied) (Fig. 293). Although callose has been detected by fluorescence microscopy in end walls of leptoids (Fig. 290), its precise location at the ultrastructural level remains problematical.

○ Their lateral walls are usually thicker than those of the neighbouring parenchyma cells. "Nacreous" walls have even been observed in some species belonging to the genus *Atrichum*, and to the section *juniperina* of the genus *Polytrichum*. However, in many cases these thickenings do not differentiate the leptoids from the contiguous parenchyma cells. Fig. 309 for instance shows a typical parenchyma cell in *Atrichum undulatum*, in which a chroroplast with large starch

(1) This description refers to the most highly differentiated leptoids only. Such leptoids are not found in all mosses, but only in some Polytrichales. Even the leptoids of certain large polytrichaceous gametophytes (e.g. various species of *Dawsonia*) do not show such a high level of specialization. The characteristics of sieve elements in vascular plants are thoroughly reviewed in Esau (1969) and Aronoff et al. (1975).

(2) In *Dendroligotrichum dendroides*, the leptoids at the periphery of the central strand possess such steeply inclined end walls that they have been interpreted as "hydroids" by Lorch (1931, p. 74). Similarly, fig. 2 in Scheirer (1972) (part of this figure reappears in Scheirer 1973, p. 44) most probably correspond to such leptoids, and not to hydroids as claimed in the caption. These cells were identified as leptoids for the first time by Schulz (1937).

grains is identifiable, and which nevertheless presents thick "nacreous" lateral walls [(1)].

Although apparently not very abundant, connections are seen between all cells of the leptom in their lateral walls. Plasmodesmata exist between parenchyma cells and leptoids *s.str.*, and between the leptoids themselves (Figs. 287 and 295). Plasmodesmata have also been observed to occur laterally between those leptom cells which possess "nacreous" thickenings (Fig. 296). ER is frequently seen in close connection with the plasmodesmata in all these different cells.

○ A nucleus persists in aged leptoids *s.str.* Developmental and cytological evidence (see above) shows that this nucleus is degenerating (Figs. 142, 301 and 303). This interpretation has been questioned (Eschrich & Steiner 1968ab), but experimentation has shown that in *P. commune* the ability of such nuclei to incorporate tritiated uridine is seriously affected (Hébant 1973e). Similar results have been obtained on *Polytrichadelphus magellanicus* (Fig. 302). Polyribosomes are of course also absent from aged leptoids.

○ As a result of the partial (controlled) autolysis of their protoplast, the contents of mature leptoids *s.str.* appear clearer than those of the adjacent cells under the light microscope (Figs. 285 and 286).

○ Mitochondria, small plastids, various vesicles and lipid droplets persist in mature leptoids *s.str.*, but no dictyosomes are seen. The structure of the plastids is striking : only vesicles, and frequently looped invaginations of their inner membrane persist in the stroma, where plastoglobuli are also recognizable (Figs. 291, 297, 299, 300 and 303) [(2)]. In aged leptoids *s.str.*, the plastids are sometimes so regressed that after conventional staining techniques some sections through them are no longer easily distinguishable from sections

(1) It must also be pointed out that if all the characteristics of the tissues are taken into consideration, the leptom of the mosses with nacreous thickenings that I have investigated up to now may not be the best differentiated of all amongst Polytrichales.

(2) Examination of a large number of plastids in "adult" leptoids however sometimes reveals the occasional persistence of small grana (Fig. 298) or small starch grains. Spiral or reticulate structures are also seen.

through the neighbouring mitochondria.

Inclusions similar in appearance to the refractive spherules of pteridophyte phloem have been seen in several species (especially *P. commune*) (Fig. 305). The exact nature and physiological significance of these structures in bryophytes and pteridophytes are not definitely known. Their ultimate fate (disappearance ?) is far from being clearly understood. Further investigation in this field is needed.

○ Most remarkable is the extensive system of endoplasmic reticulum which is encountered in the leptoids at certain stages of their active life (Figs. 291, 294, 306 and 307). This ER shows associations with the connections in the end walls (Figs. 293 and 294) and with the plasmodesmata in lateral walls. Sheets or tubules of ER are also frequently seen appressed to leptoid plastids (Figs. 298-300) and "spherules" (Fig. 305). Flattened saccules frequently occur near the plasmalemma (e.g. in Figs. 294 and 299). Stacks and various other forms of aggregated ER persist for some time in mature leptoids (for details, see Hébant 1976b).

Leptoids in setae : As far as indisputable "leptoids" of setae are concerned, their ultrastructure is known only from two rather cursory studies dealing with Polytrichales (Favali & Bassi 1974, on *Polytrichum*; Hébant 1975a, on *Dawsonia* and *Dendroligotrichum*) [(1)]. In none of these studies was the preservation of structures really satisfactory, and the question remains unanswered as to whether this was the result of poor fixation, or if the protoplast of the leptoids was degenerating at this stage. In all the forms investigated, a great many plasmodesmata were seen in the more or less oblique end walls of the leptoids. In addition to a nucleus, small plastids, mitochondria and some endoplasmic reticulum were identified in their protoplasts. The plastids of the leptoids contrast with those of the parenchyma cells of the inner cortex by their simpler structure and the frequent

(1) In the opinion of the present reviewer, the cells described as "leptoids" in the setae of *Tortula muralis* (Favali & Gianni 1973), *Mnium orthorrhynchum* (Bassi & Favali 1974) and *Funaria hygrometrica* (Schulz & Wiencke 1976) would probably be better interpreted as elements of conducting parenchyma (see pp. 46-49).

absence of starch in their stroma (plastids of parenchyma cells may accumulate large starch grains).

Further electron microscope studies of leptoids in setae are needed.

Comparison of leptoids with sieve elements of vascular plants : When comparisons between food-conducting cells of bryophytes and tracheophytes are made, it must be kept in mind that a marked diversity exists in the sieve elements of the latter (cf. Esau 1969, Aronoff et al. 1975). With reference to such structural features as the lack of P-protein, the retention of a degenerated nucleus, the organization of the "pores" and their relations with the ER, the resemblance of the leptoids *s.str.* of the most highly differentiated polytrichaceous gametophytes to the *protophloem* elements of certain angiosperms (cf. e.g. Parthasarathy 1974ab) and to the sieve elements of gymnosperms (cf. Kollmann 1964, Neuberger & Evert 1974) is particularly impressive. The resemblance of leptoids to protophloem elements is not surprising as both are involved in short- or relatively short-distance translocation. Extension of knowledge on the phloem of pteridophytes will perhaps make possible further comparisons between bryophytes and vascular plants.

□ Elements associated with leptoids

In the setae of various Polytrichales, the "leptoids" are not intermixed with cells of parenchyma (cf. p. 47). On the contrary, in the gametophytes of the same mosses, various kinds of parenchyma cells are associated with the leptoids. In the leafy stems of a highly developed strain of *Polytrichum commune*, elements showing ultrastructures reminiscent of those of the parenchyma cells associated with sieve elements in certain pteridophytes (*Polypodium vulgare*, Maxe 1965; *Lycopodium cernuum*, Warmbrodt & Evert 1974) have been found in the vicinity of the leptoids *s.str.* (Hébant 1972b); the other neighbouring elements were either ordinary parenchyma cells or elements of conducting parenchyma specialized to various extents.

Elements with structures "intermediate" between those of leptoids and of parenchyma cells were also found in the vicinity of leptoids *s.str.* in various other Polytrichales (*Polytrichadelphus magellanicus, Polytrichum spp., Dendroligotrichum dendroides,...*). When highly differentiated leaf traces are present, part of these elements establish the connection between the deuters of the traces and the leptoids of the central system (cf. pp. 42-44). However, this was studied in some detail in *Polytrichum commune* only (Hébant 1972b) and further correlated light and electron microscope studies on a wide range of species are desirable.

□ A note on structural variability in leptom

From the evidence summarized in various sections of the present review it is clear that a tremendous structural variability exists within food-conducting cells in bryophytes. This variability depends upon the forms considered, their stage of development, the phase studied (gametophytic or sporophytic), etc. Sieve element-like leptoids are encountered in certain highly differentiated polytrichaceous gametophytes only. A wide range of "intermediary" types of organization are seen in the gametophytes of many other less differentiated forms. Elements with an "intermediate" structure also accompany the leptoids *s.str.* of the most highly developed species. Finally, mosses without any leptom at all and presenting only cells of conducting parenchyma are, and by far, the most commonly encountered.

Such an impressive structural variability has important physiological implications (cf. p. 97).

6.3. THE LIGNIN PROBLEM

The occurrence in various mosses and liverworts of specialized water-conducting cells, some of which, by their development and structure, may be strikingly reminiscent of the tracheary elements of vascular plants, renders the lignin problem a particularly vivid one in bryophytes. Are some of the aromatic polymers which occur in the cell walls of many mosses and liverworts to be classified as "lignins"?

This question, which has aroused much interest and controversy since the second half of the 19th century, is perhaps close to being solved now.

■ The lignin concept

What is to be regarded as a bryophyte lignin ? A point which must be stressed at this level is the initial parallelism of those metabolic pathways which lead not only to lignins, but also to other polymeric plant phenolics (cf. e.g. Walker 1975). End products may present close structural similarities which render their identification more difficult. Moreover, the fact that lignin and flavonoid biosynthetic pathways have much in common is also of interest (both kinds of coumpounds are present in all vascular plants), since flavonoids have been identified with certainty in bryophytes (Harborne et al. 1975, Swain 1975).

Sarkanen and Ludwig (1971) point out that "the exact definition and the differentiation of lignins from other polyphenolic plant constituents has remained a matter of debate until very recently". According to them, true lignins are "polymeric natural products arising from an enzyme-initiated dehydrogenative polymerization of three primary precursors : *trans*-coniferyl 1, *trans*-sinapyl 2 and *trans-p*-coumaryl 3 alcohols.

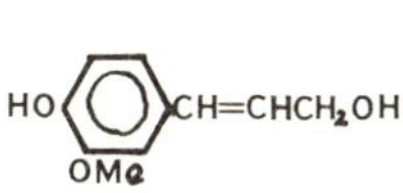

1

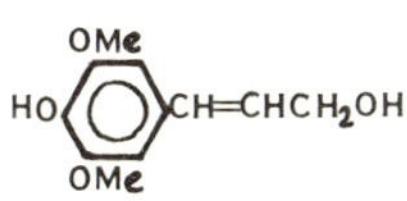

2

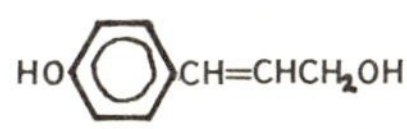

"

3

■ Methods of identification

A variety of methods are used for the identification of lignins in plant tissues. They are of unequal value and the contradictory reports found scattered in the litterature result largely from the fact that widely different methods were used. According to Sarkanen and Ludwig (1971), isolation of the so-called "Hibbert's ketones" from the ethanolysis products of the specimens is one of the most reliable criteria. A perhaps even more selective procedure, which is referred to as "oxidative degradation", has recently been devised by Erikson et al. (1973). Among histochemical tests, the phloroglucinol-HCl procedure (Wiesner 1878) is considered by Sarkanen and Ludwig (1971, p. 6) of "universal applicability to all lignins, although the reaction may be weak or even absent in lignins containing high amounts of syringyl-propane units". Since the latter are mostly "advanced" lignins, found principally in angiosperms, the Wiesner test can be considered of real value for such "primitive" land plants as the bryophytes. Other methods, especially physical ones such as U.V. or I.R. spectrophotometry, are also of interest when used in conjunction with chemical procedures. On the other hand, various classical botanical stains such as safranin, iodine green or toluidine blue have little histochemical value as far as the precise identification of lignins is concerned. At the most, they can only be indicative of the possible occurrence of such compounds.

▪ Evidence for and against the presence of lignins in bryophytes

There is a general agreement that aromatic polymers are of common occurrence in the cell walls of bryophytes (for reviews, see e.g. Pottier 1923, Sarkanen & Hergert 1971, Suire 1975), but a controversy exists as to whether or not these polymers, or part of them, may be termed "lignins".

Several early reports of the presence of lignins in bryophytes (e.g. by Tansley & Chick 1901, Cavers 1903, Mader 1929) are apparently based on inaccurate staining procedures, or on shortcomings in the application of more specific tests. In reference to this, J.L. Smith (1966) noted that "apparently positive results" could be obtained when the Wiesner test was applied to thalli of *Symphyogyna circinata*; but when sections of the same thalli were treated with the acid component of the test alone, without the phloroglucinol, red and yellow coloring was also obtained in the conducting cell walls. Smith concluded from his investigation that these walls were not lignified. Indeed, a number of workers have reported negative Wiesner reactions for a variety of bryophytes, both mosses and liverworts (Gjokić

1895, Linsbauer 1899, Finlayson 1950, Mueller 1973, Hébant 1974 and 1976a).

In spite of this, several claims for positive phloroglucinol-HCl reactions in bryophytes have appeared within recent years :

- in *Polytrichum* peristome teeth (Siegel, 1962)
- in the rhizoids of *"Jungermannia sp."* (Siegel, 1962)
- in *Dawsonia* gametophytes (Siegel, 1969)
- in *Dendroligotrichum* gametophytes (Siegel, 1969; Scheirer, 1972)
- in *Sphagnum spp.* (Bendz, Martensson & Nilsson, 1967).

In fact, these claims are contradicted by other reports, old and recent. The present reviewer has tested a very comprehensive range of bryophyte material, including the above, within the past ten years, without ever obtaining one convincing positive result. Controls were run at the same time with "primitive" vascular plants which always yielded strongly positive tests (cf. Hébant 1974). It is also to be noted that some of those authors who at first reported successful Wiesner reactions, subsequently expressed reserve. According to Siegel (pers. comm., 1973), positive results in *Dawsonia* and *Dendroligotrichum* "may have been an artifact of the history of the material", one alternative interpretation being "that there are in fact strain or race differences in lignification" in these plants (1). In a recent paper, Scheirer (1975, p. 121) states : "my conclusion is now considered precocious and the natural red coloration of the cell wall was misinterpreted as a positive Wiesner reaction". It has also been suggested (Hébant 1974, p. 584) that "since *Dawsonia* and *Dendroligotrichum* are frequently found on humus-rich forest floors, a positive lignin reaction could also be the result of contamination of their apoplastic continuums by degradation products of higher plant lignins".

Thus, no truly convincing evidence for the presence of lignin in bryophytes seems to have emerged from a large-scale application of

(1) However I have consistently obtained negative results with specimens from a variety of countries and environmental conditions, including a large bulk of New Zealand forms (especially : *Dawsonia superba* var. *superba* and *Dendroligotrichum dendroides*).

the Wiesner reaction, which is considered one of the most reliable histochemical tests presently available.

A number of attempts towards isolation and identification of bryophyte lignin, through a variety of chemical procedures, have also been made (recent reviews appear in Sarkanen & Hergert 1971, Erickson & Miksche 1974b, Suire 1975). According to the different investigators, the isolated compounds have been variously termed, – e.g. "lignins", "pseudolignins" or "non-lignin phenolic polymers". Recent reviewers on the subject remain highly cautious as to the final interpretation which should be attributed to these contradictory claims. In this respect, one can note that similar attempts towards isolation of "lignins" have even been extended to Algae (Salo 1957, Dudkin et al. 1971). If presence of "lignins" in certain algae proved to be true, then these compounds would no longer appear so important as a distinctive characteristic of vascular plants.

Where highly selective procedures have been applied to bryophytes, they have yielded negative results. Sarkanen and Latif have established the absence of Hibbert's ketones in the reaction products obtained from ethanolysis of the moss *Rhytidiadelphus loreus* (unpublished; quoted in Sarkanen & Hergert 1971, p. 81). A decisive contribution was recently made by Miksche and his co-workers, through the use of a "oxidative degradation" procedure. Erickson & Miksche (1974b) established that six moss species, including *Polytrichum commune* and *Sphagnum sp.*, and two species of liverworts, were clearly devoid of lignin, but contained other types of phenolic cell wall material. Since then, Miksche & Yasuda (in preparation) were able to show that *Dendroligotrichum dendroides* and *Polytrichadelphus magellanicus*, both from New Zealand, as well as several species of *Dawsonia* (including *D. superba* var. *superba* from New Zealand) arealso devoid of lignin. According to Miksche & Yasuda (in preparation), the giant forms (*Dawsonia superba, Dendroligotrichum*) do not appear to be fundamentally distinct from the other polytrichaceous mosses by the structure of their cell wall polymers. Thus, and as far as bryophytes are concerned, Siegel's hypothesis according to which "lignin distribution did not evolve strictly on systematic-phylogenetic lines, but included a capacity for gravitational/mechanical response as well" (Siegel et al. 1972, p. 224) has not been confirmed by recent investigations.

Chapter 7

A STRUCTURE OF PARTICULAR IMPORTANCE IN BRYOPHYTES : THE CONTACT-ZONE BETWEEN GAMETOPHYTE AND SPOROPHYTE

A number of early investigators noticed that distinctive patterns exist at the level of the interphase between gametophyte and sporophyte in bryophytes (cf. plates 34 and 74 to 76). Indeed, some of the first descriptions ever made with the light microscope of *transfer cells* refer to this region; – as pointed out by Pate & Gunning (1972) certain of these descriptions could "scarcely be bettered to-day" –. A review of these early works appears in Blaikley (1933). Special mention must be made of the publications of Rosander (1906) who investigated a large number of mosses, and of Lorch (1925ab, 1931) who produced some remarkably accurate drawings of transfer cells, in which the wall ingrowths are clearly identifiable (Figs. 120, 122 and 124). Motte (1928), in the course of a cytological investigation of mosses, also noted the great abundance of mitochondria within cells of the outer region of the "foot"(1) of *Grimmia.*

However, a precise knowledge of the structure of the foot and of surrounding gametophytic tissues, as well as the concept of the general occurrence of transfer cells (2) at this level, have come to light only within recent years, as the result of several ultrastructural investigations made on a variety of bryophytes (Monroe 1965, on *Funaria*; Eymé & Suire 1967, on *Mnium*; Maier 1967, on *Polytrichum*; Ridgway 1968 (3); Kelley 1969, on the liverwort *Sphaerocarpus*; Hébant 1975a, on *Dawsonia* and *Dendroligotrichum*; Wiencke & Schulz 1975, on *Funaria*). Transfer cells were seen in all investigated species, although variations were recorded concerning their relative development and location. It seems that some at least of these variations

(1) Foot = basal region of the sporophyte, deeply inserted within gametophytic tissues.

(2) For review of general characteristics of transfer cells, see Gunning & Pate 1969, Pate & Gunning 1972.

(3) Ridgway did not specify which bryophyte material he investigated.

may be dependent on the stage of development and on the environmental conditions; experimental evidence for this was provided by Maier & Maier (1972) who increased the volume of wall ingrowths in the haustorium of *Polytrichum* by adding monosaccharides to the culture media. In several investigated forms, transfer cells were found to be present on both sides of the interphase, viz. in the gametophyte and in the sporophyte (e.g. *Mnium*, Eymé & Suire 1967, cf. Fig. 317; *Sphaerocarpus*, Kelley 1969). In others, such as *Polytrichum* (Maier 1967), *Dawsonia* and *Dendroligotrichum* (Hébant 1975a) they seem to exist only in the foot of the sporophyte (Figs. 320 and 321) (1). The reverse situation, viz. transfer cells in the gametophytic side of the placenta only, has been reported by Gunning & Pate (1969) for *Anthoceros laevis* (see also Figs. 172-173). This seems rather unusual and would deserve further investigation.

Wiencke & Schulz (1975) state that specialization occurs in the haustorium of *Funaria hygrometrica*. According to them, the basal part plays a prominent role in the uptake of water, whereas the nutrients are absorbed preferentially in the middle part.

No plasmodesmatal connection exists between the two generations (Kienitz-Gerloff 1902, Piskernik 1914, and many authors after them). On the other hand, the wall ingrowths in the transfer cells greatly increase the surface area of the plasma membrane (cf. Fig. 318), and therefore facilitate transport of solutes from the cells of the gametophyte to the foot cells of the sporophyte (Ridgway 1968). Phenomena of active (energy requiring) transport certainly occur at these levels, as attested by the abundance of mitochondria in the cells (Fig. 319). Extensive systems of endoplasmic reticulum have also been observed in the foot cells of certain Polytrichales (Maier & Maier 1972, Hébant 1975a). The important development of a specialized ER has been similarly reported from various other cell types implicated in transport phenomena (cf. Kollmann 1964, Dörr & Kollmann 1975, etc.). In several species, the nuclei of certain cell types were

(1) Inner cells of vaginula (= part of gametophyte) in *Polytrichum* were erroneously interpreted as transfer cells by Hébant (1973b) on the basis of a light microscope investigation of enzyme activities. In fact these cells are degenerating.

found to be of large size and lobed; these nuclei may eventually prove to be endopolyploid.

In liverworts, the upper region of the foot, and the stalk (seta), consist of what appear to be cells of conducting parenchyma. In mosses, more complex patterns may be observed, especially in those species whose sporogonium possesses a well developed conducting tissue-system, comprising both hydroids and leptoids. Several detailed studies are available, which concern *Polytrichum* (Maier 1967, Maier & Maier 1972), *Funaria* (Wiencke & Schulz 1975), *Dawsonia* and *Dendroligotrichum* (Hébant 1975a). In all these mosses, the extreme tip of the foot penetrates directly into the central strand of the leafy stem. At the end of the foot an appendage (1) is seen whose cells are usually more or less necrotic (Figs. 121 and 311). The central strand of hydroids in the sporogonium originates from just above the appendage. The latter is thus the only barrier to complete continuity between the water-conducting strands of both generations. Hydroids from lower regions of the strand in the sporophyte are shorter and wider, and correspond to the "water-absorbing organ" described by Vaisey (1888) in *Polytrichum* (Fig. 121). In upper regions of the foot, as well as in the seta, they normally become more elongated. Cells of conducting parenchyma establish the connection between the transfer cells at the periphery of the foot and the leptoids of the axial system. They are also in relation with the cells of the inner cortex, which probably intervene in transport phenomena. The most complex organization was found in the foot of the Polytrichales (especially of *Dawsonia superba*), where lacunae are present, contributing to the formation of a "sponge-like" absorbing organ (Fig. 320). Electron microscope investigations on mosses simpler in structure than those studied above are needed.

The structural evidence summarized above suggests that the interphase between gametophyte and sporophyte is a particularly active

(1) This appendage corresponds to the lower part of the embryo which apparently plays an important role during the early stages of sporogonium development, especially during its implantation within gametophytic tissues (cf. Fig. 310).

zone, which plays a fundamental role in conduction in the sporophyte. Indeed, physiological data (see p. 98) show that in many bryophytes the foot acts as a "pump" driving water and substances towards the capsule during its development.

PART II

Physiology of Conduction

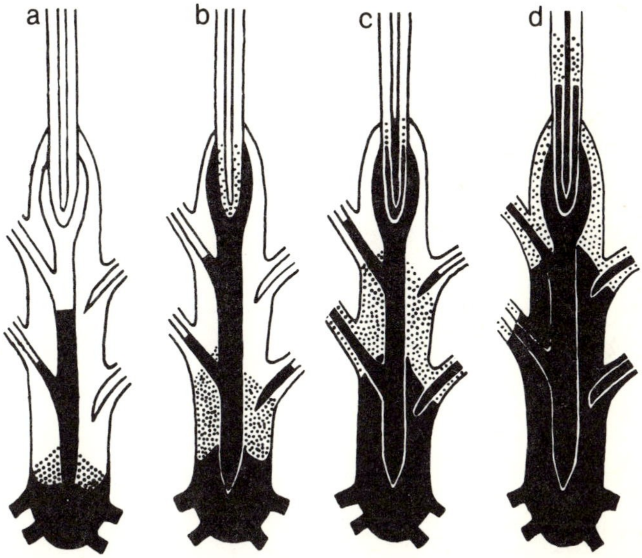

(BOPP & WENIGER 1971)

Chapter 8

WATER RELATIONS [1]

The metabolic activity of an organism is in direct relation to its hydration. It is commonly assumed that growth in bryophytes is effective only when the plants are fully turgescent, or approaching such a state (Anderson & Bourdeau 1955). However, a great many bryophytes are *poikilohydrous* [2] (Walter & Stadelmann 1968) which means that they undergo alternate periods of drought and humidity, as they are largely dependent on atmospheric conditions for their water supply. In this context one may further note that most bryophytes lack absorptive organs which penetrate deeply into the soil and that they also lack, at least in their gametophytes, a system efficient in controlling transpiration. Indeed, in these plants the dominant (gametophytic) phase behaves differently from the dominant (sporophytic) phase of vascular plants in its water relations. In addition to this, the structural and physiological aspects of drought resistance in liverworts and mosses are still incompletely understood (Abel 1956, Mahmoud 1965). Thus many uncertainties persist which must be kept in mind when the physiology of water conduction in bryophytes is studied. In the first place, several well established concepts concerning water economy in higher plants cannot be directly transposed to these plants.

Not only internal conduction, but also – and perhaps to a greater

(1) This brief review has primarily been prepared as an introduction to future investigations. It will rapidly become clear to specialists while reading this that much more work remains to be done. In particular, various modern concepts of fundamental importance, for instance that of water potential, have been very little applied to bryophytes (cf. Bayfield 1967 and 1973, Klepper 1963 and 1968). (For a modern introduction to water economy in plants, see Meidner & Sheriff 1976).

(2) This term, which parallels the well known term "poikilothermous", is easily understood and will be retained here. "Poikilohydric" has also been proposed. The term "pollacauophyte" (Buch 1947) is accepted as meaning nearly the same.

extent – external conduction play a fundamental role in the water economy of bryophytes. Although both types of conduction have been extensively investigated, their relative importance and their relationships under various environmental conditions remain a matter of debate. In addition to this, certain important aspects such as the speed of internal conduction in intact (uncut) gametophytes, or the rate of penetration of water and solutes in underground axes (if present), remain practically uninvestigated.

For all these reasons, only a very superficial and provisional account of water conduction in bryophytes can be presented here.

8.0.1. Evidence for external conduction in gametophytes

Only gametophytes are considered here since the sporophytes are not adapted to this type of conduction. However it may be noted that during the early stages of sporogonium development water can circulate by capillarity between the latter and the surrounding gametophytic tissues (especially the accrescent archegonium) (Clee 1939, 1943).

▪ Mosses

Circulation of water along capillary channels is readily apparent in mosses upon direct observation. Indeed, as early as 1787 Hedwig was able to describe the ascent of water in the stem of *Hedwigia ciliata* (op. cit., p. 109) (cf. also Fig. 336). According to K. Schimper (1857; cf. also his poem "Mooslob") external conduction is the only effective path for water supply in mosses : "Die Moose führen Feuchtigkeit nicht im Innern des Stengels in die Höhe, sondern aussen durch die Capillarräume, welche die dem Stengel anliegenden Blätter bilden". Mention must also be made of the Sphagnales, although they stand apart amongst Musci due to their many structural and biological peculiarities (cf. p. 17). The fact that they may act as "wicks" – in fact this is more than external conduction – has long been noted (W. Schimper 1857). W. Schimper (op. cit, p. 26-27) writes : "Une tige de Sphaigne, haute de plusieurs décimètres, plongée à sa partie inférieure, garnie de ses rameaux, dans un flacon d'eau, videra ce flacon en très-peu de temps en déversant l'eau par son capitule, quand on a eu soin d'incliner celui-ci un peu sur le côté. On comprend que la réunion de millions de systèmes hydrauliques de ce genre, comme cela se voit dans les grandes tourbières, doit produire des effets considérables".

Following precursory observations such as those quoted above, a number of experimental investigations have firmly established the effectiveness of the phenomena of external conduction in mosses. Oltmanns (1884, p. 13) claimed that no water movement was apparent in stems of *Dicranum* and *Hylocomium* kept in a relatively dry environment (60-70 % of relative humidity) and with their bases immersed in water, if external conduction was prevented [(1)]. Bowen (1931, 1932) employed a large number of "markers" to follow the rise of water over the stems of a variety of mosses. Among these markers were *Lycopodium* powder, stains (e.g. eosin or gentian violet), colloidal solutions (dialysed iron), sodium chloride, lithium sulfate (portions of the plants taken at various levels were subsequently incinerated, the flame being examined spectroscopically for the lithium line) and potassium nitrate (sections in the plants being subsequently mounted in diphenylamine–H_2SO_4, which gives a deep blue color reaction with KNO_3). The rates of external conduction (from a few millimeters to a few centimeters in the first hour) and the amount of salt solution risen over the external surface were determined for diverse mosses from various habitats. Although Bowen's experiments have been criticized for various reasons (cf. e.g. Mägdefrau 1935; – one may note in particular that the relative humidity in them was never quoted–), the results she obtained clearly establish the possibility of an efficient external conduction in a number of mosses. Bowen also attempted to determine the rate of internal conduction in the same species (cf. p. 89). In more precise experiments, in which relative humidity was controlled, Mägdefrau (1935-6) similarly investigated both types of conduction (cf. Fig. 145 and § 8.0.4.) and demonstrated the great importance of external

(1) In fact Mägdefrau (1935-6) subsequently showed that some internal conduction could occur in these mosses, especially in *Dicranum*.

conduction for most mosses. Using a fluorescent tracer, Bopp & Stehle (1957) showed that in *Funaria hygrometrica* the solution risen externally reaches the capillary space between the gametophyte and the base (haustorium) of the sporophyte before the fluorochrome can be identified in the central strand of the upper region of the gametophyte.

It must however be noted that, according to Buch (1945-7) and Anderson & Bourdeau (1955), in many upright-growing mosses the water travelling externally from base to tip of the gametophytic stems can reach a certain level only; this may have important implications in the water economy of the plants if they are too tall [1].

Absorption of water. The importance of the phenomena of external transport over the surface of the gametophytes implies the possibility of the absorption of water and the solutes it contains at various levels of the leafy stems, especially near growing apices. A long time ago, Necker (1775) speculated that moss surfaces were covered with minute "pores" : "ces pores des mousses sont autant de petites bouches sans cesse ouvertes en temps pluvieux, pour absorber leur nourriture...". Since then, a number of authors (e.g. Haberlandt 1886, Vaisey 1887, Rostock 1901, Anderson & Bourdeau 1955, Zacherl 1956) pointed out that water could be absorbed by the leaves. Bowen (1931, 1933) studied the matter in some more detail and showed that whereas water and solutes were readily absorbed in apical regions, penetration through the thick-walled peripheral tissues of the stems was in general much slower. Apices are protected by a mucilage which is secreted by specialized hairs (Berthier et al. 1974) and which also probably plays a role in the retention of water.

▪ Liverworts and hornworts

Foliose hepaticae. Clee (1937) investigated the rate of external conduction in the leafy liverwort *Plagiochila asplenioides* var. *major.*

(1) The very structure of the capillary channels results in the fact that the rise of water over erect stems does not usually exceed a few centimeters. It would seem that in many cases, the main function of external conduction lies in the *redistribution* over the gametophytes of the water from the precipitations.

When the plants were laid prostrate on a glass plate, eosin solutions were seen to spread rapidly towards the apices (Fig. 150). The tips of the plants were regularly reached within one minute, distances up to 3.7 cm been travelled by the stain. Leaf arrangement of the "succubous" (1) type was responsible for this rapid transport. In this plant, water conduction was far less efficient in the "incubous" (1) direction (less than one cm per minute). Clee suggested a possible relationship between the incubous or succubous organizations and the habitat of the forms which present them.

Thalloid hepaticae and hornworts. Although thalloid liverworts are less important than the leafy ones, as far as number of species is concerned, they have been more extensively investigated for external conduction.

Using various solutions (vital stains, KNO_3, $FeCl_3$) and appropriate reagents, Clee (1939) showed that in *Pellia epiphylla* (Metzgeriales) a very rapid external transport of water occurs, whereas internal conduction within parenchyma cells of the ground tissue was far slower. Less than 2 min were necessary for the stain solutions to travel externally from base to tip of the plants studied (length of plants : 2 to 3 cm approx.). According to Clee, the rhizoids on the ventral surface play an important role in this phenomenon, although some water may also pass over the upper surface. According to him, the young sporophyte takes its water supply largely from external sources (Fig. 148).

Isaac (1941) investigated *Anthoceros laevis* and showed that if the plants are kept in contact on their ventral surface with a filter-paper saturated with a vital stain solution, the liquid rapidly passes over the margins of the thallus on to the dorsal surface where it quickly spreads. Although supply of the sporophyte from external sources was here again observed, internal conduction through gametophytic tissues apparently also plays an important role in this respect. In

(1) *Incubous* : leaves so arranged that the base of each is covered by upper portion of next lower; *succubous* : with each leaf covering part of that under it (from Henderson's Dictionary of Biological Terms, 1963). For details, see Schuster (1966).

addition, Isaac found that *Anthoceros laevis* is capable of absorbing water and solutes over the whole of the surface of the gametophyte and sporophyte (Fig. 147).

Marchantiales have been subjected to a number of studies (Bowen 1935, Clee 1939, McConaha 1939 and 1941, Rifot & Barrière 1974). In *Fimbriaria bleumeana*, Bowen (1935) observed a rapid circulation of capillary films of liquids (i.e. eosin solutions), passing up the narrow channels between the scales on the ventral surface of the thallus. She also noted that the water supply to the head of the carpocephalum (reproductive structure) is almost entirely conveyed externally; this conduction is particularly effective in the region of the rhizoidal groove, where "tuberculate" rhizoids are most numerous. In *Conocephalum conicum* water similarly travels in the form of capillary films between the tuberculate rhizoids and the ventral scales (Clee 1939, McConaha 1939). Conduction from base to apex of an average thallus requires only twenty to thirty seconds (McConaha 1939). According to Clee, water is readily absorbed in the apical region. This author also observed the entry of stains via the smooth-walled rhizoids (Fig. 149). More recently, Rifot & Barrière (1974) who experimented with a ^{14}C-sodium-acetate solution similarly attributed a role in water absorption to the rhizoids. In a comparative study of several Marchantiales, McConaha (1941) found that the rate of external conduction could vary in relation to the development of the capillary systems. Whereas in the above mentioned *Conocephalum* and in *Preissia quadrata* this rate is ca. 1 mm per seconde, it is only of 0.5 mm per seconde approximately in *Lunularia cruciata* and *Reboulia hemisphaerica.*

8.0.2 Evidence for internal conduction in gametophytes and sporophytes

■ Mosses

The concept according to which the central strand in the leafy stems and setae of many species is a *conducting* strand emerged relatively early. Schacht (1856) and Unger (1861) described this strand in the gametophyte as a "Gefässbündel", whereas W. Schimper

(1848) went so far as to claim, regarding this structure in the seta : "il est facile de voir circuler la sève dans ses longues cellules cylindriques". An experimental demonstration of the truly conducting role of these strands was provided by Haberlandt (1883, 1886) who investigated the gametophytes of several species (*Mnium undulatum, Polytrichum juniperinum, Atrichum undulatum,...*), whereas Vaisey (1887a) conducted similar experiments on the setae of *Polytrichum* and *Splachnum*. Haberlandt used solutions of eosin and of lithium sulfate, and firmly established that not only the hydroids of the central strand but also those of the leaf traces (including the so-called "Zentralzellen" of the Polytrichales) do indeed conduct water (cf. Fig. 323). He also pointed out that when "false" leaf traces are present, water must pass from the latter to the central system through the cells of conducting parenchyma of the internal cortex.

Subsequent work on a wide range of species and under a variety of experimental conditions [(1)] (but nearly always with *cut stems*) confirmed that hydroids contitute a preferential pathway for the transport of water and solutes (Rostock 1902, Goebel 1906, Strunk 1914, Blaikley 1932, Zacherl 1956, Bopp & Stehle 1957, Bopp & Weniger 1971, Hébant 1974, etc.) (see also Figs. 324 to 327). Conduction in the central strand is always more rapid than in surrounding cortical tissues (Rostock 1902, Zacherl 1956). Diffusion of the stain or fluorochrome from the central strand into the parenchyma cells of the cortex was also frequently observed. "True" leaf traces take their water supply directly from the central strand (cf. e.g. Fig. 325). As suggested by Haberlandt (1886), when "false" leaf traces are present, the solutions must first cross the parenchyma cells of the internal

(1) Whereas early investigators used stains (especially eosin), or various salt solutions (KNO_3, $FeCl_3$, $LiSO_4$) together with appropriate reagents, recent workers used fluorescent dyes or radioactive tracers. Having no adequate stain at his disposal, Goebel (1906) demonstrated the conducting role of the hydroids in *Dawsonia superba* with blue ink ! Zacherl (1956) tested several fluorochromes and recommended the yellow of oxypyren as it readily penetrates into bryophytic tissues. On the other hand, it is a well known fact that the path taken by the water is not necessarily the same as that of the solutes (or stains, or markers) dissolved in it.

cortex before reaching the traces (Rostock 1902, Zacherl 1956). It also seems probable that the water absorbed by the leaves can reach the central strand via the leaf traces.

Fairly rapid rates of conduction were reported by Zacherl (1956) for cut stems of mosses with large central strands :

– 140 cm/h in *Polytrichum commune* kept under a relative humidity of 70 % (up to 200 cm/h in lower regions of stems where the central strand is smaller in diameter).

– 120 cm/h in *Mnium undulatum* kept under a relative humidity of 65 %.

These results fall well within the range of values known for a number of vascular plants. Various authors (e.g. Oltmanns 1884, Rostock 1902, Mägdefrau 1935-6, Zacherl 1956) also noted that "small" or "reduced" central strands are of little use or even of no apparent use at all for water conduction in leafy stems.

In *Polytrichum* gametophytes, the parenchyma cells which surround the central strand of hydroids (cells of the "hydrom-sheath") show enhanced enzyme activities (acid phosphatase, ATPase, peroxidases, succinate dehydrogenase, cytochrome oxidase, etc.; Figs. 333 to 335) (Hébant 1970). By comparison with what is known in higher plants (see e.g. Sauter 1972) it can be suggested that a functional association may exist between hydroids of the central strand and these parenchyma cells (e.g. intake or release of organic compounds).

Conduction in sporophytes. Not being adapted to external conduction by capillarity, the setae must necessarily conduct water internally. This fact was stressed by Oltmanns (1884) himself, who otherwise did not greatly favour the concept of internal conduction for the corresponding gametophytes. Indeed, and as pointed out earlier (p. 46), a strand is commonly seen in the setae of mosses which lack it in their leafy stems. The truly conducting role of the strand was established by Vaisey (1887a) in *Polytrichum formosum* and *Splachnum sphaericum*, with the use of eosin solutions. According to him a transpiration stream thus exists in the sporophyte of mosses. Vaisey observed a rate of transport up to 1 mm per minute in *Splachnum*. Further evidence for conduction in the setae was provided by Bopp & Stehle (1957) and Bopp & Weniger (1971) who investigated *Funaria hygrometrica* and certain Polytrichales with fluorescent dyes. Bopp & Weniger (1971) also found that the flow of water is slower in sporogonia without haustoria than in those with intact haustoria. According to them the presence of transfer cells appears to increase notably the uptake of water by the sporophyte. Unfortunately, no precise study of water economy within capsules of mosses is available.

Current concepts on water conduction in leafy stems and sporogonia of mosses with well developed strands were recently summarized by Bopp & Weniger (1971) and illustrated in the form of a diagram which is reproduced Fig. 144.

■ Liverworts

"True" water-conducting strands are not of common occurrence in liverworts (see p. 22) and conduction in them has been very little investigated.

Tansley & Chick (1901) performed the first experiments showing that the strand of *Pallavicinia lyellii* actually conducts water. They observed the rise of an eosin solution of 1 cm within 20 min "on a hot sunny day", whereas the same distance was traversed by the solution in half an hour on another "grey, damper day". A more rapid transport was recorded by Finlayson (1950) in *Symphyogyna hymenophyllum*. In her experiments, the eosin solution travelled 1.5 cm within the strand in 7-15 minutes. An even faster conduction was observed by Smith (1966) in *S. circinata* : up to 1.5 cm in four to five minutes.

In *Moerckia flotoviana*, whose strands are not as highly differentiated as those in *Pallavicinia* and allied forms, Cavers (1903) observed that "as a rule about half an hour sufficed for the staining solution to travel nearly the whole length of the thallus and to become recognizable in sections taken a little behind the apex".

In *Takakia lepidozioides* Hébant (1972a) also reported the rapid conduction of a solution of K-fluorescein (cf. Fig. 328).

Diffusion of the stain (or fluorochrome) from the central strand into the parenchyma cells of the cortex was commonly recorded in liverworts (Tansley & Chick 1901, Cavers 1903, Smith 1966, Hébant 1972a) (cf. Figs. 329-330).

Nearly all the experiments mentioned above were performed on cut thalli or stems, and no attempt towards a precise control of the relative humidity was made. However, Smith (1966) investigated the rate of penetration of eosin in thallus bases of intact (uncut) gametophytes of *Symphyogyna circinata* and found that "considerable time" was necessary for the stain to reach the central strand. In addition, Grubb (1969) conducted some precise experiments with potometers on the water relations of the gametophytes of *Haplomitrium mnioides* (a species with a central strand) and *Plagiochila asplenioides* (a species without such a strand). Two species of *Mnium* were also investigated for comparison. The potometers were kept in chambers in which relative humidity was controlled (Fig. 146). Grubb found that "for approximately the same amount of shoot exposed there is a much greater, persistent rate of water flow in *Haplomitrium* than in *Plagiochila* and, in fact, about the same rate as in *Mnium* [(1)]".

Remark. Apparently not aware of previously published work, Rifot & Barrière (1974) reinvestigated water conduction in the thallus of *Conocephalum conicum* (a species without a "true" water conducting strand). According to these authors, the nerve constitutes a preferential pathway for water transport. However, it is not clear from their experimental procedure (use of an organic tracer : ^{14}C-sodium-acetate; autoradiography of dried entire thalli) whether internal or both internal and external conductions were dealt with, nor if apoplastic alone or both apo- and symplastic transports were involved.

8.0.3. Other problems related to water economy in bryophytes

In addition to the more specific aspects of water transport reviewed above, water economy of bryophytes in general raises many

(1) Here again, the experiments were performed on cut stems.

problems which have been partly discussed in the preceding pages. In fact more questions are raised than answered, and only a brief outline of some of the major problems encountered will be presented here.

- **Uptake of water from substrate**

Very little experimental work deals with this aspect. Certain mosses (especially the Polytrichales, see p. 39) and hepatics (cf. Schuster 1966) possess underground axes. Whether or not these axes in the liverwort *Haplomitrium* should be termed "roots" has raised some discussions (Hofmeister 1854, Schacht 1859, Grubb 1970, Berthier 1971). In any case, no convincing evidence for a truly active and efficient uptake of water by these various specialized axes in bryophytes is presently available. Schulz (1937) experimented with light green solutions on intact "rhizomes" of *Polytrichum* and found that they actually absorbed the stain which was eventually found in the central conducting system of the stems. However Schulz performed long-term experiments (several hours) and the actual rate of water uptake is not known. Bopp & Stehle (1957) on *Funaria*, as well as Smith (1966) on the liverwort *Symphyogyna circinata*, noted that a long time was necessary for the solutions they used to cross the cortical tissues of the lower portions of the gametophytes and reach the central strand.

The role of the *rhizoids* is not yet completely understood. The fact that they play a role in anchorage (cf. e.g. Paul 1903), and in water conduction by capillarity is generally admitted. Their role in the active uptake of water and solutes has been less evidenced and published observations (e.g. Clee 1943, Bopp & Stehle 1957, Rifot & Barrière 1974) are not in complete agreement. Intense enzyme activities (especially phosphatase activities) have been observed in young developing rhizoids (Berthier & Hébant, unpublished). By comparison with what is known for rhizoids of fern prothalli (Smith 1972), the possibility that these rhizoids are particularly active in uptake of water when still young is to be considered.

An interesting contribution to the study of epiphyllous liverworts was recently made by Berrie & Eze (1975) who showed that when

Radula flaccida develops on a leaf, rhizoids penetrate through the cuticle and epidermal cells into the tissues of the host. Berrie & Eze established that water and dissolved ^{32}P salts can pass from the leaf into the hepatic, which thus appears to be a semi-parasite rather than a simple epiphyte.

▪ Absorption of water over the whole surface of the plant

From the evidence summarized in the previous pages, it is clear that water and solutes can frequently be absorbed at all levels of the gametophytes, although some zones of lesser penetrability (e.g. the outer cortex of certain moss stems) or even regions almost entirely impermeable (e.g. the dorsal surface of the thalli of various Marchantiales) may exists. These facts have been repeatedly underlined (cf. Oltmanns 1884, Vaisey 1887bc, Isaac 1841, Anderson & Bourdeau 1955, etc.). Although a cuticle is of widespread occurrence in bryophytes (Strunk 1914), that on the gametophyte apparently does not, at least in the majority of species, constitute a barrier to water penetration from the outside. This is noteworthy since water in the form of rain, dew, etc., represents a notable – perhaps the most important – source of water for a number of bryophytes (Anderson & Bourdeau 1955, Klepper 1963 and 1968).

▪ Retention of water

There is some resemblance between the *sporogonium* of at least certain mosses and the sporophyte of vascular plants as far as control of water loss is concerned. Both have an efficient cuticle and possess stomata [(1)]. Interestingly, Garner & Paolillo (1973) have recently established that the stomata on the still young capsule of *Funaria hygrometrica* respond in the same manner as those of higher plants to certain environmental stimuli (especially : light/darkness).

Gametophytes of both hepatics and mosses are apparently not highly adapted to the retention of water, although they may vary

(1) Not all moss capsules possess apparently functional stomata. Various specialized or reduced stomatal types exist; total lack of stomata is also known for a number of species (for extensive review, see Bierschenk 1971).

widely in this respect according to the different species. Their – usually thin – cuticles do not efficiently prevent water loss in most cases. True stomata are constantly absent, whereas the "pores" on the upper surface of a number of Marchantiales remain open even under high water stress. Indeed Wenzl (1933) found that in *Marchantia polymorpha* the water transpired per hour, under extreme transpiration, equals the total water content of the thallus and that the loss is two to three times that of leaves of *Betula pendula* under the same conditions. Blaikley (1932) similarly found that cut shoots of *Erica vagans* of approximately the same size as cut stems of *Polytrichum commune* transpire at about one-tenth of the rate of the moss.

▪ Importance of walls in water economy

Various authors have sought for an explanation of the phenomena of water absorption and retention in the wall structure and composition (especially pectic substances) (cf. e.g. Roberts & Haring 1937). An old observation (Palisot de Beauvois 1822) illustrates in a striking manner the important role played by the walls in moss gametophytes : when wet, even a dead moss takes the appearance of a living one. In addition, the necessity for conduction within parenchyma in bryophytes is readily evident from a number of their structural features (e.g. "false" leaf traces; absence of junction between strands of lateral branches and of main stem; complete absence of "true" water-conducting cells in numerous forms; etc.). There is some evidence to suggest that the walls of the parenchyma cells constitute preferential pathways for conduction of water and solutes : stains and fluorochromes are frequently identified in them (cf. e.g. Buch 1945-7). Recently Bonnot (1968) observed in the meristematic cells of *Thuidium tamariscinum* that small vacuoles appear literally to "bud off" from the walls, at the level of the plasmodesmata (Fig. 162).

Indeed, all these observations are well in accordance with current concepts according to which, in vascular plants, the water in the cell walls acts as a "buffering" system during periods of water stress, and that these walls constitute the main pathway for water transport outside of the xylem (Dainty 1963, Weatherley 1963 and 1970). With reference to their possible "buffering" role, one may recall that quite thick walls are observed in the gametophytes of many "xeric" mosses although Patterson (1964) expressed the conflicting opinion that "there are no structural modifications significantly adaptative".

■ Drought resistance

This aspect is related to the previous ones in several ways.

Nörr (1974) stresses the fact that drought resistance is one of the important factors in delimiting the habitat of a given species.

It has been pointed out that various "xeric mosses" present high osmotic values in their cells (cf. Rostock 1902; some more recent work is reviewed in Patterson 1964). Lower osmotic values are found in diverse more "mesic" species, in which soluble sugars are less abundant, whereas starch may accumulate in their plastids.

The mucilage secreted by specialized hairs in the apical region of both moss and liverwort gametophytes apparently plays a role in protecting the fragile meristematic tissues against drought (Bonnot 1968c; Berthier et al. 1974). "Mucilage-ducts" are found within the thalli of *Conocephalum conicum* and, according to Clee (1943) and others, may intervene in water retention.

A number of growth forms (especially tufts, cushions) restrain water evaporation and thus may facilitate survival in periodically dry environments.

The few facts reported above (e.g. high osmotic pressure; thick walls; existence of specialized growth forms) are not sufficient in themselves to account for the many aspects of drought resistance and revival in bryophytes. Indeed, and although these phenomena have been subjected to a number of studies, no definite understanding of the mechanisms involved has emerged up to now.

From an extensive survey of the literature in the field and in association with his own experimental results, Mahmoud (1965) concluded : "It would seem that the secret of true drought resistance will have to be sought in the molecular organization of the protoplasm".

■ **"Growth forms", "life forms", types of internal organization and microhabitats**

A certain correlation exists between "growth forms" (for review see Watson 1971, p. 139) or "life forms" (Mägdefrau 1969) and the water economy of the considered species. It is a well known fact that compact cushions are found in sites where periods of severe drought are to be endured, whereas some of the tallest forms are found only in almost constantly wet environments. A long time ago, Oltmanns (1884), Rostock (1902) and others pointed out that the tufted habit has a notable influence on water economy. Recently, Zacherl (1956) made some accurate measurements with a microhygrometer and found considerable differences in the relative humidities within a tuft (ca. 90 %) and just above it (50 % at a distance of a few centimeters). In fact, and as far as "growth forms" and "life forms" are concerned, a multitude of adaptative patterns apparently exist, which would deserve detailed investigations.

Attempts have also been made to correlate anatomical structures, especially the presence or absence of a central strand, with the ecology of the different forms. Haberlandt (1886) noted for instance that aquatic, epiphytic or epilithic species frequently lack a central strand, or possess only a reduced one (see also Buch 1947, Melikyan & Dil'Daryan 1975). However, any attempt towards systematization would present many difficulties. Indeed, it is a well known fact that mosses of widely differing internal structures may coexist in the same environment (Morin 1893). One possible – but only partial – explanation for this is that some of them are rather tolerant in their environmental requirements and may thus be found in quite distinct microhabitats notwithstanding their internal structure.

8.0.4. External versus internal conduction : towards a synthesis

All the evidence summarized above points to the fact that the specialized tissues described by the histologists do in fact constitute preferential pathways for the conduction of water and solutes. The functional value of such structural features as the lack of a protoplast and the hydrolysis of the walls (in the hydroids of many mosses) or the presence of small "pores" (in the water-conducting cells of various liverworts) is readily evident. In at least certain moss sporophytes, the central strand is apparently the most important pathway for water; as far as water economy is concerned, these sporogonia somewhat resemble the sporophytes of vascular plants. On the other hand, the efficiency of external conduction is a well established fact for the gametophytes of a number of species, both mosses and liverworts. What remains a matter of debate is the relative importance of the two systems for those gametophytes which possess a central strand, especially in natural conditions.

Although various early investigators tried to determine the respective roles of external versus internal conduction in bryophytic gametophytes (Oltmanns 1884, Rostock 1902, Grebe 1917, etc.), it is not until the nineteenthirties that serious attempts towards an experimental analysis of these phenomena were conducted. Bowen (1931, 1933) and Blaikley (1932) published conflicting results, apparently because their experiments were performed under different conditions. Bowen (1931) claims that in *Polytrichum commune* the function of the central strand is supporting and strengthening rather than conducting whereas according to Blaikley (1932), this strand plays an important part in the conduction of water in the same moss. From a more extensive investigation on a variety of species, Bowen (1933) reached the conclusion that, with the exception of the genus *Mnium*, in all forms examined the amount of water conducted over the external surface exceeded that conducted internally. Neither Bowen nor Blaikley provided quantitative data on the relative humidity in their experiments. A subsequent work by Mägdefrau (1935-6) threw some light on these problems. His experiments were performed on a variety of species with either large or small strands, or no strand at all, and the relative humidity was constantly under control (see also Fig. 145). Mägdefrau found that in *Polytrichum*, which possesses a large conducting strand, internal conduction alone is sufficient to maintain turgidity at 90 % relative humidity whereas at 70 % both internal and external conduction are necessary. According to Mägdefrau, external conduction is "good" and internal conduction reduced or absent in the great majority of species. According to him, species in which both internal and external conduction are truly effective are quite few (he gives as examples *Mnium seligeri* and to a lesser extent *Mnium undulatum* and *Polytrichum commune* [1]).

Bayfield (1973) made an exceedingly precise study of water relations in cut stems of *P. commune* using a "wind tunnel" in which environmental conditions were thoroughly controlled. He found that the phenomena are far from simple, and that water loss in this moss is controlled by a complex series of leaf arrangement changes, and by changes in water potential deficit of the shoots. According to him, this contrasts with the water relations of another moss, *Rhacomitrium lanuginosum*, in which there is apparently little control over loss. Bayfield found that water conduction in *P. commune* is predominantly internal under high evaporative flux, and external under moderate flux. He added that under many stress conditions both pathways would probably be necessary to maintain an optimum water balance [2]. Finally, he noted that in the field, and for several reasons (microecology of tussocks; resistance to water uptake from the soil; etc.) the pattern of water loss would differ from that he observed in his experiments.

From the above review, it is clear that although some excellent studies are available, many more investigations on water conduction in bryophytes, – and on their water relations in general –, are needed. In particular, very little is known about the effective role of the central strand and its relation to external conduction in moss gametophytes when in their natural habitat.

Remark. Buch (1945-7, 1947) introduced a distinction between "endohydric" forms, in which internal conduction plays the same role as in vascular plants, "mixohydric" forms where both internal

(1) All these species quoted by Mägdefrau possess a large central strand. The fact is that "reduced" strands apparently are of very limited value for conduction (see p. 81).

(2) The possibility also exists that externally conducted water plays an important role in evapo-transpiration, whereas water conducted internally is more directly involved in vital processes (assimilation, growth, etc.) (P. Cruiziat, pers. comm.).

and external conductions are important, and "ectohydric" ones whose water supply comes only from external conduction coupled with water uptake over the whole surface of the gametophyte. From the evidence summarized above, and as far as gametophytes are concerned, it would seem that a majority of contemporary bryophytes are either "mixo-ectohydric" or "ectohydric", according to this terminology, and that truly "mixohydric" ones are not very abundant. Strictly speaking, "endohydric" gametophytes apparently do not exist in bryophytes. Only the sporogonia of at least certain mosses are likely to have a truly endohydric habit.

Chapter 9

TRANSPORT OF ORGANIC COMPOUNDS

The occurrence of leptoids *s.str.* with clear sieve element-like characteristics being quite exceptional in bryophytes (cf. pp. 34 and 59-63), symplastic transport of organic substances occurs mostly through cells of conducting parenchyma. Although the latter elements may be specialized to some extent, translocation in them presumably does not reach high velocity rates. This probably is not critical since only short- or relatively short-distance transport phenomena are involved.

■ "Sources" and "sinks"

In leafy forms (mosses and liverworts) not only the leaves but also the stems, which are frequently green, can be the site of photosynthesis and therefore act as *"sources"*. In various thalloid hepatics, some internal specialization in the distribution of photosynthetic tissues may occur. Certain Marchantiales even show a complex structure which somewhat simulates, as a result of convergent evolution, that of a leaf of a phanerogam. Various kinds of photosynthetic filaments or lamellae exist in a number of species. The cells of the green lamellae on the adaxial surface of the leaves of diverse Polytrichales are typical chlorenchyma (Paolillo & Reighart 1967). Associations between plastids and peroxisomes, similar to those found in higher plants, have been seen in these cells (Hébant & Marty 1972). Efficiency of the photosynthetic apparatus in gametophytes varies with populations and age. Collins & Oechel (1974) determined the relative photosynthetic rates of successive seasonal growth increments in *Polytrichum alpinum* and found that by the 3rd year the capacity of assimilation had dropped to about 40 % of that in the youngest tissues of the same leafy shoots. A large bulk of evidence shows that photosynthetic tissues also exist in sporogonia (Haberlandt 1886, Magdeburg 1886, Bold 1938 and 1940, Paolillo &

Bazzaz 1968, etc.). No qualitative difference was found in pigments of both generations, haploid and diploid (Freeland 1957 on mosses; Taylor et al. 1972 on the liverwort *Lophocolea*). The fact that at least in certain species these sporophytes are independent to various degrees of the gametophytes is demonstrated by cultivation experiments (Haberlandt 1886, Campbell 1917, Studhalter 1938) and by certain field observations of "isolated" sporogonia (Campbell 1924, Dubois & Hébant 1968).

Among *"sinks"* one may cite growing apices, developing lateral buds, developing gemmae, underground growing axes, developing embryos and capsules, etc. Special mention must be made of the interphase between gametophyte and sporophyte which from the early stages of sporogonium development to the maturation and final emission of spores must act as a very demanding "sink" for the gametophyte.

▪ Translocation of organic compounds in bryophytes

Although the concept of transport of organic substances in bryophytes emerged relatively early (see e.g. Firtsch 1883 [1], Haberlandt 1886), experimental evidence for this is still meagre.

Interesting observations on transport of soluble sugars in the moss *Mnium hornum* were made by Marchal & Marchal (1906). They cultivated side by side an intact leafy stem, a stem from which leaves had been removed, and isolated leaves. One month later the intact leafy stem contained large amounts of starch, that without leaves contained only very small amounts of starch, whereas the isolated leaves contained much more starch than those attached to the stem. The Marchals concluded that : "sous l'influence de la lumière, l'amidon s'accumule dans les feuilles détachées; dans les feuilles attachées, il émigre partiellement dans la tige". *Mnium hornum* lacks sieve element-like food-conducting cells. This translocation must therefore take place through cells of specialized conducting parenchyma (viz. deuters of the leaf nerves; parenchyma cells of the cortex of stems).

(1) I suspect Firtsch, who was a student of Haberlandt, to be the true discoverer of "moss phloem".

Indirect evidence for transport of soluble N-compounds was also provided by Uhel (Guyomarc'h) for *Dicranum majus* and *D. scoparium* which are devoid of leptoids, and for *Polytrichum formosum*, a species with leptoids (Guyomarc'h 1969, Uhel 1975).

Mägdefrau (1935-6) attempted to demonstrate the conducting role of the leptom in leafy stems of *Polytrichum* with fluorescein. These attempts were unsuccessful, but as Mägdefrau pointed out, similar failures are known for certain vascular plants and thus they cannot be taken as a proof that the leptom in *Polytrichum* does not translocate.

More recently Thrower (1964) fed $^{14}CO_2$ to gametophytes of an antipodial polytrichaceous moss, *Dawsonia intermedia (= D. superba* var. *minor)* without observing however any movement of labelled compounds.

Successful demonstrations of an effective translocation of organic substances in the gametophytes of *Polytrichum commune* were made by Eschrich & Steiner (1967), Bayfield (1967) and Bonnemain (pers. comm.). The article by Eschrich & Steiner constitutes the first published account of rapid translocation in a moss. These authors found that after application of ^{14}C-bicarbonate on a single leaf, labelled organic material moves upward with a velocity of at least 32 cm per hour. They also made histoautoradiographs of transections in fed plants, which showed that radioactivity occurs in the leptom of the stem (Fig. 338). In his thesis, Bayfield (1967) presented evidence showing that in the same species, translocation can occur both from stem apices to lateral branches, and to connected basal shoots.

Further evidence for the translocation of labelled compounds in the tundra moss *Polytrichum alpinum* was provided by Collins & Oechel (1974). Above-ground shoots were exposed to $^{14}CO_2$ in the field. By autoradiography and subsequent counting, Collins & Oechel were able to show that preferential translocation had occurred into new below-ground shoots arising from the extensive underground system (Fig. 337). The recorded rates were quite low ("minimum rates of 3.1 mm per hour"). However in subsequent experiments on *P. alpestre* and *P. commune*, Collins (pers. comm.) found rates of translocation "at least as great as those measured by Eschrich & Steiner".

Eschrich (1975) reported that translocation also occurs in the seta of *P. commune*. According to him, applied ^{14}C-sucrose moves acropetally as well as basipetally at a velocity of around 50 cm per hour. Histoautoradiographs of transections were made, and here again the labelling was found to be most prominent in the ring of leptoids. An indirect proof of the redistribution, hence of transport, of organic compounds in the sporogonium of *Polytrichum* had also previously been given by Paolillo & Bazzaz (1968) who showed that the dry weight of the seta decreases as the weight of the capsule increases.

Finally, mention must be made of the experiments of Bopp & Knoop (1974) and of Larpent-Gourgaud (1974) who established the possibility of (slow?) transport of a number of substances (Thymidin, glucose, phenyl-alanine, etc.) in the protonemata [(1)] of mosses. These findings were related to various aspects of the morphogenesis of the filaments.

In liverworts, evidence for translocation of ^{14}C-sucrose within the thalli of *Marchantia polymorpha* was provided by Rota & Maravolo (1975). These investigators conducted elegant experiments in which transport phenomena were here again correlated with developmental patterns of the gametophytes. The isotope or its metabolites were found to move acropetally during normal growth or regeneration of the thalli. However, when IAA was supplied to the midribs of thalli without apices, both regeneration and sucrose transport were inhibited. Presumably, the translocation rate was not very high, since leptoids *s.str.* do not exist in *Marchantia.*

▪ Enzyme activities

Histochemical investigations have shown that relatively high enzyme activities (phosphatases, peroxidases, respiratory enzymes)

(1) *Protonemata* : green filaments originating from germination of spores in Mosses. They subsequently give rise to the leafy shoots. Electron microscope studies (Młodzianowski 1970, Mueller 1972) have shown that the contact walls between the cells in these filaments are traversed by numerous plasmodesmata.

exist in the leptom region of mature but still young portions of leafy stems of *Polytrichum* (Hébant 1970, 1973) (Figs. 333 to 335). These observations suggest that the translocating leptom, like the phloem of vascular plants, is quite an active tissue.

■ Exudation

Exudation phenomena have been observed in the leafy stems of *Polytrichum commune* (Hébant 1975) (Figs. 331-332) and of *Polytrichadelphus magellanicus* (unpublished observations). The amount of exudate is quite weak, probably due to the modest size of the plants. Interestingly, this exudate originates not only from the leptoids *s.str.* but also from various cells of specialized conducting parenchyma associated with them, in which leptoid characteristics are more or less well developed. However, the leptoids *sensu stricto* seem to be more efficient; – in wilted plants, exudation, if any, originates preferentially from them.

No chemical analysis of the exudate is available, although some data on the whole "sap" content of the gametophytic stems of *Polytrichum formosum* and other mosses are provided by Walland & Kinzel (1966). Eschrich & Steiner (1967) identified as sucrose one of the labelled organic substances found in the leaves of the moss (*P. commune*) they investigated.

■ Aphids

Aphids are known to feed on vascular plants, preferentially (but not exclusively) on their phloem. Interestingly, various aphids have also been found on a number of mosses, a large percentage of which were Polytrichales, viz. forms with a leptom in their gametophyte (Müller 1973).

■ Conclusion

From the above review of the few investigations which have been made on translocation of organic substances in bryophytes, it is clear that, here again, further work is needed. Divergences exist between the published results, which may concern such aspects as the direction of this translocation, its velocity or even its actual occurrence. It must however be noted that structural variability or differences in the physiological condition of the investigated plants could account for a number of these apparent discrepancies. In any case, the great interest of bryophytes for future investigations on translocation mechanisms should be underlined : the structural variability which is so impressive in those plants, especially that in the leptom of mosses, may provide a new insight into structure-function relationships. Indeed, several important questions are raised. One of these is : at which level of differentiation does the efficiency of the leptom in mosses become comparable to that of the phloem in tracheophytes ? Experimental work is also needed to specify which factors control the differentiation of the food-conducting cells in these plants. Do they respond in the same way as the vascular plants to external factors ? A wide field is open to further investigations.

Chapter 10

PHYSIOLOGY OF THE SPOROPHYTIC-GAMETOPHYTIC INTERPHASE IN BRYOPHYTES

A large bulk of evidence (structural, developmental, biological) stresses the fact that the interphase between the gametophyte and the sporophyte is a very important zone for bryophytes. However experimental work on its physiology remains remarkably scarce.

In the course of an investigation of water transport from leafy stems to sporogonia in *Funaria* and some polytrichaceous mosses, Bopp & Weniger (1971) noted that the flow of water was slower in sporophytes without haustoria than in those with intact haustoria. They interpreted this phenomenon as related to the presence of transfer cells at the interphase. The fact that these cells may occur on both sides of the contact zone or on one side only – either gametophytic or sporophytic (cf. p. 70) –, according to the species, raises some questions and would deserve further experimental work.

Histochemical investigations (Maier & Maier 1972, Hébant & Suire 1974) have shown that the transfer cells are the site of intense enzyme activities (especially : phosphatases and respiratory enzymes) (Figs. 321-322). Thus, not only do the wall ingrowths greatly increase the surfaces of exchanges, but phenomena of active (energy requiring) transport most probably also occur. It can be noted that other cell types in haustoria and surrounding gametophytic tissues may present enhanced enzyme activities, although the latter do not generally appear to be as high as in the transfer cells themselves.

Although the experimental work reviewed above is not very abundant, all the available evidence stresses the fact that the sporophytic-gametophytic interphase is a very active zone, which plays an essential role in the development and life of the sporogonia. In many forms, the haustorium presumably acts as a "pump" driving water and substances towards the capsule during its development. That the interphase constitutes a very demanding "sink" for the gametophyte is perhaps illustrated by the fact that seçondary branches which develop in its vicinity in *Polytrichum alpinum* are sometimes quite weak at their base by comparison to those which develop near abortive archegonia (Hébant & Berthier 1972). Further studies however are needed for a better understanding of the gametophyte-sporophyte relationships. The fact that in various forms the sporogonium may become largely independent raises the problem of the duration of the active life of the haustorium. According to Wiencke & Schulz (1975), in *Funaria* transport of nutrients stops after two thirds of the total development of the sporogonium. It has also been suggested that reverse transport, – from sporophyte to gametophyte –, may sometimes occur (Studhalter 1942, in *Riella*; Collins, pers. comm., in *Polytrichum*).

The physiology of such forms as *Anthoceros* where transfer cells exist on the gametophytic side of the interphase only (see p. 70) should also be investigated.

PART III

Evolutionary Problems

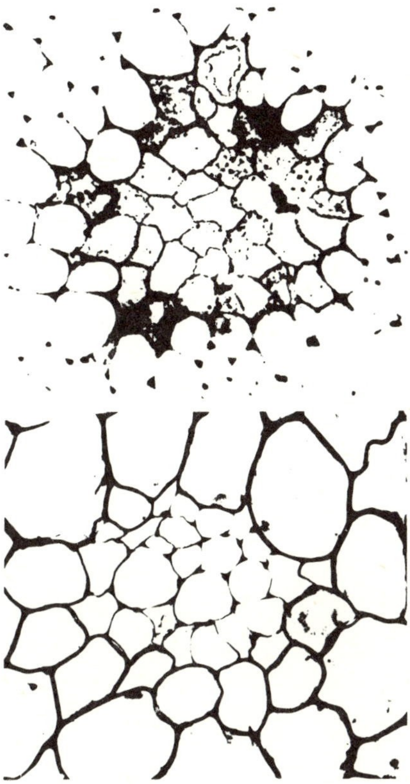

Chapter 11

SOME EVOLUTIONARY PROBLEMS RELATED TO CONDUCTING TISSUES IN BRYOPHYTES

Bryophytes are fascinating plants. Indubitably, they are authentic non-vascular [(1)] land plants. However, the occurrence in a number of them of conducting tissues strikingly similar to those of tracheophytes raises interesting evolutionary problems. Are these conducting tissues analogous or homologous to xylem and phloem ? Did bryophytes originate from a stock of primitive archegoniatae from which contemporary vascular plants also derived ? Do they represent totally independent lines of descent ? Answers to such questions have important implications when comparative studies on conducting systems are undertaken. Although presently available evidence does not provide clear answers, it is nevertheless possible to distinguish among the diverse interpretations which have been advanced those which can be held as reasonable working hypotheses, and those which should be discarded.

11.0.1. Position of bryophytes within the plant kingdom

Two outstanding questions are raised here :

1 - Do bryophytes represent a natural evolutionary group, or do the various groups within them (viz. mosses, liverworts, anthocerotes, and perhaps also Sphagnales and Andreales) represent unrelated lines of descent, not originating from a common ancestral stock ?

2 - What are the relationships of bryophytes with the tracheophytes on the one hand, and with green algae on the other ?

(1) Although the term "vascular" has sometimes been applied to bryophytes within recent years, it would certainly be preferable to restrict this term to those plants which possess "true" tracheary elements, in which the presence of lignified secondary thickenings ("ornamentations") is a distinctive characteristic. Although hydroids of mosses share many features with tracheids, they lack the lignified "ornamentations".

Only the second of these questions will be considered in some detail here as several reviews deal with the bryophytes alone (e.g. Schuster 1966, Steere 1969, Watson 1971). A certain tendency exists to revive the ancient idea that contemporary bryophytes mostly correspond to a "level of evolution and structure" and are polyphyletic in nature (Chopra 1967, Steere 1969). Without entering into details, I will retain the view that, whatever the degree of independence in the various lines, they most probably derive from a remote common ancestral stock. The reasons for such an assertion lie in the fact that bryophytes share many common characteristics [(1)] – especially in their life cycle – and in the existence of "synthetic forms". Among the latter, the anthocerotes are noteworthy, since their sporophyte may present a columella and stomata as in mosses, and elaters as in hepaticae.

Concerning the situation of the bryophytes within the plant kingdom, only a multidisciplinary approach can be expected to throw some light on this problem which is still largely a matter of conjecture. Some aspects of the presently available evidence are summarized below.

■ Biochemical evidence

○ ***Photosynthetic pigments :*** The same photosynthetic pigments are found together with "true" intraplastidial starch in the green algae *sensu lato*, bryophytes and tracheophytes. Most authors have taken this fact as a decisive argument for the derivation of the latter (bryophytes + tracheophytes) from green algae. Indeed nobody to-day would hold certain of the earlier views according to which brown algae for instance may have been ancestral to archegoniatae.

○ ***"Informative" molecules :*** Great hopes arose some ten years ago, especially after the publication of the paper "molecules as documents of evolutionary history" by Zuckerkandl and Pauling (1965), that decisive phylogenetic trees could be built from the study of

(1) A large proportion of these common characteristics are also encountered in pteridophytes and belong to what we might call the "archegoniate pattern". These characteristics are reviewed in the following pages (cf. also Fig. 151).

amino acid sequences of essential proteins. Unfortunately, no published data of this kind are available yet for bryophytes and primitive tracheophytes (determinations of amino acid sequences are not easily made). In addition, some limitations of these methods have just become apparent from the first attempts towards their applications to the plant kingdom (especially that of Boulter and his collaborators with cytochrome c). As pointed out by Cronquist (1976), not only can evolution at the molecular level proceed either notably faster or notably slower than evolution of other features [(1)], but the result of studies in this field apparently depends on "how one manipulates the data". However the possibility that base sequences of DNA [(2)] could be determined in the near future (cf. Kolata 1976) could drastically alter the present status of "molecular trees".

○ ***Enzymes and biosynthetic pathways :*** Biosynthetic pathways, and to a lesser extent enzymes, can also yield precious information provided that a sufficient number of different organisms are sampled. Here again, studies are not always easily conducted, and available information on lower archegoniatae remains extremely scarce.

Phenylalanine ammonia lyase is the first enzyme of the phenylpropanoid pathway, which can lead to both lignins and flavonoids. Its presence in vascular plants is well substantiated. Meyer & Angerman (1973) showed not only that this enzyme occurs in the moss *Funaria hygrometrica*, but that its photoregulation in this plant presents striking similarities with that known in vascular plants.

The enzyme *glycolate oxidase* was found by Frederick et al. (1973) to occur in all the bryophytes and tracheophytes they investigated,

(1) Two possible reasons for these discrepancies apparently are : 1) that only a small proportion of the nuclear DNA is "translated" into amino acid sequences of structural proteins (cf. Britten & Kohne 1968) and 2) that "phenotypic" evolution proceeds from "epiphenomena" such as gene rearrangements rather than from point mutations (Wilson et al. 1974).

(2) According to many molecular biologists attempts to derive evolutionary considerations from DNA base compositions (viz. percentages of G+C) must be taken with reserve.

whereas its absence was recorded from a number of green algae, where *glycolate dehydrogenase* occurs instead. Only certain green algae (viz. those belonging to the "Charophyceae" *sensu lato* as defined by Stewart & Mattox 1975) possess glycolate oxidase.

Urea can be degraded in plants by the enzymes *urease* and *urea amidolyase*. Up to now, the latter has been recorded from lower plants only (fungi, green algae). In an investigation of 5 liverworts Jones et al. (1973) found high urease activities, but no urea amidolyase. They concluded that "from the standpoint of enzymology of urea degradation, liverworts and flowering plants possess comparable biochemical potential".

In the opinion of the present reviewer, the above mentioned data, although of outstanding interest, must be taken with some caution, at least certain of them. Indeed, the number of investigated organisms remains in general rather small (one exception concerns glycolate oxidase). It may well happen that enzymes first reported absent from certain groups will be subsequently found in them when more extensive investigations are conducted [1].

○ ***Other biochemical evidence :*** "secondary" molecules or "end-products" can be of great help for the construction of phylogenetic trees, provided they originate from homologous synthetic pathways. Unfortunately, this is not always proven. Several reviews deal in detail with these compounds (Hegnauer 1962, Swain 1974). Among them, *flavonoids* are of particular interest, as they have been most widely investigated (Harborne et al. 1975). Although abundant and diversified in tracheophytes, flavonoids have been very seldom encountered in lower plants (algae and fungi). Interestingly, their presence has been reported from a number of mosses and liverworts (for review, see Harborne et al. 1975). Their occurrence in the charophyte *Nitella hookeri* (Markham & Porter 1969) and in certain fungi (?) is also worthy of note. The existence of fungal flavonoids raises

(1) Note for instance that a thylakoid-bound L-phenylalanine ammonia-lyase has recently (and apparently for the first time) been reported from an unicellular green alga, *Dunaliella marina* (Löffelhard et al. 1973).

the question of knowing whether or not they originate from biosynthetic pathways homologous to those of green plants.

Triterpene hydrocarbons provide another typical example of the kind of problems encountered with "secondary" products, since they are known to occur in ferns, mosses and also (if confirmed) in one lichen (for review see Marsili & Morelli 1968).

Certain little investigated compounds, such as non-protein amino acids could also prove in the future to be of great help in clarifying the position of the bryophytes within the plant kingdom (J.R.L. Walker, pers. comm.).

- **Morphological evidence** (cf. Fig. 151 and Plate 80)

The present scarcity of biochemical data, together with their possible limitations, tends to preserve the full value of "classical" morphological evidence. Here again, the question of distinguishing between those structures which are truly homologous, and those which result from parallel (homoplastic) development is not a simple one since in general no decisive criterion exists for distinguishing between the two possibilities. When a larger bulk of developmental and physiological evidence has become available, it will perhaps be possible to be more confident. Nevertheless it is of outstanding interest to note that such complex and elaborated structures as archegonia, antheridia and stomata do occur in bryophytes since no other organisms than tracheophytes share these characteristics [1] with them. Although many bryophytes (especially the liverworts) lack stomata or possess reduced or aberrant ones (cf. Bierschenk 1971), the fact that in *Funaria hygrometrica* these stomata do respond in the same way as those of vascular plants to external stimuli (Garner & Paolillo 1973) further strengthens the hypothesis of a true homology.

(1) Occurrence of "protected" oogonia in a few thallophytes is not of critical importance since both in development and final structure these reproductive organs differ from archegonia (notwithstanding the fact that the general "tendency" towards protection of oogonia in Charophyceae *sensu* Stewart & Mattox (1975) might truly have an evolutionary significance).

From an anatomical point of view, it is of interest to note that the fundamental pattern of organization of many moss stems (viz. a supportive tissue at the periphery, a "simple" strand in the middle) is strongly reminiscent of that seen in various primitive vascular plants, especially amongst the oldest known tracheophytes (cf. e.g. Dianne Edwards 1970).

The morphological similarity between the central strands of mosses and primitive tracheophytes has been further strengthened by recent histological studies (cf. e.g. Figs. 4-5, 263-264 and 343-344). Strands in both groups also have a comparable role and present a similar degeneration of their protoplasts with a peak of strong acid phosphatase activity at the last step of their development (this even occurs in the rare liverworts which possess a strand, cf. p. 27).

Lack of roots, which has long (Necker 1775 !) been considered a distinctive characteristic of bryophytes appears less important to-day since a number of primitive tracheophytes are devoid of true roots, whereas some bryophytes possess below-ground axes which may present some "root" characteristics (cf. e.g. Grubb 1970).

▪ Life cycles

The existence of many basic similarities between the life cycles of bryophytes and vascular cryptogams has been emphasized repeatedly (see e.g. Doyle 1970; cf. also Fig. 151). In both groups, the embryo starts its development within the archegonium ("embryophytic habit") and several authors, including Proskauer (1960), suggested that this may have been the case for their putative common ancestors. Parasitism being a secondary acquisition, the suggestion has also been made that the fact that the sporogonium of bryophytes almost never reaches total independence is derived rather than primitive. Moreover, the same morphogenetic plasticity [(1)] exists in the alternation of generations in both groups (Bristow 1962, Wetmore et al. 1963, Bauer 1963, Menon & Lal 1974).

(1) For instance, the possibility to express gametophytic characters from diploid cells, and vice versa.

■ **Ultrastructural evidence**

The sperm cells of bryophytes present ultrastructural similarities to those of vascular cryptogams (Duckett 1975), whereas they are distinct from the motile cells of most algae. Only the motile cells of *Chara, Nitella* and some other forms (belonging to the Charophyceae *sensu* Stewart & Mattox 1975) are known to possess characteristics reminiscent of those found in sperm cells of land plants. Similarly, the fine structure of the mitotic spindle and phragmoplast is comparable in the bryophytes, tracheophytes and in the Charophyceae *sensu* Stewart & Mattox (cf. Pickett-Heaps & Marchant 1972). In addition, comparative studies on spore wall structure of liverworts and pteridophytes do not preclude the idea of an ancient link between these different groups (Lugardon 1975; Denizot, pers. comm. 1976).

■ **Physiological evidence**

A few examples of physiological similarities have been outlined in the preceding pages (e.g. the response of stomates of *Funaria* and vascular plants to external stimuli, or the presence of glycolate oxidase in photosynthetic tissues of bryophytes, tracheophytes and Charophyceae *sensu* Stewart & Mattox). Progress of knowledge in this area will certainly contribute to a better understanding of the topics discussed here and thus allow a cross-check with the similarities found at the structural level alone.

■ **Palaeontological evidence** (cf. Plate 80)

The presently available fossil record is of little help as far as the origin of bryophytes is concerned (Lacey 1969). However, it may be noted that the earliest known tracheophytes were more similar in general appearance to certain present-day bryophytes than to most contemporary vascular plants. This concerns more particularly certain Rhyniophytina such as *Steganotheca* or *Cooksonia* (see Dianne Edwards 1970) : their tiny dichotomously branched axes with terminal sporangia are somewhat reminiscent of moss sporogonia (the latter are unbranched, but some examples of teratological branching are known; in these cases, they are frequently dichotomous ! – cf. Chadefaud 1960). As pointed out earlier, the internal organization of these early tracheophytes, with a ring of outer supportive cells and a "simple" central strand, is also more or less comparable to that of moss stems. Following their first description of the Rhynie flora, Kidston & Lang (1921) indeed claimed that "the simplicity of the independent sporophytes of the Rhyniaceae makes comparison between the asexual generation in Bryophyta and Pteridophyta easier, and to this extent narrows the gap between these two great groups" (loc. cit. p. 842).

With particular reference to *Rhynia*, it must also be noted that Kidston & Lang (1920) failed to identify lignified secondary thickenings in the tracheids of *Rhynia major*. Similar observations were made on *Nothia aphylla* by A.G. Lyon. The latter investigator writes : "I consider it unlikely that the absence of wall features from the xylem elements of *Rhynia major* & *Nothia* is merely due to degenerative changes" (pers. comm.). Similarly, Proskauer (1960) suggested that the water-conducting cells of *R. major* may have been unlignified. These elements in *R. major* and *Nothia* would then be very similar to the hydroids of mosses. Indeed the stele of *Rhynia* as a whole may show an extraordinary resemblance to certain moss steles (Figs. 339 to 341).

▪ Conclusion

Bryophytes and tracheophytes share an impressive assemblage of characteristics, although part (but part only) of these characteristics is also encountered in certain thallophytes. The following list recapitulates the most noteworthy of them (cf. also Fig. 151) :

– same photosynthetic pigments (also shared with green algae and euglenophytes)
– "true" intraplastidial starch (also shared with green algae)
– specific enzymes (phenylalanine ammonia lyase, glycolate oxidase; the latter also occurs in Charophyceae *sensu* Stewart & Mattox)
– presence of flavonoids (also found in the Charophyte *Nitella*)
– same fundamental pattern of life cycle (including the "embryophytic habit" of the young developing sporophyte)

– epidermis with cuticle and stomata (in addition to tracheophytes, the latter occur in the sporogonia of anthocerotes, Sphagnales and mosses only; cf. Fig. 345)

– "cutinized" spores (cf. Fig. 346)

– archegonia and antheridia

– resemblance in ultrastructures of sperm cells

– same structure of the mitotic spindle and phragmoplast (shared with Charophyceae *sensu* Stewart & Mattox)

– morphological and anatomical resemblances between the sporophytes of certain of the oldest known vascular plants and those of certain bryophytes (cf. e.g. Fig. 348)

– physiological resemblances (response of stomata to external stimuli; "lysosomal" degeneration of central strands; etc).

As pointed out by Stewart & Mattox (1975), the fact that apparently unrelated characteristics occur together within different organisms is a good indication of their affinity. The above list shows that this is precisely the case for bryophytes and tracheophytes. In addition, cytological and biochemical evidence further suggests that the archegoniatae also have affinities with green algae, and among the latter, especially with the Charophyceae *sensu* Stewart & Mattox.

Decisive evidence concerning the relationships of bryophytes with other organisms will perhaps emerge in the forthcoming years, especially from biochemical and palaeontological studies. This might alter the contemporary view of the question, since in the present state of knowledge a relative amount of uncertainty persists. However, it is clear from the reviewed evidence that the basic idea according to which *the Archegoniatae constitute a natural evolutionary group* can still be retained as a reasonable working hypothesis. The affinities of the Archegoniatae with the Charophyceae also seem well established. These concepts are hardly new, and indeed an impressive number of authors have expressed similar viewpoints; specialized terms have

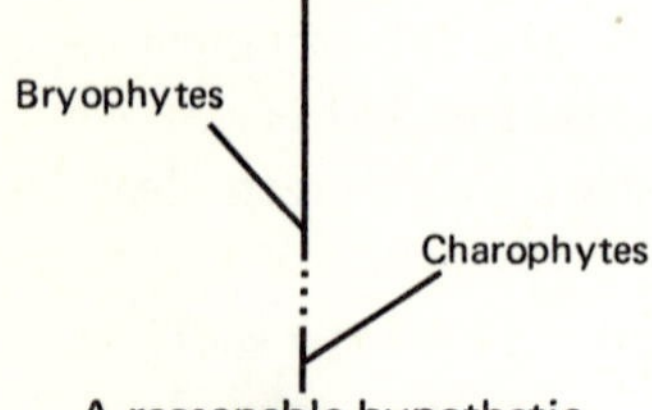

A reasonable hypothetic phylogenetic tree

even been proposed to designate the putative ancestral stock from which modern bryophytes and tracheophytes would have derived ("prohépatiques", Lignier 1903 and 1908; "Antho-rhyniaceae", Mehra & Handoo 1953).

11.0.2. Evolutionary trends in conducting tissues of bryophytes

- **Mosses**

To present the facts in the most simplified form, one can roughly state that there are mosses with well developed conducting tissues, mosses with poorly developed conducting tissues, and mosses without any conducting tissues at all (for details, see pp. 32 *sqq.*). Three differing types of interpretation have been proposed for these facts : progressive evolution, regressive evolution and evolution in both directions, starting from a "median type" :

Progressive evolution

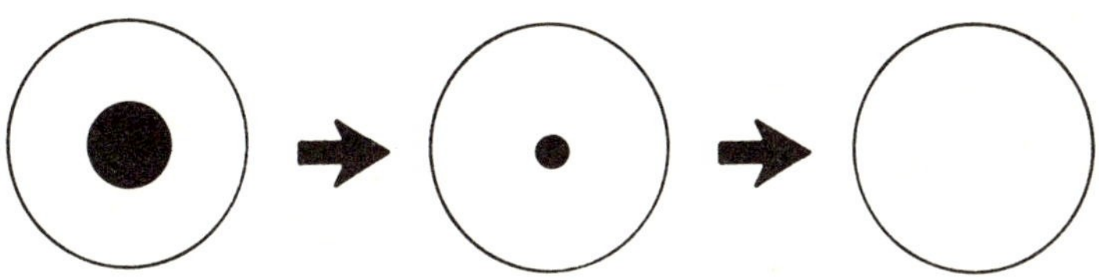

Regressive evolution

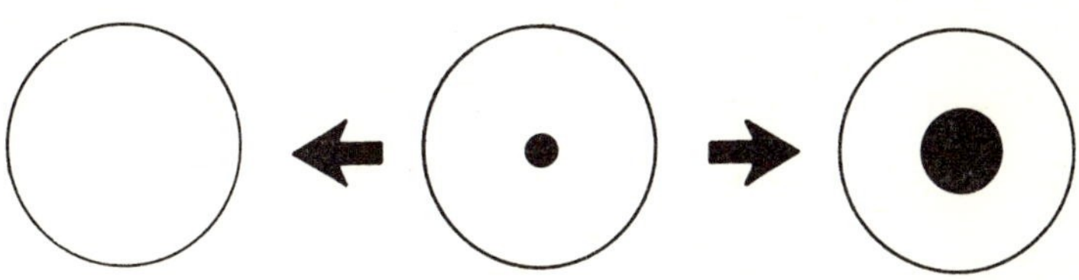

Evolution in both directions starting from a "median" type

(Schematic transections of stems; central strands dark)

It would be tedious to discuss at length these various hypotheses. However one important fact is worthy of note : forms with well developed strands, poorly developed strands and no strand at all coexist in varying proportions within many of the "natural" groups (viz. orders, families, ...) of mosses recognized by taxonomists. To admit, as is most commonly done, that the possession of conducting tissues might correspond to a secondarily acquired organization implies a multitude of cases of parallel evolution which resulted, along many lines, in the independent development of conducting tissues.

An alternative and perhaps more likely interpretation is to consider the absence of conducting tissues a *derived state.*

Certain evidence gives further support to this interpretation : mosses in which the gametophyte has no conducting strand, or only a rudimentary one, generally belong to taxa considered by most authors as displaying a group of advanced characteristics. The others correspond to specialized forms of which the relationship with groups normally having conducting tissues is generally incontestable (Hébant 1966b); various aquatic forms, for instance fall into this category (cf. Haberlandt 1886).

The interpretation in terms of regressive evolution of the diversity of structures found in moss stems is hardly new and many classic authors have suscribed to it. Lorentz (1867-8) who conducted the first extensive investigations described many "regressive" series. He described them as such for the sake of "convenience", being reluctant to take this as reality because, he claimed, Nature cannot move backwards ("Ich sage depauperirt, obwohl ich weiss, dass die Natur den umgekehrten Weg einschlägt und fortbildend das Aermere bereichert, nicht aber das Reiche verarmt"; loc. cit. p. 441). On the other hand, Haberlandt (1886), Lorch (1931) and many others clearly admitted the possible existence of a regressive evolution of the conducting tissues in the leafy stems of mosses.

■ Liverworts

Hepaticae which are known to possess "true" water-conducting strands [1] are quite rare (see p. 22). While recently contemplating the revised version of Mehra's phylogenetic tree of hepaticae (Mehra & Sokhi 1972), I have been impressed by the fact that in the liverworts Mehra places at the base of his tree (Takakiales, Calobryales), a "true" water-conducting strand is constantly (?) [2] present, whereas such a strand occurs only occasionally in those he places in the median region (the Anacrogynae), and is totally absent in those he considers the most advanced (cf. Fig. 152). This again fits well with the hypothesis of a regressive evolution of the conducting tissues.

■ At this point, it must be borne in mind that the concept of a regressive evolution of the conducting tissues in bryophytes is not inconsistent with what is known of the general evolutionary trends in the group as a whole. Indeed, many facts pertaining to the comparative morphology of these plants are better explained in terms of simplification and regression (cf. e.g. Richards 1951, Chadefaud 1960, Fulford 1964, etc.).

It may also be noted that comparable regressions of their conducting system have occurred in the smaller sporophytic axes of certain vascular plants whose way of life approaches that of bryophytes (e.g. certain species of *Trichomanes*; cf. Boodle 1900, Bell 1960).

A last remark will concern stomata : a long time ago, De Candolle (1813) noted that a very strong correlation exists between the presence of stomata and of conducting tissues in vascular plants. Could then the occurrence (persistence ?) of stomates in certain bryophytic sporophytes otherwise devoid of conducting tissues (e.g. *Anthoceros, Sphagnum* [3] – cf. Fig. 345) be an indication that their ancestors indeed possessed conducting tissues ?

(1) For the concept of water-conducting strand, see p. 3.

(2) Disappearance of this strand under certain circumstances is possible (cf. p. 9). Species of *Haplomitrium* without a strand perhaps also exist.

(3) In *Sphagnum*, these stomata are claimed to be non-functional (Haberlandt 1914).

11.0.3. Speculation

Taking into account as much as I can of the available evidence I reach the conclusion that two reasonable working hypotheses, neither of which is new, may be retained for the moment :

– First hypothesis (well substantiated) : *Bryophytes and tracheophytes have originated from a common stock of primitive Archegoniatae.*

– Second hypothesis (reasonable but less substantiated than the previous one) : *Bryophytes have undergone a regressive evolution of their conducting tissues. This evolution has occurred to various degrees, resulting in the observed diversity in internal structures of these plants.*

I am then tempted to further speculate that those primitive Archegoniatae from which bryophytes and tracheophytes are derived either already had conducting tissues (at least : water-conducting cells), or had the morphogenetic potentiality to develop such tissues [1,2]. The first of these two alternatives in no way implies that bryophytes are derived from authentic vascular plants (viz. with "true" tracheary elements, see note 1 p. 102); there is at present not one serious argument supporting such an assertion. Nevertheless it may be noted that the water-conducting cells of the most primitive tracheophytes (= tracheids), of mosses (= hydroids) and of liverworts share common characteristics which are possibly inherited from common ancestral progenitors. These characteristics are :

– occurrence of the water-conducting cells in the median region of stems or thalli, usually in the form of a single solid protostele

– elongated form of the conducting cells

– a lysosomal degeneration of their protoplast with a "peak" of

(1) This second alternative raises the complex problem of parallel evolution within closely related groups as opposed to parallel evolution within unrelated groups.

(2) Recent paleobotanical evidence supports the view that a diversified flora of non-vascular land plants existed in the Silurio-Devonian and possibly earlier (Chaloner et al. 1974, Kevan et al. 1975).

strong acid phosphatase activity constitutes the last step in their development.

However these cells also show important differences, especially in the structure of their walls, which further strengthen the separation of the three groups of contemporary land plants which present them :

– *tracheids of vascular plants* (Fig. 5) : hydrolysed primary walls + lignified secondary thickenings ("ornamentations")

– *hydroids of mosses* (Fig. 4) : hydrolysed primary walls may occur but lignified "ornamentations" are absent

– *water-conducting cells of liverworts* (Figs. 2 and 3) : no hydrolysed primary walls [(1)] nor lignified "ornamentations", but presence of numerous small plasmodesmata-derived pores in their walls.

Food-conducting cells also raise a problem. Up to now, a leptom with a level of differentiation truly comparable with that of the phloem in vascular plants has been found in the gametophytic phase [(2)] of certain polytrichaceous mosses only. Then what ? Homology or parallel development ? If the interpretation sustained here (pp. 48-49) of the persistence of a "residual" leptom in the setae of certain non-polytrichaceous mosses is substantiated, this might provide some support for the (fragile !) interpretation that a true homology is at hand.

In many mosses, conducting tissues exist in both generations. In the Polytrichales, this concerns not only hydroids but also the leptoids. However my suggestion (1964) that this situation parallels that found in certain Rhyniophytina where both generations might have been vascularized still needs substantiation. In spite of a number of reports (Merker 1958, Pant 1962, Lemoigne 1968) no decisive

(1) One exception is known : *Moerckia flotoviana*, see p. 27.

(2) The leptom in setae of Polytrichales is less highly differentiated (see § 6.2.).

and truly convincing evidence for the existence of vascularized gametophytes in these early land plants has emerged as yet [(1)]. Indeed, evidence for an alternation of "homologous" – and vascularized – generations in the early tracheophytes, although not completely lacking, is still weak (cf. Holloway 1939; see also the recent comments by Bierhorst 1971).

(1) The interpretation of *Rhynia gwynne-vaughanii* as the gametophyte of *Rhynia major* has been seriously shaken by the discovery of sporangia attached to typical *R. gwynne-vaughanii* axes (David S. Edwards 1973). That some of the axes preserved in the Rhynie chert might be vascular gametophytes nevertheless remains an interesting possibility according to A.G. Lyon (pers. comm. 1976).

LITERATURE CITED

ABEL, W.O. 1956. Die Austrocknungsresistenz der Laubmoose. Sitzgsber. Wien Akad. Wiss., math.-nat. Kl., Abt I, 165 : 619-707

ANDERSON, L.E., BOURDEAU, P.F. 1955. Water relations in two species of terrestrial mosses. Ecology 36 : 206-212

ARONOFF, S., DAINTY, J., GORHAM, P.R., SRIVASTAVA, L.M., SWANSON C.A. Eds. 1975. *Phloem Transport*, Plenum Press, New York

BAKER, J.R. 1946. The histochemical recognition of lipine. Quart. J. Micr. Sc., N.S., 87 : 441-470

BASSI, M., FAVALI, M.A. 1974. Seta ultrastructure in *Mnium orthorhynchum.* Nova Hedwigia 25 : 337-346

BASTIT, E. 1891. Recherches anatomiques et physiologiques sur la tige et la feuille des mousses. Rev. gén. Bot. 3 : 255-271, 306-316, 341-360, 373-388, 406-426, 462-489, 521-530, Pl. 12-13

BAUER, L. 1963. On the physiology of sporogonium development in mosses. J. Linn. Soc. (Bot.) 58 : 343-351

BAYFIELD, N.G. 1967. *Experimental ecology of Polytrichum commune and other heathland bryophytes*, Ph.D. Thesis, Aberdeen

BAYFIELD, N.G. 1973. Notes on water relations of *Polytrichum commune* Hedw. J. Bryol. 7 : 607-617

BELL, P.R. 1960. The morphology and cytology of sporogenesis of *Trichomanes proliferum*. New Phytol. 59 : 53-59

BENDZ, G., MÅRTENSSON, O., NILSSON, E. 1967. Moss Pigments. 6. On the pigmentation of *Sphagnum* species. Bot. Not. 120 : 345-354

BERRIE, G.K., EZE, J.M.O. 1975. The relationship between an epiphyllous liverwort and host leaves. Ann. Bot. 39 : 955-963

BERTHIER, J. 1971. Ramification du *Takakia* Hatt. et Inoué et morphogénèse des Cormophytes. Bull. Soc. Bot. France, Mémoires 1971 : 29-39

BERTHIER, J. 1972. Recherches sur la structure et le développement de l'apex du gamétophyte feuillé des mousses. Rev. Bryol. Lichénol. 38 : 421-551, Pl. 1-30

BERTHIER, J., BONNOT, E.-J., FABRE, M.-C., HÉBANT, C. 1974. L'appare sécréteur des Bryales : données morphologiques, ultrastructurales et cytochimiques. Bull. Soc. Bot. Fr. 121, Suppl. Coll. Bryol. : 97-100

BERTHIER, J., GALTIER, J., HÉBANT, C., HÉBANT-MAURI, R. 1971. Remarques sur la ramification de la stèle chez les Mousses *s.str.* (Bryopsida). Comparaison avec certains végétaux vasculaires archaiques. C.R. Acad. Sc. 272D : 2868-2871

BIERHORST, D.W. 1971. *Morphology of vascular plants*, MacMillan, New York

BIERSCHENK, K. 1971. *Morphologie und Entwicklungsgeschichte der Laubmoos-Spaltöffnungen*, Thesis, Tübingen

BLAIKLEY, N.M. 1932. Absorption and conduction of water and transpiration in *Polytrichum commune*. Ann. Bot. 46 : 289-300

BLAIKLEY, N.M. 1933. The structure of the foot in certain mosses and in *Anthoceros laevis*. Trans. Roy. Soc. Edinburgh 57 : 699-709

BOLD, H.C. 1938. The nutrition of the sporophyte in the Hepaticae. Amer. J. Bot. 25 : 551-557

BOLD, H.C. 1940. The nutrition of the sporophyte in the Musci. Amer. J. Bot. 27 : 318-322

BONNOT, E.-J. 1967a. L'infrastructure des plamodesmes de deux Bryales. C.R. Acad. Sci. (Paris) 264D : 2276-2279, Pl. 1-2

BONNOT, E.-J. 1967b. Sur la structure de l'apex du gamétophyte feuillé de la mousse *Anomodon viticulosus* (L.) Hook. et Tayl. Bull. Soc. Bot. Fr. 114 : 4-11

BONNOT, E.-J. 1968a. La formation des vacuoles dans les cellules méristématiques du gamétophyte feuillé de *Thuidium tamariscinum* (Hedw.) B.S.G. Proc. Fourth European Region. Conf. Elect. Microscopy, Rome : 409-410

BONNOT, E.-J. 1968b. Sur la structure et les propriétés de la cellule apicale du gamétophyte feuillé des Bryales. Bull. Soc. Bot. Fr. 115, Mém. : 208-222

BONNOT, E.-J. 1968c. *Contributions à la connaissance structurale de l'apex du gamétophyte feuillé des bryales*, Thèse, Dijon

BOODLE, L.A. 1900. Comparative anatomy of the Hymenophyllaceae, Schizaeaceae and Gleicheniaceae. I. On the anatomy of the Hymenophyllaceae. Ann. Bot. 14 : 455-496

BOPP, M., KNOOP, B. 1974. Régulation de la différentiation chez le protonéma des mousses. Bull. Soc. Bot. Fr., Coll. Bryologie : 145-152

BOPP, M., STEHLE, E. 1957. Zur Frage der Wasserleitung im Gametophyten und Sporophyten der Laubmoose. Z. Bot. 45 : 161-174

BOPP, M., WENIGER, H.-P. 1971. Wassertransport vom Gametophyten zum Sporophyten bei Laubmoosen. Z. Pflanzenphysiol. 64 : 190-198

BOWEN, E.J. 1931. Water conduction in *Polytrichum commune*. Ann. Bot. 45 : 175-200, Pl. 7

BOWEN, E.J. 1933. The mechanism of water conduction in the musci considered in relation to habitat. Ann. Bot. 47 : 401-422, 635-661, 889-912

BOWEN, E.J. 1935. A note on the conduction of water in *Fimbriaria bleumeana*. Ann. Bot. 49 : 844-848

BRISTOW, J.M. 1962. The controlled *in vitro* differentiation of callus derived from a fern, *Pteris cretica* L., into gametophytic and sporophytic tissues. Develop. Biol. 4 : 361-375

BRITTEN, R.J., KOHNE, D.E. 1968. Repeated sequences in DNA. Science 161 : 529-540

BROTHERUS, V.F. 1924-5. Musci, in Engler, A., Prantl, K., *Die natürlichen Pflanzenfamilien*, vol. 10 (1924) : 143-478; vol. 11 (1925) : 1-552

BRUCH, Th., SCHIMPER, W.Ph., GUMBEL, Th. 1836-55. *Bryologia europaea, seu Genera Muscorum europaeorum monographice illustrata*; 6 vols.; E. Scheizerbart, Stuttgart

BUCH, H. 1945-7. Ueber die Wasser- und Mineralstoffversorgung der Moose. Soc. Scient. Fenn., Comm. Biol. IX, 16 (1945) : 1-44; ibid. IX, 20 (1947) : 1-66

BUCH, H. 1947. De l'adaptation des Bryophytes, des Algues et des Lichens aux conditions d'humidité et de la concentration de la solution nutritive dans les milieux naturels. Rev. bryol. lichénol. 16 : 54-59

BURR, R.J., BUTTERFIELD, B.G., HEBANT, C. 1974. A correlated scanning and transmission electron microscope study of the water-conducting elements in the gametophytes of *Haplomitrium gibbsiae* and *Hymenophyton flabellatum*. Bryologist 77 : 612-617

CAMPBELL, D.H. 1917. Growth of isolated sporophytes of *Anthoceros*. Proc. Nat. Acad. Sci. 3 : 494-496

CAMPBELL, D.H. 1920. Studies on some East Indian Hepaticae. *Calobryum blumei* N. ab E. Ann. Bot. 34 : 1-12, Pl. 1

CAMPBELL, D.H. 1924. A remarkable development of the sporophyte in *Anthoceros fusiformis* Aust. Ann. Bot. 38 : 473-483

CAMPBELL, D.H. 1925. The relationships of the Anthocerotaceae. Flora 118-9 : 62-74

CAMPBELL, D.H., WILLIAMS, F. 1914. A morphological study of some members of the genus *Pallavicinia*. Stanford Univ. Publ., Univ. Ser. : 1-44

CAMPBELL, E.O., MARKHAM, K.R., PORTER, L.J. 1975. Dendroid liverworts of the order Metzgeriales in New Zealand. N.Z.J.Bot. 13 : 593-600

CANDOLLE, A.P. de 1813. *Théorie élémentaire de la Botanique*, Paris

CAVERS, F. 1903. Notes on Yorkshire Bryophytes. II. *Pallavicinia flotowiana.* The Naturalist 1903 : 441-444, 451-455, Pl. 16

CHADEFAUD, M. 1960. Les végétaux non vasculaires. In Chadefaud, M., Emberger, L., *Traité de Botanique*, Tome 1, Masson, Paris

CHALONER, W.G., MENSAH, M.K., CRANE, M.D. 1974. Non-vascular land plants from the devonian of Ghana. Palaeontology 17 : 925-947, Pl. 120-124

CHOPRA, R.S. 1967. Relationship between Liverworts and Mosses. Phytomorphology 17 : 70-78

CLEE, D.A. 1937. Leaf arrangement in relation to water conduction in the foliose hepaticae. Ann. Bot. N.S. 1 : 325-328

CLEE, D.A. 1939. The morphology and anatomy of *Pellia epiphylla* considered in relation to the mechanism of absorption and conduction of water. Ann. Bot. N.S. 3 : 105-111

CLEE, D.A. 1943. The morphology and anatomy of *Fegatella conica* in relation to the mechanism of absorption and conduction of water. Ann. Bot. N.S. 7 : 185-193

COESFELD, R. 1892. Beiträge zur Anatomie und Physiologie der Laubmoose. Bot. Z. 50 : 153-164, 169-176, 185-193, Pl. 4

COLLINS, N.J., OECHEL, W.C. 1974. The pattern of growth and translocation of photosynthate in a tundra moss, *Polytrichum alpinum*. Can. J. Bot. 52 : 355-363

CRIBBS, J.E. 1918. A columella in *Marchantia polymorpha*. Bot. Gaz. 65 : 91-96, Pl. 1-2

CRONQUIST, A. 1976. The taxonomic significance of the structure of plant proteins : A classical taxonomist's view. Brittonia 28 : 1-27

CROW, W.B. 1946. The uniformity of structure in the sporophyte of land plants. J. Indian Bot. Soc., M.O.P. Iyengar Commemoration vol. : 25-28

DAINTY, J. 1963. Water relations of plant cells. Adv. Bot. Res. 1 : 279-326

DAVY de VIRVILLE, A. 1927-8. L'action du milieu sur les mousses. Rev. Gén. Bot. 39 (1927) : 364-383, 449-457, 515-522, 560-586, 638-662, 711-726, 767-783; ibid. 40 (1928) : 30-44, 95-110, 156-173, Pl. 1-18

DIXON, H.N. 1932. Classification of Mosses, in Verdoorn, Fr., *Manual of Bryology*, pp. 397-412, Martinus Nijhoff, The Hague

DORR, I., KOLLMANN, R. 1975. Strukturelle Grundlagen des Parasitismus bei *Orobanche*. II. Die Differenzierung der Assimilatleitungsbahn im Haustorialgewebe. Protoplasma 83 : 185-199

DOUIN, R., DAVY de VIRVILLE, A. 1924. Action du milieu sur le *Fegatella conica* Corda. Rev. gén. Bot. 36 : 513-530, Pl. 11-12

DOYLE, W.T. 1970. *The Biology of Higher Cryptogams*, Macmillan, London

DUBOIS, A., HEBANT, C. 1968. Un *Riella* nouveau pour la France. Naturalia monspeliensia, Bot., 19 : 43-46, Pl. 1

DUCKETT, J.G. 1975. Spermatogenesis in pteridophytes. In Duckett, J.G., Racey, P.A., *The biology of the male gamete*, Biol. J. Linn. Soc. 7, suppl. 1 : 97-127

DUDKIN, M.S., ARESHIDZE, I.V., CHERNO, N.K. 1971. On studying chemical composition of two *Chara* species (in Ukrainian; English summary). Rev. botanique ukrainienne 28 : 762-765

EDWARDS, David S. 1973. *Studies on the Flora of the Rhynie Chert*, Ph.D. Thesis, Cardiff

EDWARDS, Dianne 1970. Fertile Rhyniophytina from the lower devonian of Britain. Paleontology 13 : 451-461

ERICKSON, M., LARSSON, S., MIKSCHE. G.E. 1973. Gaschromatographische Analyse von Ligninoxydationsprodukten. VII. Ein verbessertes Verfahren zur Charakterisierung von Ligninen durch Methylierung und oxydativen Abbau. Acta Chem. Scand. 27 : 127-140

ERIKSON, M., MIKSCHE, G.E. 1974a. On the occurrence of lignin or polyphenols in some mosses and liverworts. Phytochemistry 13 : 2295-2299

ERIKSON, M., MIKSCHE, G.E. 1974b. Two dibenzofurans obtained on oxidative degradation of the moss *Polytrichum commune*. Acta chem. Scand. B28 : 109-113

ESAU, K. 1969. The Phloem, in *Handbuch der Pflanzenanatomie* V/2, Berlin-Stuttgart

ESAU, K., CHEADLE, V.I., GIFFORD, E.M., Jr. 1973. Comparative structure and possible trends of specialization of the phloem. Amer. J. Bot. 40 : 9-19

ESAU, K., GILL, R.H. 1973. Correlations in differentiation of protophloem sieve elements of *Allium cepa* root. J. Ultrastructure Research 44 : 310-328

ESCHRICH, W. 1970. Biochemistry and fine structure of phloem in relation to transport. Ann. Rev. Plant Physiol. 21 : 193-214

ESCHRICH, W. 1975. Bidirectional transport, pp. 401-416, in Aronoff, S. et al. Eds., *Phloem Transport*, Plenum Press, New York

ESCHRICH, W., STEINER, M. 1967. Autoradiographische Untersuchungen zum Stofftransport bei *Polytrichum commune*. Planta 74 : 330-349

ESCHRICH, W., STEINER, M. 1968a. Die Struktur des Leitgewebesystems von *Polytrichum commune*. Planta 82 : 33-49

ESCHRICH, W., STEINER, M. 1968b. Die submikroskopische Struktur der Assimilatleitbahnen von *Polytrichum commune*. Planta 82 : 321-336

EVANS, A.W. 1925. The lobate species of *Symphyogyna*. Trans. Connecticut Acad. Arts Sci. 27 : 1-50

EYME, J., SUIRE, Cl. 1967. Au sujet de l'infrastructure de la zone placentaire de *Mnium cuspidatum* Hedw. (Mousse bryale acrocarpe). C.R. Acad. Sc. 265D : 1788-1791, Pl. 1-6

FABRE, M.-C. 1973. Etude en microscopie électronique à balayage de la porosité des hyalocystes de quelques Sphaines (*Sphagnum div.sp.*, Bryophytes, Sphagnales). C.R. Acad. Sc. (Paris) 276D : 513-516, Pl. 1-2

FABRE, M.-C. 1974a. Etude en microscopie électronique à balayage de la porosité du hyaloderme des tiges de deux Sphaignes (*Sphagnum palustre* L. et *S. centrale* Jens.). Bull. Soc. Bot. Fr. 121, Suppl. Coll. Bryologie : 137-142

FABRE, M.-C. 1974b. *Contribution à l'étude structurale et cytochimique des Sphaignes de la section Palustria L.*, Thèse Doctorat Spécialité, Lille

FARMER, J.B. 1894. Studies in Hepaticae : on *Pallavicinia decipiens* Mitten. Ann. Bot. 8 : 35-52, Pl. 6-7

FAVALI, M.A., BASSI, M. 1974. Seta ultrastructure in *Polytrichum commune* L. Nova Hedwigia 25 : 451-463

FAVALI, M.A., GIANNI, F. 1973. Sporophyte ultrastructure in *Tortula muralis* Hedw. Oesterr. Bot. Z. 122 : 323-331

FINLAYSON, M.K. 1950. Some aspects of the life history of the liverwort *Symphyogyna hymenophyllum* (Hook.) M. et N. Trans. Roy. Soc. N.Z. 78 : 280-290, Pl. 32-35

FINOCCHIO, A.F. 1963. *Simple pits and other anatomical aspects of certain moss gametophores*, Ph.D. Thesis, Columbia University (University Microfilms, Ann Arbor, no. 64-2745)

FINOCCHIO, A.F. 1967. Pitting of cells in moss gametophores. Bull. Torrey Bot. Club 94 : 18-20

FIRTSCH, G. 1883. Ueber einige mechanische Einrichtungen im anatomischen Bau von *Polytrichum juniperinum* Willd. Ber. Deutsch. Bot. Ges. 1 : 83-97, Pl. 2

FLEISCHER, M. 1920. Ueber die Entwicklung der Zwergmännchen aus sexuell differenzierten Sporen bei den Laubmoosen. Ber. Deutsch. Bot. Ges. 38 : 84-92, Pl. 2

FREDERICK, S.E., GRUBER, P.J., TOLBERT, N.E. 1973. The occurrence of glycolate dehydrogenase and glycolate oxidase in green plants. An evolutionary survey. Plant Physiol. 52 : 318-323

FREELAND, R.O. 1957. Plastid pigments of gametophytes and sporophytes of Musci. Plant Physiol. 32 : 64-66

FREY, W. 1971. Blattentwicklung bei Laubmoosen. Nova Hedwigia 20 : 463-556, Pl. 1-25

FREY, W. 1974a. Bau und Entstehung der Zentralstrang-verzweigung bei Laubmoosen (Musci). Nova Hedwigia 25 : 211-223, Pl. 1-3

FREY, W. 1974b. Entwicklungsgeschichtliche Untersuchungen an *Hypnodendron dendroides* (Brid.) Touw (Hypnodendraceae, Musci). Nova Hedwigia 25 : 229-249, Pl. 1-5

FULFORD, M. 1965. Evolutionary trends and convergence in the Hepaticae. Bryologist 68 : 1-31

GARNER, D.L., PAOLILLO, D.J., Jr. 1973. On the functioning of stomates in *Funaria*. Bryologist 76 : 423-427

GJOKIĆ, G. 1895. Ueber die chemische Beschaffenheit der Zellhäute bei den Moosen. Oesterr. Bot. Z. 45 : 330-334

GOEBEL, K. 1895. Archegoniatenstudien. 6. Ueber Function und Anlegung der Lebermoos-Elateren. Flora 80 : 1-37, Pl. 1

GOEBEL, K. 1900. *Organographie der Pflanzen*. Zweiter Teil. *Spezielle Organographie.* Jena

GOEBEL, K. 1905. *Organography of Plants.* Part II. *Special Organography.* Translated by I.B. Balfour. Oxford University Press. (Reprint edition : Hafner, New York & London, 1969)

GOEBEL, K. 1906. Archegoniatenstudien. 10. Beiträge zur Kenntnis australischer und neuseeländischer Bryophyten. Flora 96 : 1-202

GOTTSCHE, C.M. 1843. Anatomisch-physiologische Untersuchungen über *Haplomitrium hookeri* N. v.E. mit Vergleichung anderer Lebermoose. Nova Acta Acad. Caes. Leop.-Carol. Nat. Cur. 20 : 267-398, pl. 13-20

GREBE, C. 1917. Studien zur Biologie und Geographie der Laubmoose. I. Biologie und Oecologie der Laubmoose. Hedwigia 59 : 1-208

GROLLE, R. 1972. Die Namen der Familien und Unterfamilien der Lebermoose (Hepaticopsida). J. Bryol. 7 : 201-236

GRUBB, P.J. 1970. Observations on the structure and biology of *Haplomitrium* and *Takakia*, hepatics with roots. New Phytol. 69 : 303-326

GUNNING, B.S.E., PATE, J.S. 1969. "Transfer cells". Plant cells with wall ingrowths, specialized in relation to short distance transport of solutes – their occurrence, structure and development. Protoplasma 68 : 107-133

GUTTENBERG, H.V. 1943. Die physiologischen Scheiden, in Linsbauer, K., Tischler, G., Pascher, A., Eds., *Handbuch der Pflanzenanatomie* I/2, bd. V, Berlin

GUYOMARC'H, C. 1969. Sur les variations de l'azote soluble au cours du développement du sporophyte chez deux Mousses Bryales. C.R. Acad. Sc. 268D : 2339-2342

HABERLANDT, G. 1879. *Die Entwickelungsgeschichte des mechanischen Gewebesystems der Pflanzen*, Engelmann, Leipzig

HABERLANDT, G. 1883. Ueber die physiologische Funktion des Centralstranges im Laubmoosstämmchen. Ber. Deutsch. Bot. Ges. 1 : 263-268

HABERLANDT, G. 1886. Beiträge zur Anatomie und Physiologie der Laubmoose. Jb. wiss. Bot. 17 : 359-498, Pl. 21-27

HABERLANDT, G. 1914. *Physiological Plant Anatomy* (trad. Drummond, M.), Macmillan, London

HALLET, J.-N. 1972. Morphogénèse du gamétophyte feuillé du *Polytrichum formosum* Hedw. I. Etude histochimique, histoautoradiographique et cytophotométrique du point végétatif. Ann. Sci. Nat., Bot., Sér. *12*, 13 : 19-118

HARBORNE, J.B., MABRY, T.J., MABRY, H., Eds. 1975. *The Flavonoids*, Chapman & Hall, London

HASSEL de MENENDEZ, G. 1961. Las especies argentinas del género *Symphyogyna*. Bol. Soc. Argentina Bot. 9 : 233-260

HATTORI, S., IWATSUKI, Z., MIZUTANI, M., INOUE, S. 1974. Speciation of *Takakia*. J. Hattori Bot. Lab. 38 : 115-121

HÉBANT, C. 1964. Sur les leptoïdes des bryophytes. C.R. Acad. Sci. (Paris) 258 : 3339-3341, Pl. 1-2

HÉBANT, C. 1964-5. Signification et évolution des tissus conducteurs chez les bryophytes. Naturalia monspeliensia, sér. Bot., 16 : 79-86

HÉBANT, C. 1966a. Différenciation des tissus conducteurs dans la tige feuillée de *Polytrichum juniperinum* Willd. (Mousses Polytrichales). C.R. Acad. Sci. (Paris) 262D : 2585-2588, Pl. 1-2

HÉBANT, C. 1966b. Précisions nouvelles sur la signification et la répartition des tissus conducteurs dans la tige feuillée des mousses. C.R. Acad. Sci. 263D : 1065-1068, Pl. 1

HÉBANT, C. 1967a. Sur la comparaison des tissus conducteurs des bryophytes et des plantes vasculaires. C.R. Acad. Sci. 264D : 901-903, Pl. 1-2

HÉBANT, C. 1967b. Structure et différenciation des tissus conducteurs dans le gamétophyte des *Polytrichum*. Naturalia monspeliensia, sér. Bot., 18 : 293-318, Pl. 1-15

HÉBANT, C. 1968. Observations en microscopie à fluorescence sur la répartition de la callose chez les mousses (Bryopsida). Naturalia monspeliensia, sér. Bot., 19 : 75-78, Pl. 1

HÉBANT, C. 1969a. Nouvelles observations sur le leptome de la tige feuillée des *Polytrichum*. C.R. Acad. Sci. 269D : 2530-2533, Pl. 1-4

HÉBANT, C. 1969b. Observations sur les traces foliaires des mousses *s.str.* (Bryopsida). I. Les hydroïdes et leurs relations avec le cylindre central. Rev. bryol. lichénol. 36 : 721-728, Pl. 1-5

HÉBANT, C. 1970a. Aspects infrastructuraux observés au cours de la différenciation du phloème (leptome) dans la tige feuillée de quelques Polytrichales. C.R. Acad. Sci. (Paris) 271D : 1361-1364, Pl. 1-4

HÉBANT, C. 1970b. Etude histochimique, histoenzymologique et infrastructurale de la différenciation des tissus conducteurs dans la tige aérienne feuillée de quelques Mousses Polytrichales. C.R. Acad. Sci. (Paris) 271D : 1986-1989, Pl. 1-4

HÉBANT, C. 1970c (1971). A new look at the conducting tissues of mosses (Bryopsida) : their structure, distribution and significance. Phytomorphology 20 : 390-410

HÉBANT, C. 1972a. Précisions nouvelles sur la structure et la signification du faisceau conducteur irriguant le gamétophyte de *Takakia lepidozioides* Hattori et Inoue. C.R. Acad. Sci. (Paris) 275D : 189-192, Pl. 1-2

HÉBANT, C. 1972b. Observations sur les traces foliaires des mousses *s.str.* (Bryopsida). II. Etude, chez quelques Polytrichales, des éléments à caractères "phloémiens". Nova Hedwigia 23 : 735-758, Pl. 1-11

HÉBANT, C. 1972c. Les tissus conducteurs des bryophytes. L'Information Scientifique 1972 (5) : 207-219

HÉBANT, C. 1973a. Observations sur le leptome de *Dawsonia* et *Dendroligotrichum*. C.R. Acad. Sci. 276D : 3131-3134, Pl. 1-2

HÉBANT, C. 1973b. Acid phosphomonoesterase activities (β-glycerophosphatase and naphtol AS-MX phosphatase) in conducting tissues of bryophytes. Protoplasma 77 : 231-241

HÉBANT, C. 1973c. Studies on the development of the conducting tissue-system in the gametophytes of some Polytrichales. I. Miscellaneous notes on apical segmentation, growth of gametophytes, and diversity in histo-anatomical structures. J. Hattori Bot. Lab. 37 : 211-227

HÉBANT, C. 1973d. Diversity of structure of the water-conducting elements in Liverworts and Mosses. J. Hattori Bot. Lab. 37 : 229-234

HÉBANT, C. 1973e. Le noyau des cellules "phloémiennes" (leptoïdes) des Bryophytes est-il dégénérescent ? C.R. Acad. Sci. 277D : 1445-1447, Pl. 1-2

HÉBANT, C. 1974. Studies on the development of the conducting tissue-system in the gametophytes of some Polytrichales. II. Development and structure at maturity of the hydroids of the central strand. J. Hattori Bot. Lab. 38 : 565-607, Pl. 1-23

HÉBANT, C. 1975a. Organization of the conducting tissue-system in the sporophytes of *Dawsonia* and *Dendroligotrichum*. J. Hattori Bot. Lab. 39 : 235-254, Pl. 1-11

HÉBANT, C. 1975b. On the occurrence of lysosomal acid phosphatase activity in the differentiating water-conducting strand of *Takakia* and its evolutionary significance. Phytomorphology 25 : 279-282

HÉBANT, C. 1975c. The phloem (leptom) of bryophytes, pp. 211-215, in Aronoff, S. et al. Eds., *Phloem Transport*, Plenum Press, New York

HÉBANT, C. 1976a. Comparative anatomy of the gametophytes in *Dawsonia* (Polytrichales, Musci). J. Hattori Bot. Lab. 40 : 221-246, Pl. 1-13

HÉBANT, C. 1976b. Studies on the development of the conducting tissue-system in the gametophytes of some Polytrichales. III. Further observations on leptoids, with particular reference to their endoplasmic reticulum. Protoplasma 87 : 79-90

HÉBANT, C. 1976c. Evidence for the presence of sieve elements in the vascularized gametophytes of *Psilotum* from Holloway's collections. New Zealand J. Bot. 14 : 187-191

HÉBANT, C., BERTHIER, J. 1972. La ramification et ses conséquences anatomiques dans la tige aérienne feuillée des Polytrichales (étude morphogénétique et histologique de quelques espèces appartenant aux genres *Polytrichum, Pogonatum* et *Dendroligotrichum*). Rev. Bryol. Lichénol. 38 : 177-240, Pl. 1-9

HÉBANT, C., JOHNSON, R.P.C. 1976. Ultrastructural features of freeze-etched water-conducting cells in *Polytrichum* (Polytrichales, Musci). Cytobiologie 13 : 354-363

HÉBANT, C., MARTY, F. 1972. Fine structural identification of peroxisomes in the cells of the photosynthetic lamellae of the leaf of *Polytrichum commune* gametophytes. J. Bryol. 7 : 195-199

HÉBANT, C., SCHÄFER, P. 1974. Aperçu historique sur les premiers travaux relatifs aux tissus conducteurs des bryophytes. Bull. Soc. Bot. France, Coll. Bryologie : 49-59

HÉBANT, C., SUIRE, Cl. 1974. Mise en évidence d'activités enzymatiques au niveau de la zone de transfert gamétophyte-sporophyte chez quelques bryophytes. Rev. Bryol. Lichénol. 40 : 171-181

HECKMAN, C.A. 1973. *Exine structure and formation in the Jungermanniae, with reference to spore and elater development*, Ph.D. Thesis, University of Massachusetts (University Microfilms, Ann Arbor, no. 73-14645)

HEDWIG, J. 1782. *Fundamentum Historiae Naturalis Muscorum Frondosorum*, Vols. 1-2, Leipzig

HEDWIG, J. 1784. *Theoria Generationis et Fructificationis Plantarum Cryptogamicarum Linnaei*

HEDWIG, J. 1787-97. *Descriptio et Adumbratio Microscopico-Analytica Muscorum Frondosorum*, Vols. 1-4, Leipzig

HEGNAUER, R. 1962. *Chemotaxonomie der Pflanzen.* Band 1. *Thallophyten, Bryophyten, Pteridophyten und Gymnospermen*. Birkhäuser Verlag, Basel und Stuttgart

HENDERSON, I.F., HENDERSON, W.D. 1963. *A dictionary of biological terms*, 8th edition by Kenneth, J.H; Oliver & Boyd, Edinburg and London

HERZOG, Th. 1925. Anatomie der Lebermoose. In Linsbauer, K., *Handbuch der Pflanzenanatomie*, bd VII, Berlin

HOFMEISTER, W. 1854. Zur Morphologie der Moose. Ber. Kön. Sächs. Ges. Wiss. Leipzig, math.-phys. Kl. : 93-106

HOLLOWAY, J.E. 1939. The gametophyte, embryo, and young rhizome of *Psilotum triquetrum* Swartz. Ann. Bot. N.S. 3 : 313-336, Pl. 8-9

HOOKER, W.J. 1812-16. *British Jungermanniae*, Longmans, London

HY, F. 1880. De la structure de la tige des mousses de la famille des Polytrics. Bull. Soc. Bot. Fr. 27 : 106-112

ISAAC, I. 1941. The structure of *Anthoceros laevis* in relation to its water supply. Ann. Bot. N.S. 5 : 339-351

JONES, R.A., MONTAGUE, M.J., TAYLOR, J. 1973. Urease and urea amidolyase : determination of activity in liverworts. Phytochemistry 12 : 1675-1676

KAHLEM, G., HEBANT, C. 1973. Electrophoretic study of acid phosphatases in *Polytrichum commune*. Bryologist 76 : 554-556

KAWAI, I. 1965. Studies on the genus *Grimmia*, with reference to the affinity of gametophyte. Sci. Rep. Kanazawa Univ. 10 : 79-132, Pl. 1-64

KAWAI, I. 1969. Studies on the affinity regarding the conducting tissue of the sporophyte in the Musci. (2). On the seta in some species of the family Polytrichaceae. Science Reports Kanazawa University 14 : 39-57

KELLEY, C. 1969. Wall projections in the sporophyte and gametophyte of *Sphaerocarpos*. J. Cell Biol. 41 : 910-914

KEVAN, P.G., CHALONER, W.G., SAVILE, D.B.O. 1975. Interrelationships of early terrestrial arthropods and plants. Palaeontology 18 : 391-417, Pl. 54-56

KIDSTON, R., LANG, W.H. 1920. On old red sandstone plants showing structure, from the Rhynie Chert Bed, Aberdeenshire. Part II. Additional notes on *Rhynia gwynne-vaughani* Kidston and Lang; with descriptions of *Rhynia major*, n.sp., and *Hornea lignieri*, n.g., n.sp. Trans. Roy. Soc. Edinburgh 52 : 603-627, Pl. 1-10

KIDSTON, R., LANG, W.H. 1921. On old red sandstone plants showing structure, from the Rhynie Chert Bed, Aberdeenshire. Part IV. Restoration of the vascular cryptogams and discussion of their bearing on the general morphology of the pteridophyta and the origin of the organization of land-plants. Trans. Roy. Soc. Edinburgh 52 : 831-854, Pl. 1-5

KIENITZ-GERLOFF, F. 1902. Neue Studien über Plasmodesmen. Ber. deutschen Bot. Ges. 20 : 93-117, Pl. 4

KLEPPER, Betty. 1963. Water relations of *Dicranum scoparium*. Bryologist 66 : 41-54

KLEPPER, Betty. 1968. A comparison of the water relations of a moss and a lichen. Nova Hedwigia 15 : 13-20, Pl. 3

KOLATA, G.B. 1976. DNA sequencing : a new era in molecular biology. Science 192 : 645-647

KOLLMANN, R. 1964. On the fine structure of the sieve element protoplast. Phytomorphology 14 : 247-264

KOLLMANN, R. 1965. Zur Lokalisierung der funktionstüchtigen Siebzellen im sekundären Phloem von *Metasequoia glyptostroboides*. Planta 65 : 173-179

KRATZL, K., EIBL, J. 1951. Ueber den chemischen und botanischen Nachweis der Verholzung (Zur Phylogenese des Lignins). Mitt. Oesterr. Ges. Holzforschung 3 : 77-85

LACEY, W.S. 1969. Fossil bryophytes. Biol. Rev. 44 : 189-205

LAMBERT, A.M. 1968. Différenciation des spores et élatères chez quelques hépatiques. Bull. Soc. Bot. Nord Fr. 21 : 79-86, Pl. 1-4

LARPENT-GOURGAUD, M. 1974. Problèmes posés par les échanges intercellulaires dans le protonéma des Bryophytes : mise en évidence des phénomènes de transport. Bull. Soc. Bot. France, Coll. Bryologie : 161-167

LEACH, W. 1930. Note on the effect of growing mosses in a moisture-saturated atmosphere and under conditions of darkness. New Phytol. 29 : 276-284

LECLERC du SABLON, M. 1885. Recherches sur le développement du sporogone des hépatiques. Ann. Sc. Nat., Bot., Sér. 7, 2 : 126-180, Pl. 7-11

LEITGEB, H. 1868. Beiträge zur Entwicklungsgeschichte der Pflanzenorgane. I. Wachsthum des Stämmchens von *Fontinalis antipyretica*. Sber. Akad. Wiss., math.-nat. Cl., 57 : 308-342, Pl. 1-4

LEITGEB, H. 1877. *Untersuchungen ueber die Lebermoose*, Vol. 3, O. Deistung, Jena

LEMOIGNE, Y. 1968. Observation d'archégones portés par des axes de type *Rhynia gwynne-vaughanii* Kidston et Lang. Existence de gamétophytes vascularisés au Dévonien. C.R. Acad. Sci. (Paris) 266D : 1655-1657, Pl. 1

LIGNIER, O. 1903. Equisétales et Sphénophyllales. Leur origine filicinéenne commune. Bull. Soc. Linn. Normandie, sér. *5*, 7 : 93-137

LIGNIER, O. 1908. Essai sur l'évolution morphologique du règne végétal. C.R. Ass. Fr. Av. Sci., Clermont-Ferrand : 530-542 (also in Bull. Soc. Linn. Normandie, sér. *6*, 3 (1911) : 35-62)

LINSBAUER, K. 1899. Zur Verbreitung des Lignins bei Gefässkryptogamen. Oesterr. Bot. Z. 49 : 317-323

LIST, A. 1963. Some observations on DNA content and cell and nuclear volume growth in the developing xylem cells of certain higher plants. Amer. J. Bot. 50 : 320-329

LÖFFELHARDT, W., LUDWIG, B., KINDL, H. 1973. Thylakoid-gebundene L-Phenylalanin-Ammoniak-Lyase. Hope-Seyler's Z. Physiol. Chem. 354 : 1006-1012

LONGTON, R.E. 1974. Genecological differentiation in Bryophytes. J. Hattori Bot. Lab. 38 : 49-65

LORCH, W. 1894. Beiträge zur Anatomie und Biologie der Laubmoose. Flora 78 : 424-465

LORCH, W. 1909. Die Polytrichaceen. Eine biologische Monographie. Abh. bayer. Akad. Wiss., math.-phys. Kl., 23 : 445-546

LORCH, W. 1925a. Ueber die Saugzellen im Fusse und in der Vaginula bei den Laubmoosen. Ber. Deutsch. Bot. Ges. 43 : 120-127

LORCH, W. 1925b. Ueber die Haustorialschläuche am Fusse der Laubmoose. Ber. Deutsch. Bot. Ges. 43 : 262-270

LORCH, W. 1931. Anatomie der Laubmoose, in Linsbauer, K., *Handbuch der Pflanzenanatomie*, Bd. VII, Gebrüder Borntraeger, Berlin

LORENTZ, P.G. 1864. *Moosstudien*, Engelman, Leipzig

LORENTZ, P.G. 1867. Studien zur vergleichenden Anatomie der Laubmoose. Flora 50 : 241-248, 257-264, 289-297, 305-313, 526-528, 529-540, 544-558, Pl. 8-12

LORENTZ, P.G. 1867-8. Grundlinien zu einer vergleichenden Anatomie der Laubmoose. Jb. wiss. Bot. 6 : 363-466, Pl. 21-28

LORENTZ, P.G. 1869. Studien zur Anatomie des Querschnittes der Laubmoose. Flora 52 : 161-173, 193-208, 209-219, 225-233, 241-250, Pl. 2-6

LUGARDON, B. 1975. Sur la structure fine de l'exospore dans les divers groupes de Ptéridophytes actuelles (microspores et isospores). In Ferguson, I.K., Muller, J., *The evolutionary significance of the exine*, Linn. Soc. Symp. series, no. 1 : 231-250, Pl. 1-6

MADER, A. 1929. Untersuchungen über die Gattung *Moerckia*. Planta 8 : 742-790

MAGDEBURG, F. 1886. *Die Laubmooskapsel als Assimilationsorgan*, Thesis, Berlin

MÄGDEFRAU, K. 1935-6. Untersuchungen über die Wasserversorgung des Gametophyten und Sporophyten der Laubmoose. Z. Bot. 29 : 337-375

MÄGDEFRAU, K. 1969. Die Lebensformen der Laubmoose. Vegetatio 16 : 285-297

MÄGDEFRAU, K. 1971. Niedere Pflanzen, pp. 425-583, in *Lehrbuch der Botanik*, 30. Auflage, Gustav Fischer Verlag, Stuttgart

MAHMOUD, M.I. 1965. *Protoplasmatics and drought resistance in mosses*, Ph. D. Thesis, University of California, Davis (University Microfilms, Ann Arbor, no. 66-2387)

MAIER, K. 1967. Wandlabyrinthe im Sporophyten von *Polytrichum*. Planta 77 : 108-126

MAIER, K., MAIER, U. 1972. Localization of beta-glycerophosphatase and Mg^{++}-activated adenosine triphosphatase in a moss haustorium, and the relation of these enzymes to the cell wall labyrinth. Protoplasma 75 : 91-112

MARCHAL, El., MARCHAL, Em. 1906. Recherches physiologiques sur l'amidon chez les bryophytes. Bull. Soc. Roy. Bot. Belgique 43 : 115-214

MARKHAM, K.R., PORTER, L.J. 1969. Flavonoids in the green algae (Chlorophyta). Phytochemistry 8 : 1777-1781

MARSILI, A., MORELLI, I. 1968. Triterpenes from mosses. I. The occurrence of 22(29)-hopene in *Thamnium alopecurum* (L.) Br. Eur. ssp. *eu-alopecurum* Giac. Phytochemistry 7 : 1705-1706

MAXE, M. 1965. Aspects infrastructuraux des cellules du parenchyme phloémien de *Polypodium vulgare*. C.R. Acad. Sc. 260 : 5609-5612, Pl. 1-4

McCONAHA, M. 1939. Ventral surface specializations of *Conocephalum conicum*. Amer. J. Bot. 26 : 353-355

McCONAHA, M. 1941. Ventral structures effecting capillarity in the Marchantiales. Amer. J. Bot. 28 : 301-306

McCORMICK, F.A. 1914. A study of *Symphyogyna aspera*. Bot. Gaz. 58 : 401-418, Pl. 30-32

MEHRA, P.N. 1968. Phyletic evolution in the Hepaticae. Phytomorphology 17 : 47-58

MEHRA, P.N., HANDOO, O.N. 1953. Morphology of *Anthoceros erectus* and *A. himalayensis* and the phylogeny of Anthocerotales. Bot. Gaz. 114 : 371-382

MEHRA, P.N., SOKHI, J.K. 1972. Embryology of *Cyathodium flabellatum* Mehra. J.Hattori Bot. Lab. 36 : 17-53

MEIDNER, H., SHERIFF, D.W. 1976. *Water and Plants*, Blackie, Glasgow and London

MELIKYAN, A.P., DIL'DARYAN, B.I. 1975. Les types de structure anatomique de la tige du gamétophyte chez quelques représentants de divers groupes écologiques de mousses feuillées d'Arménie. Biol. Zh. Armenii S.S.R. 28 : 19-24

MENDER, G. 1938. Protoplasmatische Anatomie des Laubmooses *Bryum capillare*. l. Protoplasma 30 : 373-400

MENON, M.K.C., LAL, M. 1974. Morphogenetic role of kinetin and abscisic acid in the moss *Physcomitrium*. Planta 115 : 319-328

MERKER, H. 1958. Zum fehlenden Gliede der Rhynienflora. Bot. Not. 111 : 608-618

MEYER, M.W., ANGERMAN, C. 1973. Enzymes in the moss *Funaria hygrometrica* : phenylalanine ammonia lyase and its control. Bryologist 76 : 361-365

MŁODZIANOWSKI, F. 1970. Fine structure of protonema in *Funaria hygrometrica*. Acta Soc. Bot. Pol. 39 : 45-50

MONROE, J.H. 1965. *Some aspects of ultrastructure in Funaria*, Ph.D. Thesis, University of Texas, Austin (University Microfilms, Ann Arbor, no. 65-10754)

MORIN, F. 1893. *Anatomie comparée et expérimentale de la feuille des Muscinées. Anatomie de la nervure appliquée à la classification*. 139 pp., Pl. 1-24, Thèse, Rennes, France

MOTTE, J. 1928. Contributions à la connaissance cytologique des Muscinées. Ann. Sc. Nat., Bot., *10em* Sér., 10 : 293-543, Pl. 1-11

MOZINGO, H.N., KLEIN, P., ZEEVI, Y., LEWIS, E.R. 1969. Scanning electron microscope studies on *Sphagnum imbricatum*. Bryologist 72 : 484-488

MUELLER, D.M. 1972. Observations on the ultrastructure of *Buxbaumia* protomena. Plasmodesmata in the cross walls. Bryologist 75 : 63-68

MUELLER, D.M. 1973. The peristome of *Fissidens limbatus* Sullivant. Univ. California Publ. Bot. 63 : 1-34, Pl. 1-35

MÜLLER, F.P. 1973. Aphiden an Moosen. Entomologische Abhandlungen 39: 205-242

NACHLAS, M.M., TSOW, K.-C., SOUZA, E. de, CHENG, C.S., SELIGMAN, A.M. 1957. Cytochemical demonstration of succinic dehydrogenase by the use of a new p-nitrophenyl substituted ditetrazole. J. Histochem. Cytochem. 5 : 420-436

NECKER, N. de 1775. *Physiologie des corps organisés* (French translation of *Physiologia Muscorum*, published in 1774 in Mannheim)

NEUBERGER, D.S., EVERT, R.F. 1974. Structure and development of the sieve-element protoplast in the hypocotyl of *Pinus resinosa*. Amer. J. Bot. 61 : 360-374

NOAILLES, M.-C. 1974. Etude ultrastructurale comparée de la différenciation du parenchyme et des cellules périphériques de la tige d'une Mousse pleurocarpe : *Pleurozium schreberi* (Willd.) Mitt. (Hypnobryale). C.R. Acad. Sci. (Paris) 278D : 441-444, Pl. 1-4

NOGUCHI, A. 1974. Musci Japonici. XI. The families Disceliaceae, Ephemeraceae, Oedipodiaceae, Splachnaceae, and Schistostegaceae. J. Hattori Bot. Lab. 38 : 387-404

NÖRR, M. 1974. Trockenresistenz bei Moosen. Flora 163 : 371-387

O'BRIEN, T.P. 1970. Further observations on hydrolysis of the cell wall in the xylem. Protoplasma 69 : 1-14

OLTMANNS, F. 1884. *Ueber die Wasserbewegung in der Moospflanze und ihren Einfluss auf die Wasservertheilung im Boden*, Thesis, Strassburg

PALISOT de BEAUVOIS, A. 1822. *Muscologie*, ou *Traité sur les Mousses*, Paris

PANT, D.D. 1962. The gametophyte of the Psilophytales. In Maheshwari, P., Johri, B.M., Vasil, I.K., *Proc. Summer School Bot., Darjeeling*. Min. Sci. Res. Cult. Affairs, New Delhi. Pp. 276-301

PAOLILLO, D.J., Jr., BAZZAZ, F.A. 1968. Photosynthesis in sporophytes of *Polytrichum* and *Funaria*. Bryologist 71 : 335-343

PAOLILLO, D.J., Jr., REIGHARD, J.A. 1967. Ultrastructural features of some polytrichaceous moss leaves. Bryologist 70 : 61-69

PARTHASARATHY, M.V. 1974a. Ultrastructure of phloem in Palms. II. Structural changes, and fate of the organelles in differentiating sieve elements. Protoplasma 79 : 93-125

PARTHASARATHY, M.V. 1974b. Ultrastructure of phloem in Palms. III. Mature phloem. Protoplasma 79 : 265-315

PATE, J.S., GUNNING, B.E.S. 1972. Transfer cells. Ann. Rev. Plant Physiol. 23 : 173-196

PATON, J.A., PEARCE, J.V. 1957. The occurrence, structure and functions of the stomata in British bryophytes. Trans. Brit. Bryol. Soc. 3 : 228-259

PATTERSON, P.M. 1964. Problems presented by bryophytic xerophytism. Bryologist 67 : 390-396

PAUL, H. 1903. Beiträge zur Biologie der Laubmoosrhizoiden. Bot. Jb. 32 : 231-274

PICKETT-HEAPS, J.D., MARCHANT, H. 1972. The phylogeny of the green algae : a new proposal. Cytobios 6 : 255-264

PISKERNIK, A. 1914. Die Plasmaverbindungen bei Moosen. Oesterr. Bot. Z. 64 : 107-120, Pl. 5-6

POTONIÉ, H. 1883. Ueber die Zusammensetzung der Leitbündel bei den Gefässkryptogamen. Jb. Königlichen bot. Gart. 2 : 233-278

POTTIER, J. 1923. *La Chimie des Mousses.* Imp. P. Tranchart, Saint Vit.

PROSKAUER, J. 1960. Studies on Anthocerotales VI. Phytomorphology 10 : 1-19

RICHARDS, P.W. 1951. Evolutionary trends and classification in the bryophyta. Proc. Leeds Phil. Lit. Soc. (Sci. Sect.) 6 : 58-60

RIDGWAY, J.E. 1968. Ultrastructural features of the sporophytic-gametophytic interphase in bryophytes. 26th annual EMSA Meeting : 86-87

RIFOT, M., BARRIERE, G. 1974. La conduction dans le thalle de l'hépatique *Conocephalum conicum* (L.) Dum. I. Etude du transit de l'eau à l'aide d'une solution d'acétate de sodium ^{14}C. Rev. Bryol. Lichénol. 40 : 45-52

ROBERTS, E.A., HARING, I.M. 1937. The water relations of the cell walls of certain mosses as a determining factor in their distribution. Ann. Bryol. 10 : 97-114

ROSANDER, H.A. 1906. *Studier ofver Bladmossornas Organisation. Mössa, Vaginula och Sporogon*. Akad. Afh., Uppsala

ROSTOCK, R. 1902. *Ueber Aufnahme und Leitung des Wassers in der Laubmoospflanze*, Thesis (1901), Jena

ROTA, J.A., MARAVOLO, N.C. 1975. Transport and mobilization of ^{14}C-sucrose during regeneration in the hepatic, *Marchantia polymorpha*. Bot. Gaz. 136 : 184-188

ROTH, D. 1969. Embryo und Embryotheca bei den Laubmoosen. Eine histogenetische und morphologische Untersuchung. Bibliotheca Botanica 129 : 1-49, Pl. 1-9

RUHLAND, W. 1924. Musci, Allgemeiner Teil, in Engler, A., Prantl. K., *Die natürlichen Pflanzenfamilien*, II Aufl., 10. Leipzig

SAITO, K. 1975. A Monograph of Japanese Pottiaceae (Musci). J. Hattori Bot. Lab. 39 : 373-537

SALO, M.L. 1957. Lignin studies. II. The lignin content and properties of lignin in different materials. Maataloustieteellinen Aikakauskirja 29 : 202-210

SARKANEN, K.V., HERGERT, H.L. 1971. Classification and distribution, p. 43-94, in Sarkanen, K.V., Ludwig, C.H., *Lignins. Occurrence, formation, structure and reactions*. Wiley-Interscience, New York

SARKANEN, K.V., LUDWIG, C.H. 1971. *Lignins. Occurrence, formation, structure and reactions.* Wiley-Interscience, New York

SAUTER, J.J. 1972. Respiratory and phosphatase activities in contact cells of wood rays and their possible role in sugar secretion. Z. Pflanzenphysiol. 67 : 135-145

SCHACHT, H. 1856. *Lehrbuch der Anatomie und Physiologie der Gewächse*, erster Theil, Berlin

SCHACHT, H. 1859. *Lehrbuch der Anatomie und Physiologie der Gewächse*, zweiter Theil, Berlin

SCHAUER, T. 1967. Anatomische und systematische Studien über die mitteleuropäischen Arten der Gattung *Seligeria* (Musci). Nova Hedwigia 14 : 313-325

SCHEIRER, D.C. 1972. Anatomical studies in the Polytrichaceae. I. The gametophore of *Dendroligotrichum dendroides* (Hedw.) Broth. Bryologist 75 : 305-314

SCHEIRER, D.C. 1973. Hydrolysed walls in the water-conducting cells of *Dendroligotrichum* (Bryophyta) : histochemistry and ultrastructure Planta 115 : 37-46

SCHEIRER, D.C. 1974. *The anatomy, histochemistry and ultrastructure of the gametophore of Dendroligotrichum dendroides (Hedw.) Broth. (Bryopsida : Polytrichaceae)*, Ph.D. Thesis, Pennsylvania State University (University Microfilms, Ann Arbor, no. 75-09836)

SCHEIRER, D.C. 1975. Anatomical studies in the Polytrichaceae. II. Histochemical observations on thickened lateral walls of hydroids of *Dendroligotrichum*. Bryologist 78 : 113-123

SCHIMPER, K.F. 1857a. Bot. Zeitung 15 : 769

SCHIMPER, K.F. 1857b. Auszug, Stücke aus dem noch ungedruckten : ***Mooslob***, oder die schönsten Geschichten der Moose, alte und neue, in Versen für eine junge Dame zu einer eleganten Moossammlung von Dr. Karl F. Schimper. Festgabe für Bonn. Mainz, Sept. 1857

SCHIMPER, W.Ph. 1848. *Recherches anatomiques et morphologiques sur les mousses*, Thèse, Strasbourg

SCHIMPER, W.Ph. 1857. *Mémoire pour servir à l'histoire naturelle des Sphaignes*, 96 pp., Pl. 1-24, Paris

SCHNEPF, E. 1973. Mikrotubulus-Anordnung und -Umordnung, Wandbildung und Zellmorphogenese in jungen *Sphagnum*-Blättchen. Protoplasma 78 : 145-173

SCHOENAU, K. von 1912. Zur Verzweigung der Laubmoose. Hedwigia 51 : 1-56

SCHÖNHERR, J., ZIEGLER, H. 1975. Hydrophobic cuticular ledges prevent water entering the air pores of liverwort thalli. Planta 124 : 51-60

SCHULZ, C. 1937. *Beiträge zur Anatomie und Biologie der Polytrichaceen*, Thesis, Hamburg

SCHULZ, D., WIENCKE, C. 1976. Sporophytenentwicklung von *Funaria hygrometrica* Sibth. II. Differenzierung des Wasser- und Stoffleitungssystems in der Seta. Flora 165 : 47-60

SCHUSTER, R.M. 1966a. Studies on Hepaticae XV. Calobryales. Nova Hedwigia 13 : 1-76

SCHUSTER, R.M. 1966b. *The Hepaticae and Anthocerotae of North America* Vol. I, Columbia University Press, New York and London

SIEGEL, S.M. 1962. *The Plant Cell Wall*, Pergamon Press, New York

SIEGEL, S.M. 1969. Evidence for the presence of lignin in moss gametophytes. Amer. J. Bot. 56 : 175-179

SIEGEL, S.M., CARROL, P., UMENO, I., CORN, C. 1972. The evolution of lignin : experiments and observations. Recent Adv. Phytochem. 4 : 223-238

SIESSEGGER, B. 1966. *Die Elateren der Lebermoose, ihr Bau und ihre taxonomische Verwendbarkeit*, Thesis, Tübingen

SLOOVER, J. de 1959. Considération sur la valeur spécifique de *Moerckia flotoviana* (Nees) Schiffn., Dilaenacée nouvelle pour la flore belge. Bull. Jardin Bot. Etat (Bruxelles) 29 : 157-181

SMITH, D.L. 1972. Localization of phosphatases in young gametophytes of *Polypodium vulgare* L. Protoplasma 74 : 133-148

SMITH, G.L. 1971. A Conspectus of the genera of Polytrichaceae. Mem. New York Bot. Gard. 21(3) : 1-83

SMITH, G.M. 1955. *Cryptogamic Botany,* vol. 2, *Bryophytes and Pteridophytes*, 2nd ed., McGraw-Hill, New York and London

SMITH, J.L. 1964. Water-conducting system of *Symphyogyna*. Nature 202 : 617

SMITH, J.L. 1966. The liverworts *Pallavicinia* and *Symphyogyna* and their conducting system. Univ. Calif. Publ. Bot. 39 : 1-83, Pl. 1-18

SPRUCE, R. 1895. Hepaticae Elliottianae. J. Linn. Soc. London, Bot., 30 : 331-372, Pl. 20-30

SRIVASTAVA, L.M. 1975. Structure and differentiation of sieve elements in Angiosperms and Gymnosperms, pp. 33-62, in Aronoff, S. et al. Eds., *Phloem Transport*, Plenum Press, New York

STEERE, W.C. 1969. A new look at evolution and phylogeny in Bryophytes. In Gunckel, J.E., *Current Topics in Plant Science*, Academic Press, New York and London, 134-143

STEVENSON, D.W. 1974. Ultrastructure of the nacreous leptoids (sieve elements) in the polytrichaceous moss *Atrichum undulatum*. Amer. J. Bot. 61 : 414-421

STEWART, K.D., MATTOX, K. 1975. Comparative cytology, evolution and classification of the green algae with some consideration of the origin of other organisms with chlorophylls a and b. Bot. Rev. 41 : 104-135

STRASBURGER, E. 1891. *Histologische Beiträge*. Heft III. *Ueber den Bau und die Verrichtungen der Leitungsbahnen in den Pflanzen*. Gustav Fischer, Jena

STRUNK, R. 1914. *Beiträge zur Kenntnis der Organisation der Moose*, Thesis Bonn

STUDHALTER, R.A. 1938. Independence of the sporophyte in *Riella* and *Sphaerocarpus*. Ann. Bryol. 15 : 1-38

STUDHALTER, R.A. 1942. The foot of *Riella americana* and its relation to nutrition of the sporophyte. Bot. Gaz. 103 : 633-650

SUIRE, Cl. 1975. Les données actuelles sur la chimie des bryophytes. Rev. Bryol. Lichénol. 41 : 105-256

SWAIN, T. 1974. Biochemical evolution in plants, pp. 125-302, in Florkin, M., Stotz, E.H. Edts, *Comprehensive Biochemistry* 29A, Elsevier, Amsterdam

TANSLEY, A.G., CHICK, E. 1901. Notes on the conducting tissue-system in Bryophyta. Ann. Bot. 15 : 1-38, Pl. 1-2

TAYLOR, J., THOMAS, R.J., OTERO, J.G. 1972. Chromatographic and spectrophotometric analyses of the photosynthetic pigments of the gametophyte and the sporophyte of *Lophocolea heterophylla*. Bryologist 75 : 36-42

THIERY, J.P. 1967. Mise en évidence des polysaccharides sur coupes fines en microscopie électronique. J. Microscopie 6 : 987-1018, Pl. 1-12

THROWER, S.L. 1964. An investigation of translocation in *Dawsonia intermedia* C.M. Trans. British. Bryol. Soc. 4 : 664-667, Pl. 1

TROTTER, L.B.C. 1948. *Andreaea hartmanii* Thed. – A histological study. Trans. British Bryol. Soc. 1 : 101-103, Pl. 1-2

UHEL (GUYOMARC'H), C. 1975. Quantitative changes in free amino acids in *Polytrichum formosum* during development. Phytochemistry 14 : 2337-2340

UNGER, F. 1861. Beiträge zur Physiologie der Pflanzen. VII. Ueber den anatomischen Bau des Moostammes. Sitzber. math.-nat. Cl. Akad. Wiss. Wien 43 : 497-518, Pl. 1-4

VAISEY, J.R. 1887a. The transpiration of the sporophore of the musci. Ann. Bot. 1 : 73-74

VAISEY, J.R. 1887b. On the absorption of water and its relation to the consti tution of the cell-wall in mosses. Ann. Bot. 1 : 147-152

VAISEY, J.R. 1887c. On the constitution of cell-walls and its relation t absorption in Mosses. Rep. Brit. Ass. Adv. Sc., Manchester, 1887 : 77

VAISEY, J. R. 1888. On the anatomy and development of the sporogonium o the mosses. J. Linn. Soc., Bot., 24 : 262-285, Pl. 9-12

VAISEY, J.R. 1890a. Alternation of generations in green plants. Ann. Bot. 4 371-378

VAISEY, J.R. 1890b. On the morphology of the sporophyte of *Splachnum luteum*. Ann. Bot. 5 : 1-10, Pl. 1-2

WAENKER von DANKENSCHWEIL, H. 1915. *Beiträge zur Anatomie de Laubmoose*, Thesis, Freiburg

WALKER, J.R.L. 1975. *The biology of plant phenolics*, Arnold, London

WALLAND, A., KINZEL, H. 1966. Ueber die Zusammensetzung der Zellsäft bei Archegoniaten. Flora, Abt. A, 156 : 597-633

WALTER, H., STADELMANN, E. 1968. The physiological prerequisites fo the transition of autotrophic plants from water to terrestrial life. Bio Science 18 : 694-701

WARDLAW, C.W. 1952. *Phylogeny and Morphogenesis*, MacMillan, Londo

WARDLAW, C.W. 1968. *Morphogenesis in Plants*, Methuen, London

WARMBRODT, R.D., EVERT, R.F. 1974. Structure of the vascular paren chyma in the stem of *Lycopodium lucidulum*. Amer. J. Bot. 61 : 437-44

WATSON, E.V. 1971. *The Structure and Life of Bryophytes*, 2nd edn., Hut chinson University Library, London

WEATHERLEY, P.E. 1963. In *The water relations of plants*, Rutter, A.J. Whitehead, F.H., eds., pp. 85-100, Blackwell, Oxford

WEATHERLEY, P.E. 1970. Some aspects of water relations. Adv. Bot. Res 3 : 171-206

WENIGER, H.P. 1962. *Versuche über die Wasserleitung bei Laubmoosen, insbesondere die Wasserversorgung des Sporophyten*, Thesis, Freiburg

WENZL, H. 1933. Untersuchungen über den Wasserhaushalt von *Marchantia polymorpha*. Jb. wiss. Bot. 79 : 311-352

WETMORE, R.H., De MAGGIO, A.E., MOREL, G. 1963. A morphogenetic look at the alternation of generations. J. Indian Bot. Soc. 52A : 306-320, Pl. 1-2

WIENCKE, C., SCHULZ, D. 1975. Sporophytenentwicklung von *Funaria hygrometrica* Sibth. I. Strukturelle Grundlagen der Wasser- und Nährstoffaufnahme im Haustorium. Protoplasma 86 : 107-117

WIESNER, J. 1878. Note über das Verhalten des Phloroglucins und einiger verwandter Körper zur verholzten Zellmembran. Sitzber. math.-nat. Cl. Akad. Wiss. Wien 77 : 60-66

WIGGLESWORTH, G. 1956. Further notes on *Polytrichum commune* L. Trans. Brit. Bryol. Soc. 3 : 115-120

WILSON, A.C., SARICH, V.M., MAXSON, L.R. 1974. The importance of gene rearrangement in evolution : evidence from studies on rates of chromosomal, protein, and anatomical evolution. Proc. Nat. Acad. Sci. U.S.A. 71 : 3028-3030

WINKLER, S. 1969. Anatomische Untersuchungen über Leitstränge der Gametophyten bei den Lebermoosen. Oesterr. Bot. Z. 117 : 348-364

ZACHERL, H. 1956. Physiologische und ökologische Untersuchungen über die innere Wasserleitung bei Laubmoosen. Z. Bot. 44 : 409-436

ZANTEN, B.O. van 1973. A taxonomic revision of the genus *Dawsonia* R. Brown. Lindbergia 2 : 1-48

ZUCKERKANDL, E., PAULING, L. 1965. Molecules as documents of evolutionary history. J. Theoret. Biol. 8 : 357-366

ADDENDUM

KAWAI, I. 1976. Systematic studies on the conducting tissues of the gametophyte in musci. (6). On the essential coordination among the anatomical characteristics of the stem in some species of Hypnaceae. Sci. Rep. Kanazawa Univ. 21 : 47-124, pl. 1-51

MARAVOLO, N.C. 1976. Polarity and localization of auxin movement in the hepatic, *Marchantia polymorpha.* Amer. J. Bot. 63 : 526-531

NIKLAS, K.J. 1976. The role of morphological biochemical reciprocity in early land plant evolution. Ann. Bot. 40 : 1239-1254

SCHEIRER, D.C. 1976. Some fine structural observations on the rhizome of *Dendroligotrichum* (Bryophyta). Protoplasma 89 : 323-337

INDEX

PLATES

I should like to express my sincere thanks to the many Authors, Editors, Publishing Companies and Societies, who generously granted the permission to reproduce photographs or drawings in these plates*:

Prof. E.-J. Bonnot (161, 162)
Prof.-Dr. M. Bopp (144)
Dr. E.J. Bowen (Mrs Brunsden) (28)
Robin Burr (188)
Dr. B.G. Butterfield (188)
Prof. D.A. Clee (148 to 150)
Dr. N.J. Collins (337)
Prof.-Dr. W. Eschrich (338)
Prof. J. Eymé (317)
Dr. M.-C. Fabre (167 to 169)
Prof. A.F. Finocchio (160)
Dr. W. Frey (239, 240)
Dr. P.J. Grubb (146)
Dra. G. Hässel de Menéndez (57)
Prof. I. Isaac (147)
Prof. G. Kahlem (143)
Dr. A.-M. Lambert (176, 177)
Prof.-Dr. K. Mägdefrau (30, 145)
Prof. P.N. Mehra (152)
Mme Marthe Penot (325)
Prof. M. Penot (325)
Dr. B. Siessegger (48)
Dr. J.L. Smith (56)
Dr. Cl. Suire (317)

* Numbers in parentheses refer to the figures on the plates. References in the corresponding captions refer to full quotations of the original publications in the bibliography (pp. 117-142). Absence of reference in a caption means that the figure is original.

The American Journal of Botany (27)
The Annals of Botany (28, 44, 45, 50, 58, 117, 147 to 150)
J.B. Baillière : *L'Information Scientifique* (33, 43)
Berichte der deutschen botanischen Gesellschaft (123, 124)
Blackwell Scientific Publications : *Transactions of the British Bryological Society* (70 to 72)
The Bryologist (143, 188)
Bulletin de la Société Botanique de France (167, 168)
Bulletin de la Société Botanique du Nord de la France (176, 177)
Bulletin of the Torrey Botanical Club (160)
J. Cramer (Engelmann) : *Jahrbücher für wissenschaftliche Botanik* (6, 9, 10, 67, 92, 95, 97, 98, 118, 119); *Nova Hedwigia* (14, 101 to 103, 108 to 111, 239, 240, 249, 252, 253)
Gustav Fischer Verlag : (26, 30); *Flora* (37 to 39, 78 to 84, 91, 96); *Zeitschrift für Botanik* (145); *Zeitschrift für Pflanzenphysiologie* (144)
Gauthier-Villars : *Comptes Rendus de l'Académie des Sciences* (125, 161, 169, 181, 220 to 223, 232, 263, 264, 276, 285, 289, 301, 305, 308, 317, 328 to 330, 333, 334, 344)
Gebrüder Borntraeger : (36, 122); *Jahrbücher für wissenschaftliche Botanik* (93, 94, 116)
Journal of the Hattori Botanical Laboratory (13, 15, 16, 17 to 25, 69, 100, 152, 178, 182, 210, 211, 224 to 226, 230, 231, 236, 247, 248, 251, 254, 259 to 262, 265, 267, 268, 270 to 272, 274, 275, 281, 282, 311, 318 to 320, 326)
Journal of the Linnean Society of London, Botany (52, 113 to 115, 121)
National Research Council of Canada : *Canadian Journal of Botany* (337)
Naturalia monspeliensia, Série Botanique (31, 55, 85 to 89, 126, 131 to 141, 163 to 165, 290)
The Naturalist (59, 60)
The New Phytologist (146)
New Zealand Journal of Botany (343)
Phytomorphology (32, 46, 61 to 63, 127, 128, 205)

Plenum Publishing Corporation (331, 332)

Revue Bryologique et Lichénologique (73, 74, 76, 77, 104 to 107, 112, 255, 321, 322)

Revue Générale de Botanique (7, 8, 11, 12)

Springer-Verlag, Heidelberg : *Planta* (338)

Springer-Verlag, Wien : *Oesterreichische botanische Zeitschrift* (35); *Protoplasma* (193, 277 to 279, 293, 294, 304)

Stanford University Press (53, 54)

University of California Press (56)

The University of Chicago Press : *Botanical Gazette* (47)

LIST OF PLATES

LINE DRAWINGS

PHOTOGRAPHS

PLATE 1

Water-conducting strands in Bryophytes

Fig. 1. When present a "true" water-conducting strand consists of dead empty cells. It usually appears as a solid "protostele" in the median region of the leafy stem, or the seta, or the prothallus. A slight protrusion of the living cells of the internal cortex within the central strand is commonly observed – this being a consequence of differences in turgor pressure between the living (cortical) and dead (conducting) cells –. The present figure illustrates the general appearance of such a "true" water-conducting strand in a transverse section of a robust erect female leafy stem of the liverwort *Takakia lepidozioides* (specimen from British Columbia). x 280.

C, cortex; CS, central strand.

Figs. 2 to 5 (Diagrams). The structure of the water-conducting cells shows important variations according to the different groups. In hepatics (Figs. 2 and 3), they possess small plasmodesmata-derived pores in their walls [1] (Figs. 2B and 3B). In *Takakia* and *Haplomitrium* these conducting elements retain the general appearance of elongated parenchyma cells – from which they differ, however in lacking a protoplast when mature (Fig. 2A). In *Pallavicinia* and allied forms, they are more specialized, being tracheid-like in appearance (Fig. 3), although the structure of their walls is quite distinct from that of the tracheids of vascular plants (compare with Fig. 5). In mosses too, the water-conducting cells (*hydroids*) (Fig. 4) are reminiscent of tracheids (Fig. 5) with their elongated shape and tapering ends. The walls of the hydroids, especially their end walls, may show partial hydrolysis (Fig. 4B) and then are very similar in appearance to the non-lignified hydrolysed walls of the tracheary elements of vascular plants (Fig. 5B). In the latter, lignified secondary thickenings also occur (Fig. 5). Such lignified "ornamentations" have not been reported as yet from the water-conducting cells of Bryophytes.

EW, end wall; HW, hydrolysed wall; LST, lignified secondary thickenings; LW, lateral wall; PR, small "pore" in the wall.

(1) *Moerckia* stands apart from the other hepatics. See figs.59, 60 and 208 to 21…

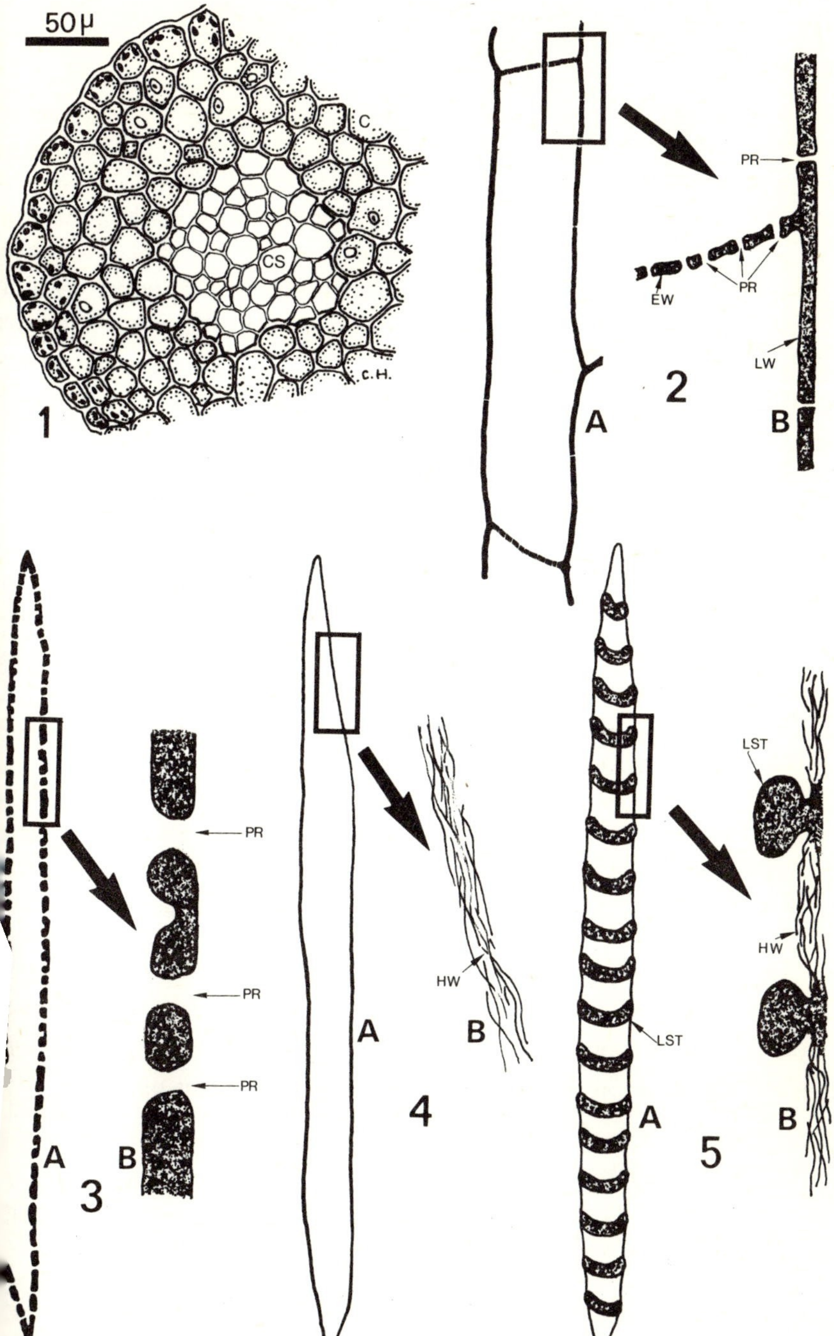

50 μ
C
CS
C.H.
1
PR
EW
LW
A
B
2
A
B
3
HW
A
B
4
LST
HW
A
B
5

PLATE 2

Some aspects of structural variability in *Mnium* (Eubryales, Musci)

Fig. 6. *Mnium undulatum.* Leafy gametophyte. Transverse section of a robust stem collected in the field. A conspicuous water-conducting strand (CS) is present. It consists of hydroids. The inner cortex (IC) comprises thin-walled cells of conducting parenchyma. The elements of the outer cortex (OC) possess thick walls. A "false" leaf trace (LT) is also identifiable. x 160 approx.
Redrawn from Lorentz (1867-8).

Fig. 7. *Mnium undulatum.* Gametophyte. Transverse section in creeping "rhizome"-like portion of stem, from a specimen collected in the field. The overall organization is simpler than that of the erect green-leaved portion of the stem. The central strand of hydroids (CS), in particular, is much smaller. x 230 approx.
Adapted from Davy de Virville (1927-8).

Fig. 8. *Mnium undulatum.* Transverse section of an erect leafy stem grown under high humidity. A marked regression of the internal structures is evident (compare with fig. 6). CS, central strand; x 260 approx.
Adapted from Davy de Virville (1927-8).

Note : Comparisons have been made with fresh specimens, which showed patterns very similar to those illustrated in this plate.

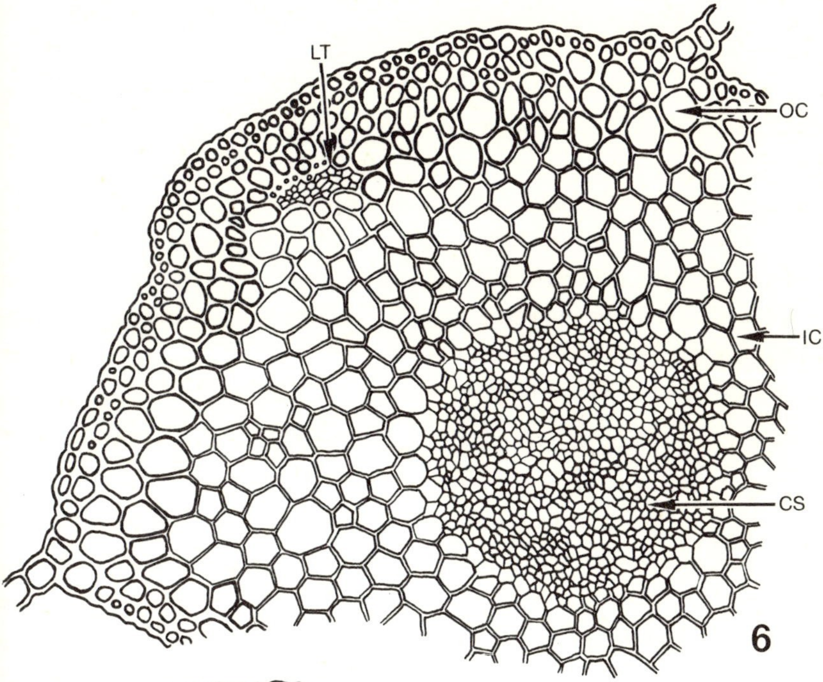

6

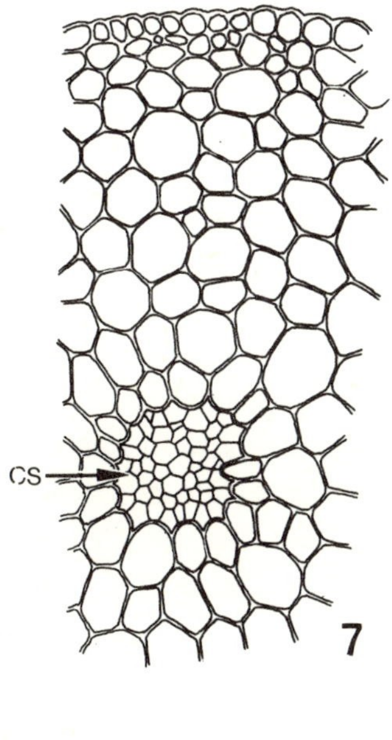

7

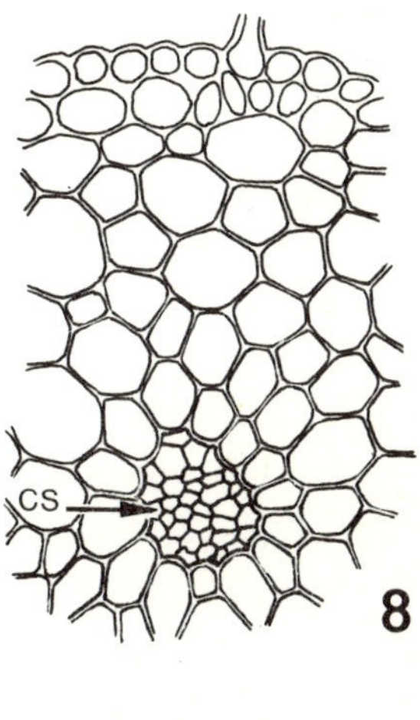

8

PLATE 3

Structural variability.

Figs. 9 and 10. *Rhynchostegium rusciforme* (Hypnobryales, Musci). Gametophyte. Fig. 9, transection of a well developed branch in which a central strand (CS) is identifiable. Fig. 10, transection of a small branch in the same species, without a central strand. x 380 approx.
Redrawn from Lorentz (1867-8).

Note : In a reinvestigation of a number of specimens of *Rhynchostegium rusciforme* from various sites in France (herbarium material), I could not find patterns similar to those described by Lorentz and reproduced here (viz. presence of a strand in the largest axes and its disappearance from small branches); indeed, all the stems I could investigate were actually devoid of a strand. Thus a problem is raised here (identification of the material ? differences according to strains ?). However the facts illustrated here by Lorentz have since been observed in several other species (e.g. *Thuidium tamariscinum*, Hypnobryales, Musci).

Figs. 11 and 12. *Mnium undulatum* (Eubryales, Musci). Variations in leaf structure in plants collected in the field. Fig. 11 shows the complex internal organization of the nerve of a fully developed green leaf from the upper region of an erect stem. Cells of specialised conducting parenchyma (deuters, D), water-conducting cells (hydroids, H) and supportive cells (stereids, ST) can be distinguished. Fig. 12 illustrates the simpler organization of the nerve of a scale-like leaf from the base of an erect stem. Both figs. are of transverse sections. Figs. 11 and 12, x 330 approx.
Redrawn from Davy de Virville (1927-8).

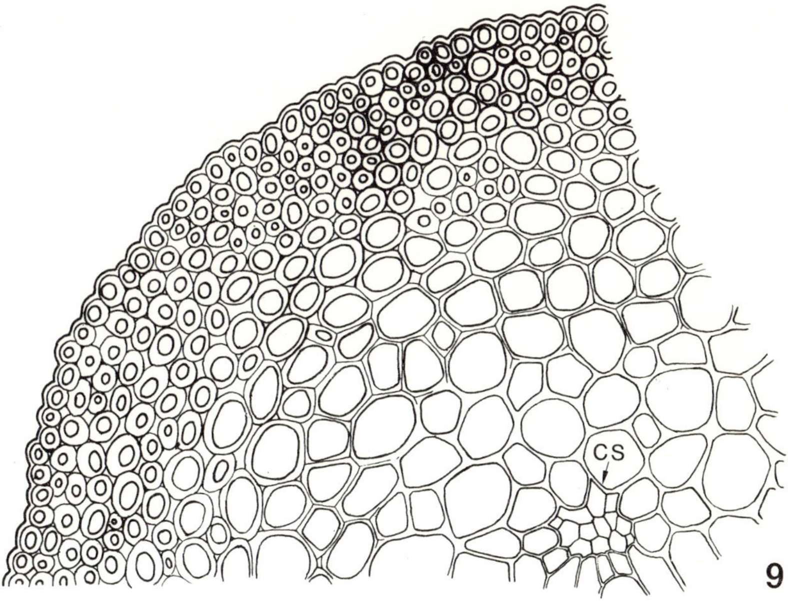

9

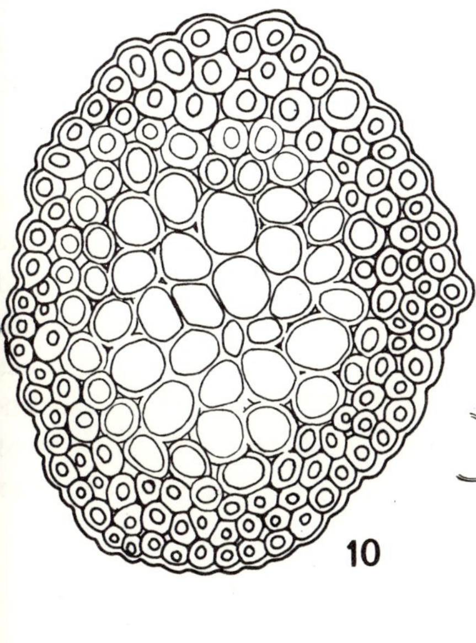

10

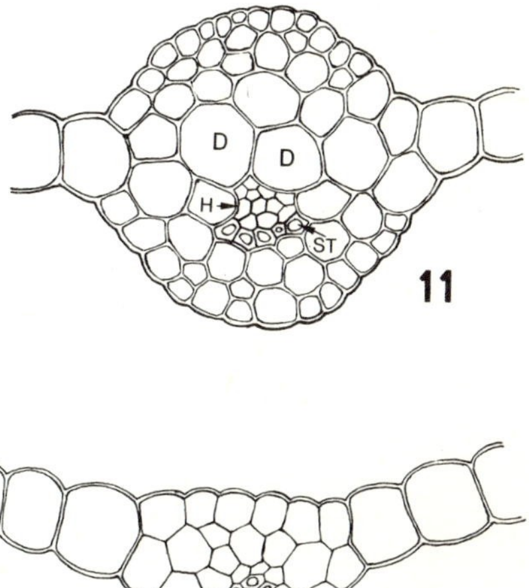

11

12

PLATE 4

Some aspects of structural variability in *Polytrichum commune* (Polytrichales, Musci). I.

Fig. 13. Transection of a very young plantlet grown under sterile conditions; no internal differentiation.

Fig. 14. Transverse section of a young erect leafy stem which originated from a previously inhibited lateral bud after degeneration of the bearing axis (specimen collected in a peat bog). The section was cut at 1.5 cm above the base of the stem. Note the relative simplicity of its organization, which is reminiscent of that found in various mosses of the Arthrodontae. The leaf traces seen at this level are still "false" leaf traces, and the deuters in them are apparently more or less ordinary parenchyma cells.

Fig. 15. Portion of transection in a fully developed specimen collected from a peat bog. A complex internal organization is seen. "True" leaf traces in particular are present (viz. they join the central strand).

Fig. 16. *idem*, specimen collected on pseudo-alpine grassland. The leaf traces here are exceptionnally well developed.

CS, central strand; D, deuters; GT, ground tissue; H, HYD, hydroid(s); HS, hydrom-sheath; IC, inner cortex; L, leaf; LEP, leptoid; LT, leaf trace; MH, mucilage hair; N, nucleus; P, plastid; ST, stereid. Camera-lucida drawings. All figs. are at the same scale. Figs. 13, 15 and 16 after Hébant (1973c) and fig. 14 after Hébant (1972b).

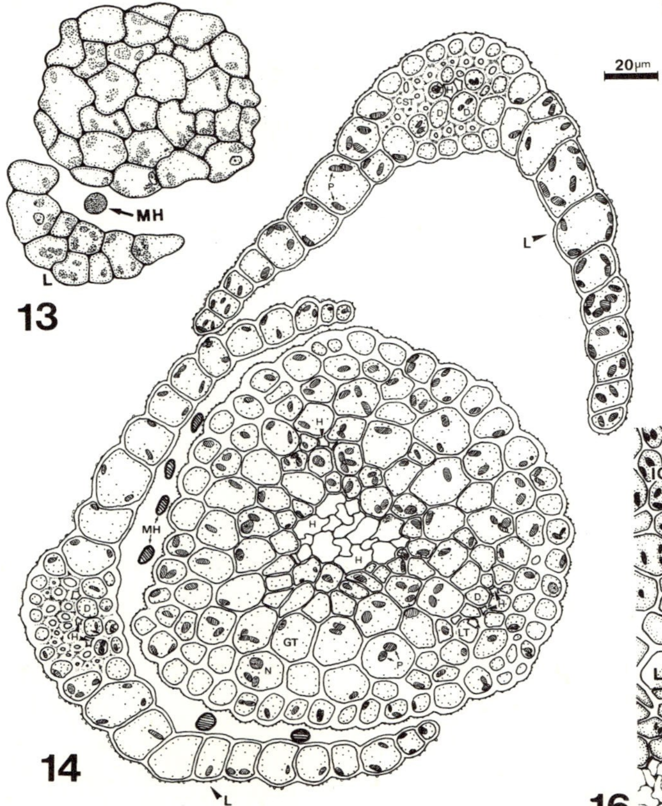
MH
L
20µm
P
CST
D
H
L
MH
H
GT
N
P
LT
D
L
LT
LEP
IC
LT
HYD
HS
LEP
CS
HYD
IC
LT
LEP
HS
LT
CS
HYD
13

14

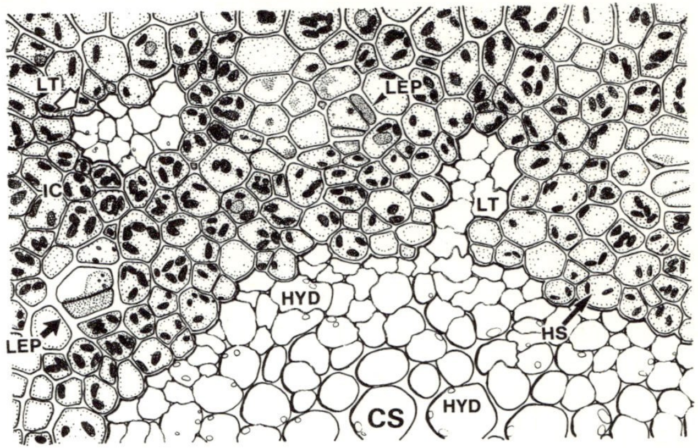
15

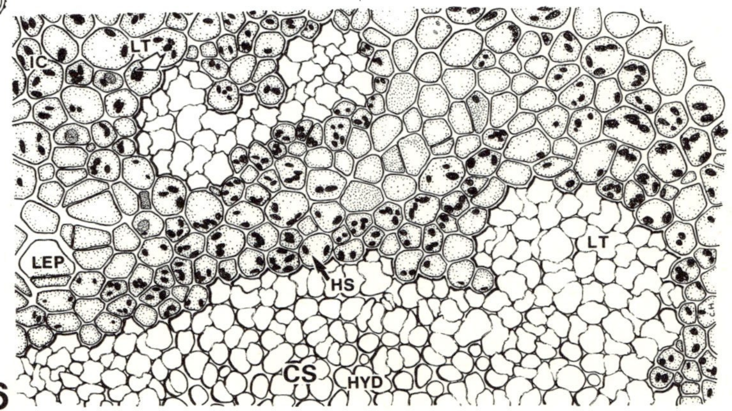
16

PLATE 5

Some aspects of structural variability in *Polytrichum commune.* II.

Fig. 17. Lateral branch developed from a previously inhibited lateral bud after apex of main aerial leafy stem has been severed.

Fig. 18. Typical "rhizome-like" organization of lower part of the lateral branch, as seen in transection (schematic; see also figs. 85 to 89 and 228 to 233).

Fig. 19. Transection of green-leaved portion of the lateral branch (schematic).

Fig. 20. Part of transection of lateral branch near its insertion on the main stem; no central strand.

Fig. 21. Portion of transection in "rhizome-like" region of lateral branch. A complex histological structure can be seen.

Figs. 22 and 23. Apical organization of young lateral branches, at the level where rhizome-like structure is produced (longitudinal sections). Young hydroids of central strands can be identified from young stereids at a very early stage of development because of their exceptionnally large nucleus and nucleolus. Young leptoids are also visible.

A, apical cell; HYD, hydroid; IC, inner cortex; LEP, leptoid; MH, mucilage hair; OC, outer cortex; ST, stereid.

Figs. 20 to 23 are camera-lucida drawings of paraffin sections of material fixed in Craf I and stained with haematoxylin-fast green. After Hébant (1973c).

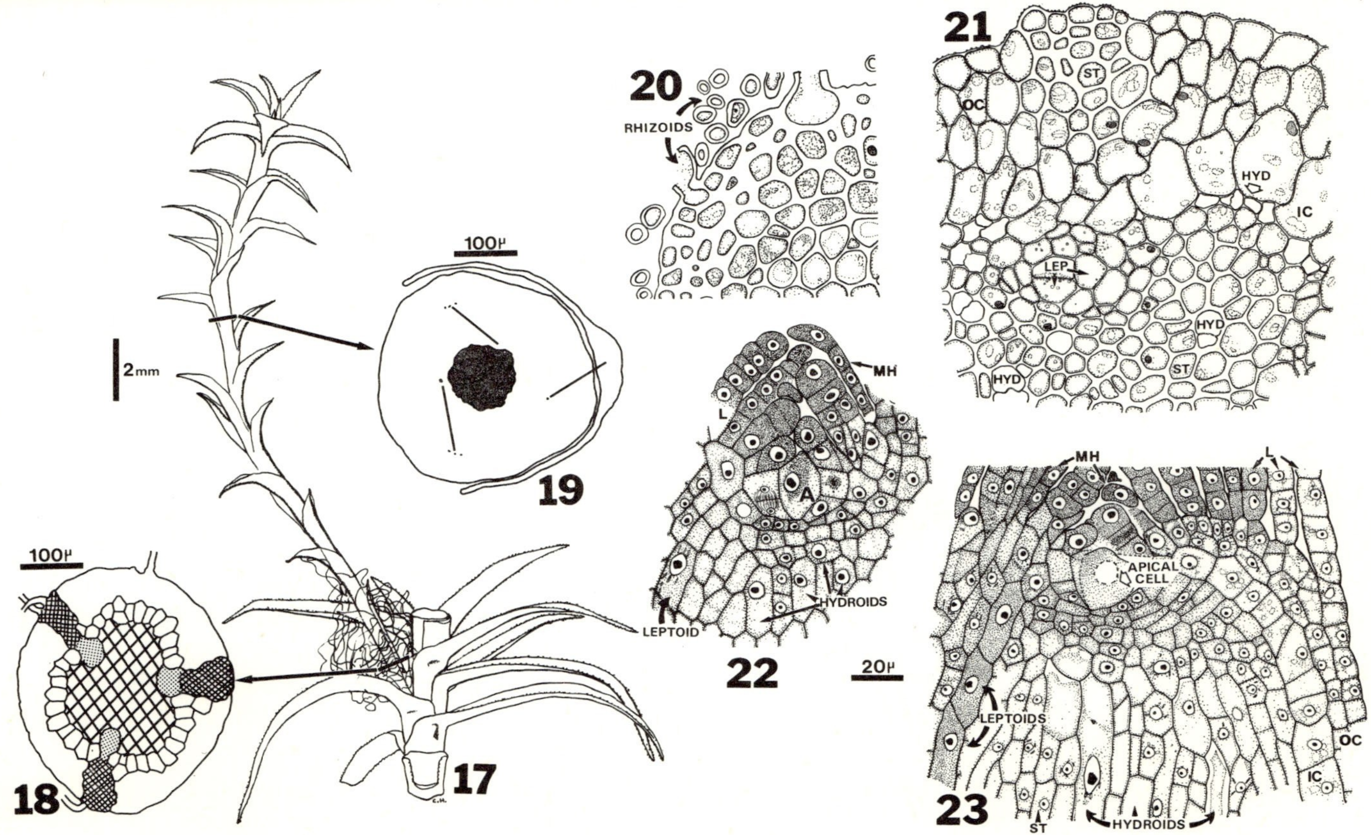
17
18
100µ
19
100µ
2mm
20
RHIZOIDS
21
OC
ST
HYD
IC
LEP
22
MH
L
A
HYDROIDS
LEPTOID
20µ
23
MH
L
APICAL CELL
LEPTOIDS
OC
IC
ST
HYDROIDS

PLATE 6

Atrichum undulatum (Polytrichales, Musci)

Fig. 24. Transection of upright (aerial) leafy stem at a level where central strand is devoid of stereids.

Fig. 25. Transection in upper part of fertile specimen bearing sporogonium; hydroids of central strand are intermixed with parenchyma-like cells (which may or may not have thickened walls) (see also Haberlandt 1886, fig. 7 on plate 23).

GP, ground parenchyma; HYD, hydroid; LEP, leptoid; LT, leaf trace. After Hébant (1973c).

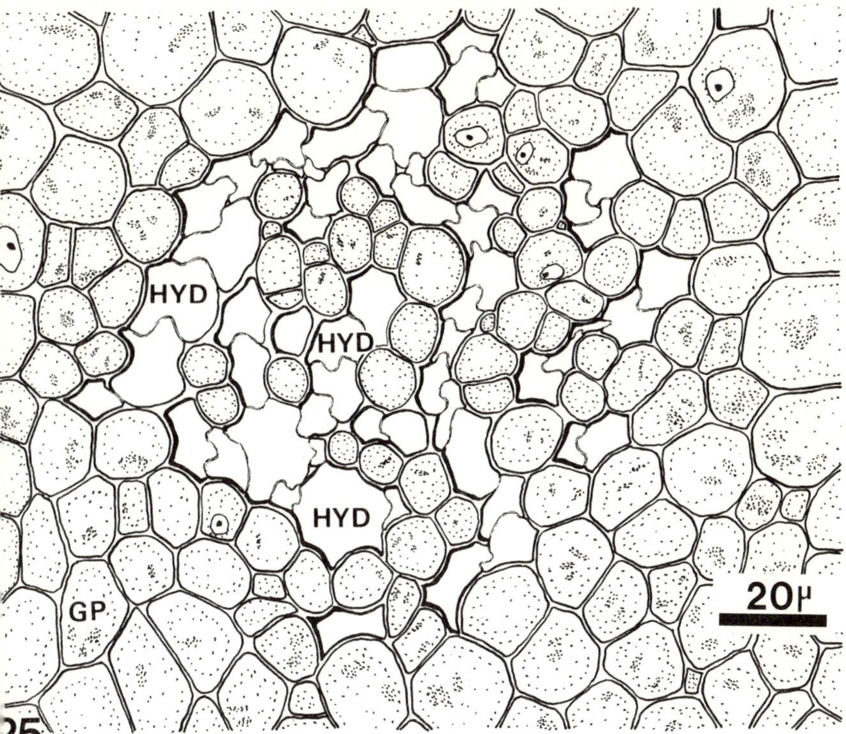

25

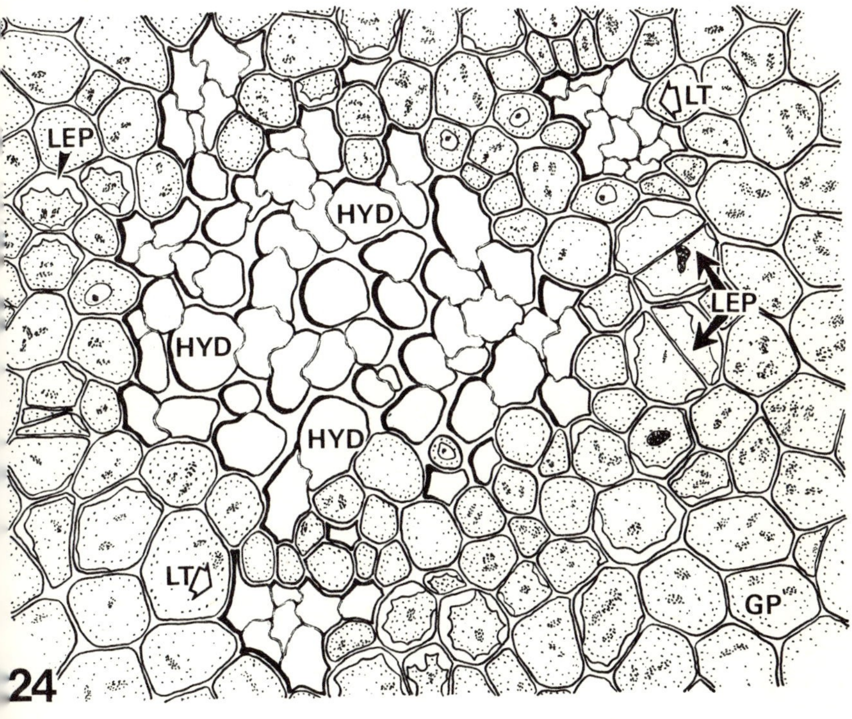

24

PLATE 7

Structures involved in "external" conduction by capillarity

Fig. 26. *Marchantia polymorpha* (Marchantiales, Hepaticae). Ventral surface of a thallus showing the complex pattern of rhizoids and scales which greatly facilitates conduction by capillarity. x 4 approx.
Redrawn from Goebel (1900).

Fig. 27. *Conocephalum conicum* (Marchantiales). Transverse section of thallus. A great many rhizoids run along ventral surface. Together with scales, they constitute very efficient capillary channels.
Redrawn from McConaha (1939).

Fig. 28. *Fimbriaria bleumeana* (Marchantiales). Section of thallus and base of stalk of the carpocephalum (=female reproductive structure). A "rhizoidal groove" (RG) exists in the latter, which permits the rise of water and solutes towards reproductive structures.
Redrawn from Bowen (1935).

Fig. 29. *Encalypta streptocarpa* (Pottiales, Musci). General appearance of leafy gametophyte (n) bearing sporogonia (2n). The dense imbrication of leaves and the presence of numerous rhizoids at base of stems provide capillary channels for external conduction. Approx. natural size.
Redrawn from Bruch *et al.*, *Bryologia europaea* (1836-55).

Fig. 30. *Leucobryum glaucum* (Dicranales, Musci). Structure of leaf. Small cells with a living protoplast and well developed plastids are intercalated with large and empty dead cells. Large "pores" are seen in the walls of the latter. x 290.
Redrawn from Mägdefrau (1971).

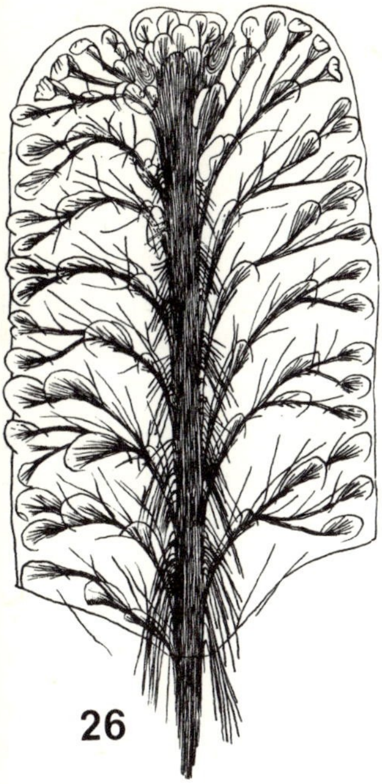

26

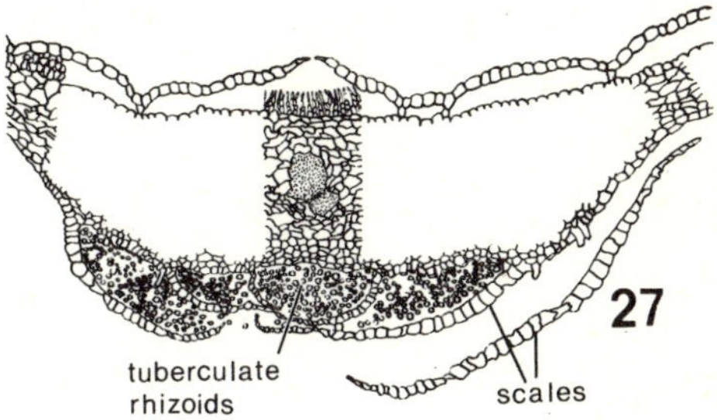

27

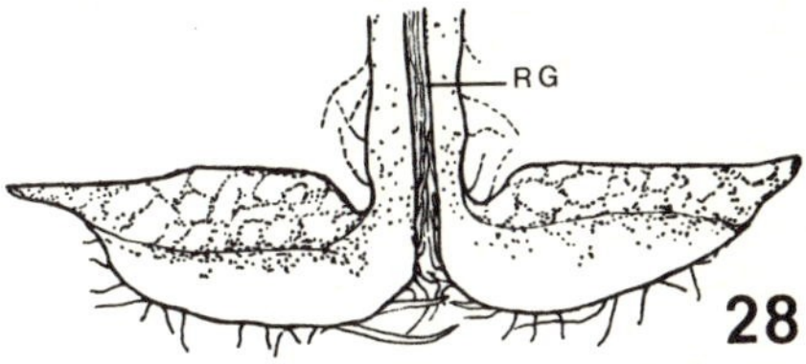

28

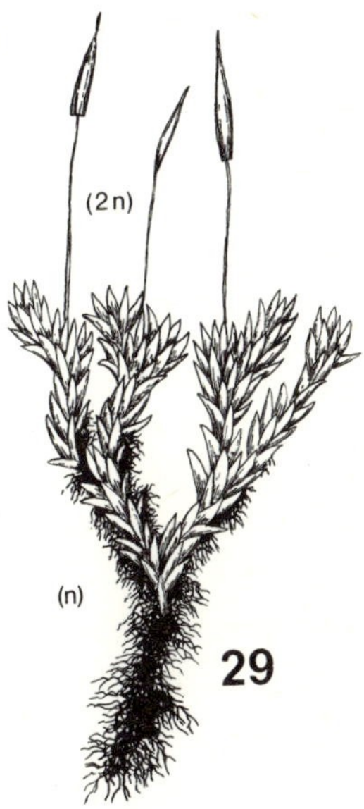

29

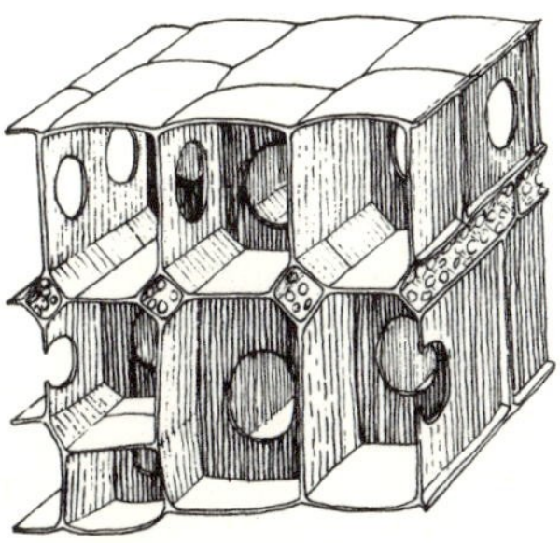

30

PLATE 8

Fig. 31. *Marchantia polymorpha* (Marchantiales, Hepaticae). General appearance of thallus in transection (schematic) and structure of some of the most highly differentiated cell types from the nerve and from the ventral parenchyma of the wings (the cells shown isolated by maceration are drawn from actual photographs). Section of thallus x 37; macerated cells x 185.
After Hébant (1964-5).

Fig. 32. *Dendroceros crispus* (Anthocerotales, Anthocerotae). Gametophyte. Cells from thallus with large primary pit fields. Drawn from part of figure 10 in Proskauer (1960). Reproduced with permission. x 650.

Fig. 33. *Scapania sp.* (Jungermanniales, Hepaticae). Portion of transection of gametophytic stem. The arrows establish the relationship with the corresponding cell types as they appear in maceration. Note that the median portion of the stem ("medulla") consists of elongated cells of conducting parenchyma, whereas supportive cells are seen at the periphery. Inset : macerated cell from the limb of the leaf. x 180.
After Hébant (1972c).

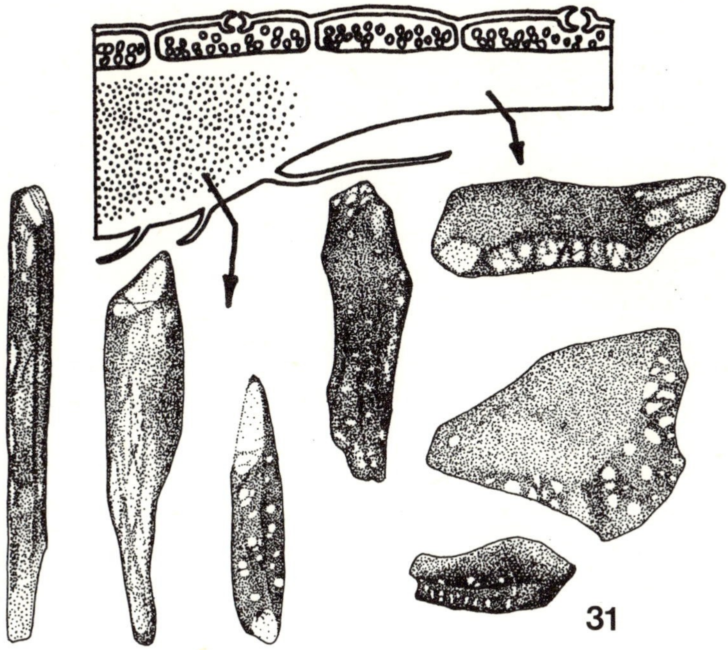

31

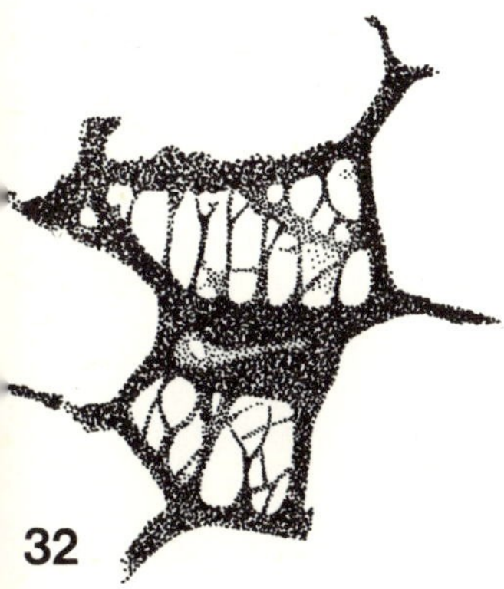

32

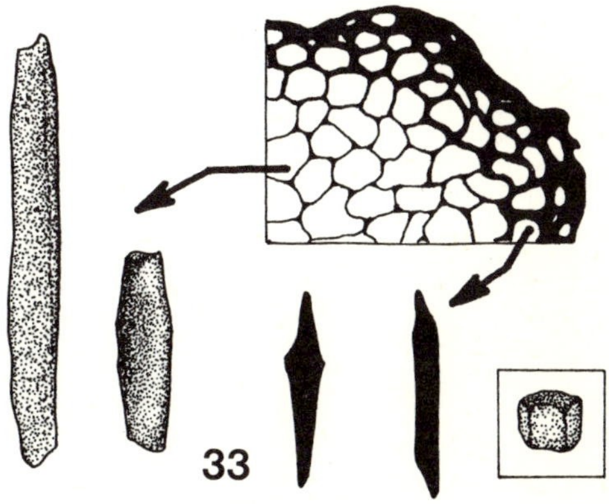

33

PLATE 9

Fig. 34. *Bryum pseudotriquetrum* (Eubryales, Musci). Gametophyte. Parenchyma cells with large primary pit fields. x 450.
After Coesfeld (1892); redrawn.

Fig. 35. *Fontinalis antipyretica* (Isobryales, Musci). Leaf. A considerable number of plasmodesmata are seen to cross the walls of each cell. x 620.
Redrawn from Piskernik (1914).

Fig. 36. *Andreaea petrophylla* (Andreaeales). Transverse section of gametophytic stem. The cells of the median region (dotted) possess a large number of "pits". They probably correspond to a conducting parenchyma.
Redrawn from Lorch (1931).

Fig. 37. *Syrrhopodon lycopodioides* (Syrrhopodontales, Musci). Part of the lower portion of a leaf. Empty cells with pores can be seen.
Redrawn from Lorch (1894).

Figs. 38 and 39. *Calymperes fasciculatum* (Syrrhopodontales, Musci). Fig. 38, part of the lower portion of a leaf in surface view. Perforated cells are present. Fig. 39, the same in transverse section.
NV, nerve of the leaf.
Redrawn from Lorch (1894).

34

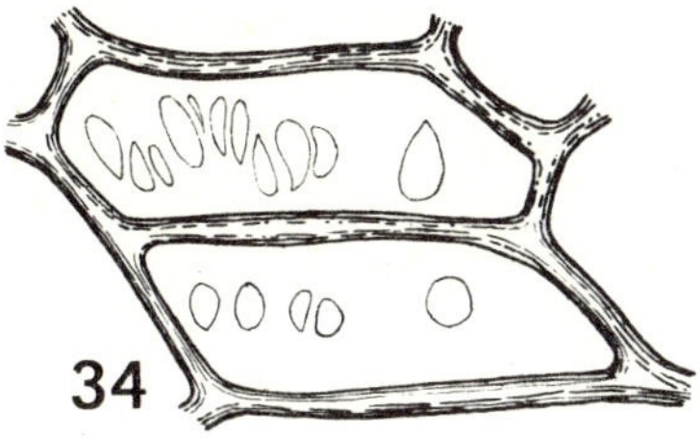

35

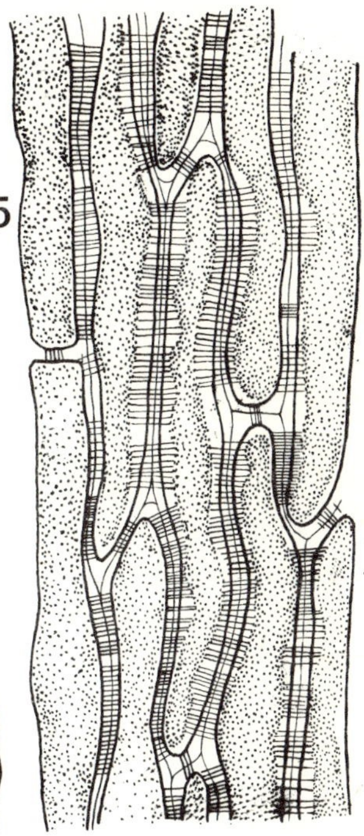

36

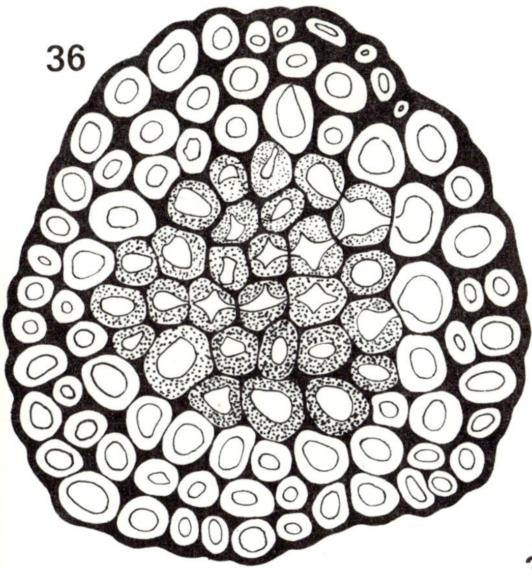

37

38

39

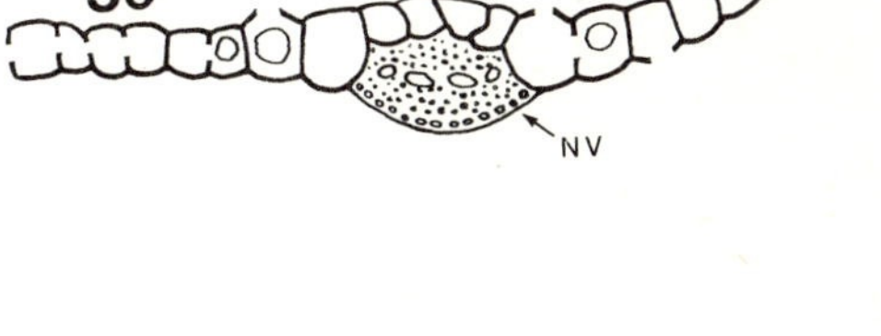

PLATE 10

Sphagnum palustre (Sphagnales)

Fig. 40. General organization of a mature stem. The outer cortex consists of hyaline cells. The insertion of two leaves (L) and of a bundle of small lateral branches (LB) are also shown. x 290.

Fig. 41. Structure of a hyaline cell from the outer cortex of the stem. Thickenings (called "fibers") strenghthen the walls which are perforated.

Fig. 42. Detail of wall structure of a hyaline cell. x 1,500 approx.

All redrawn from W.Ph. Schimper (1857).

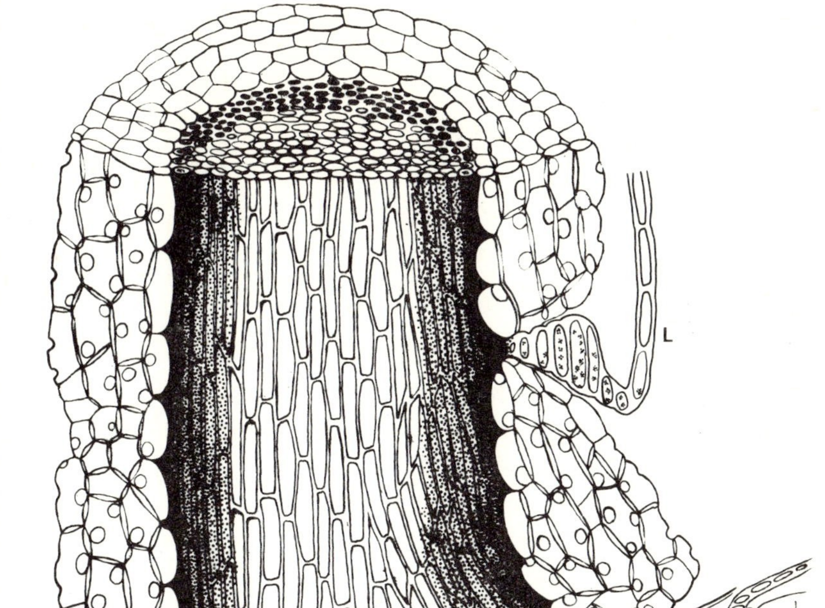

40

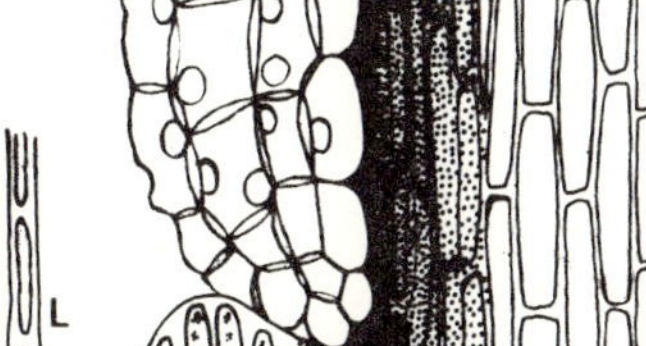

41

42

PLATE 11

Fig. 43. *Anthoceros sp.* (Anthocerotales, Anthocerotae). Transverse section of sporophyte. A columella (COL) is present in its median region. x 75. SP, spores.
After Hébant (1972c).

Figs. 44 and 45. *Anthoceros fusiformis.* Transverse (fig. 44) and longitudinal (fig. 45) sections of an exceptionally large sporophyte in which the columella (COL) exhibits a notable development such as to superficially resemble a stele, whereas the sporogenous tissue (SP) is abortive. x 55 approx.
Redrawn from Campbell (1924).

Fig. 46. *Megaceros sp.* (Anthocerotales). Portion of an abnormal columella in which ordinary columella cells are replaced by spirally thickened elements. x 220.
Redrawn from Proskauer (1960).

Fig. 47. *Marchantia polymorpha* (Marchantiales, Hepaticae). Longisection of a sporangium approaching maturity, in which a structure reminiscent of a columella is present (a columella is normally totally absent from sporogonia of this species). x 55 approx.
Redrawn from Cribbs (1918).

Fig. 48. *Ptilidium pulcherum* (Jungermanniales, Hepaticae). Portion of a mature elater. Its general appearance suggests that of a tracheid. x 1250.
Redrawn from Siessegger (1966).

43

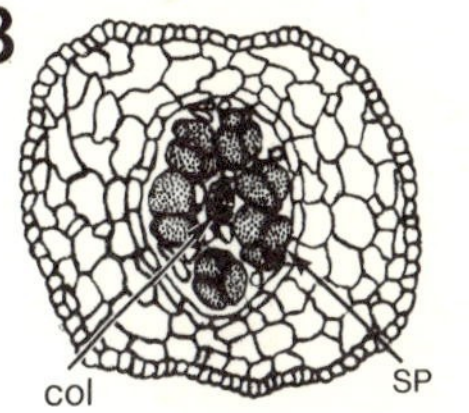

44

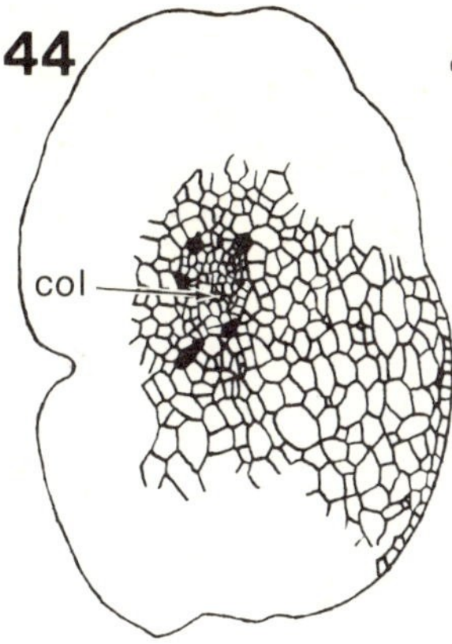

45

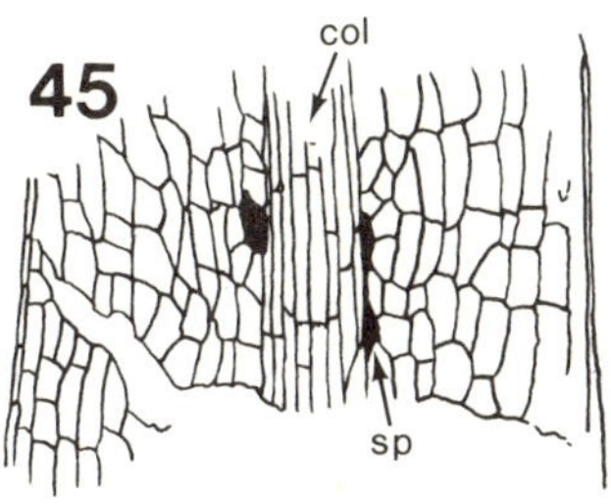

46

47

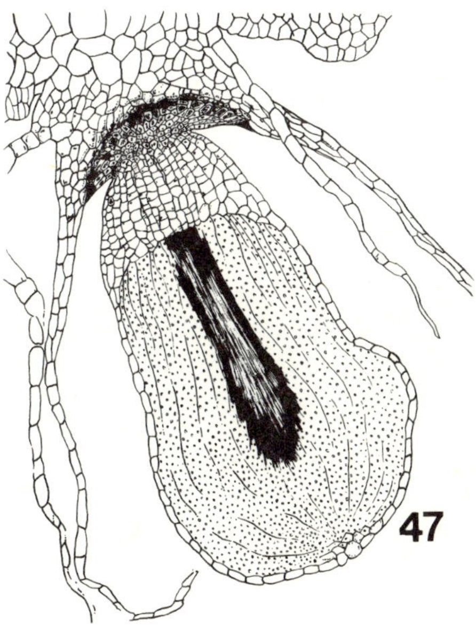

48

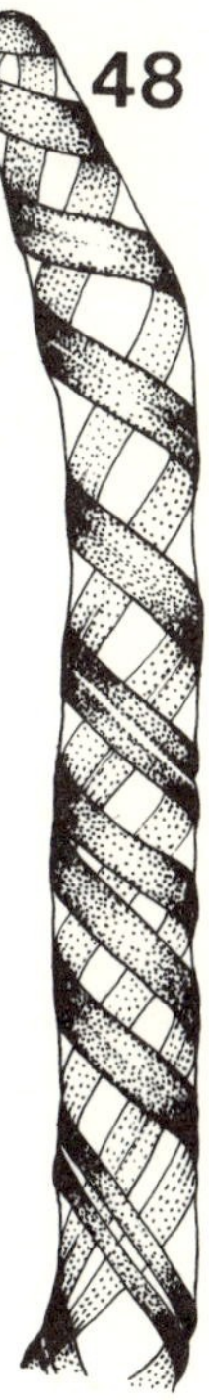

PLATE 12

Fig. 49. *Haplomitrium hookerii* (Calobryales, Hepaticae). Longisection of an erect fertile female stem bearing sporophyte. A well developed conducting strand (CS) is present in the gametophytic stem.
Redrawn from Gottsche (1843).

Fig. 50. *Haplomitrium blumei.* Transverse section of leafy shoot showing central strand (CS). x 46.
Redrawn from Campbell (1920).

Fig. 51. *Pallavicinia lyellii* (Metzgeriales, Hepaticae). An early representation of the thallus in which the central strand (CS) is identifiable.
Redrawn from Hooker (1816).

Fig. 52. *Symphyogyna trivittata* (Metzgeriales). An early illustration of the multistrand condition which occurs in several South American species of *Symphyogyna.* Three strands are identifiable in this transection of the thallus. x 23.
Redrawn from Spruce (1895).

Figs. 53 and 54. *Pallavicinia radiculosa* (Metzgeriales). Development of the central strand within apical derivatives (fig. 53; longisection), and structure of the young differentiating conducting strand as seen in transverse section (fig. 54).
A, apical cell, CS, central strand.
Redrawn from Campbell & Williams (1914).

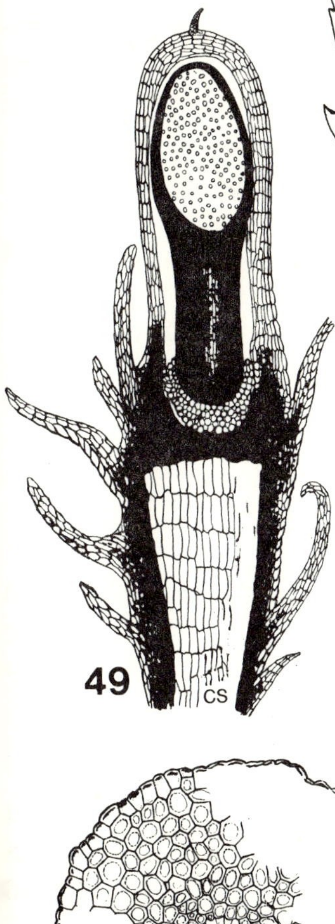

49

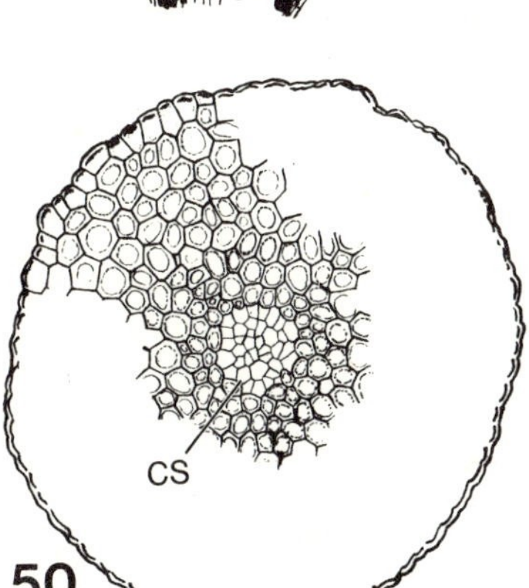

50

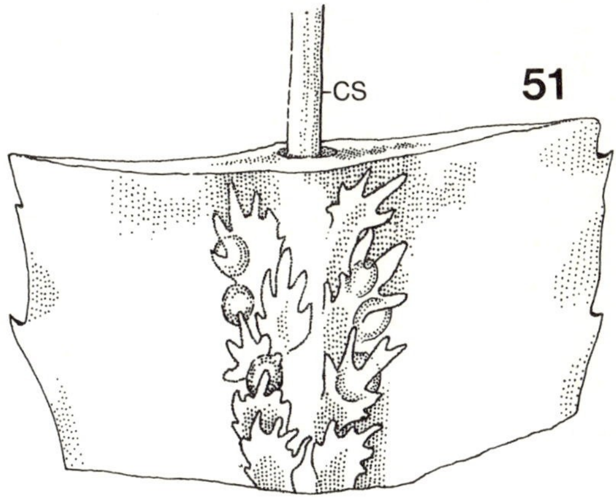

51

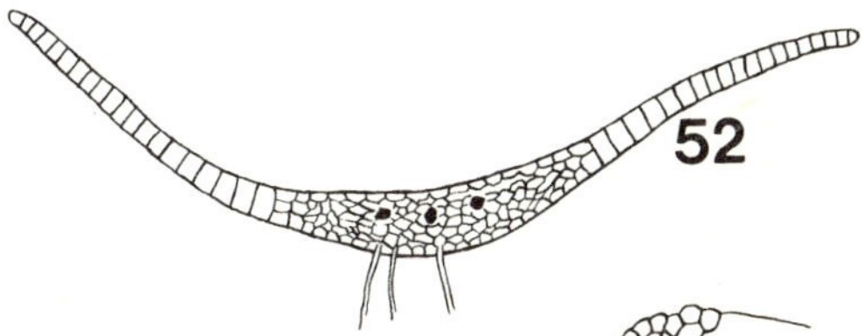

52

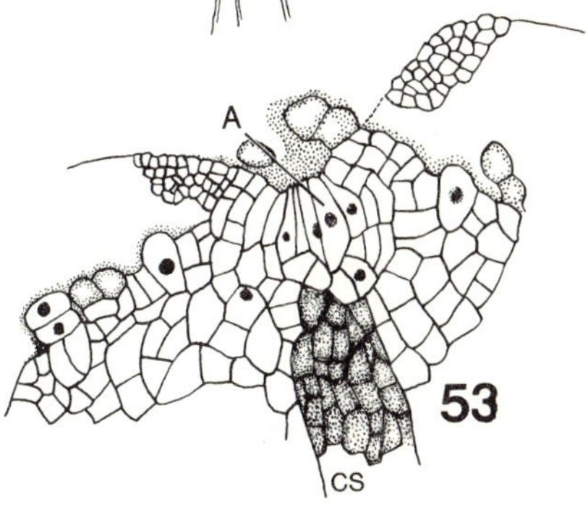

53

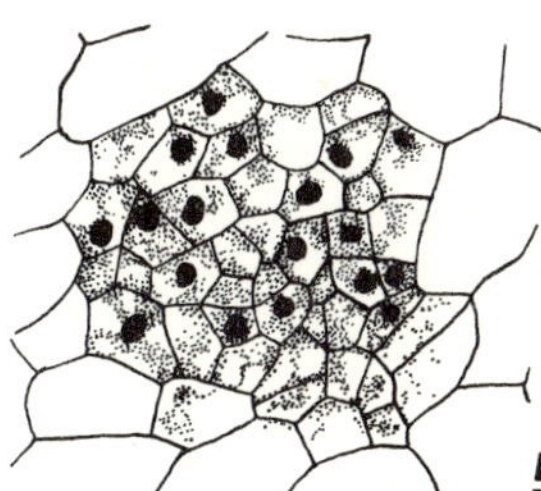

54

PLATE 13

Fig. 55. *Pallavicinia lyellii* (Metzgeriales, Hepaticae). Diagram showing the general organization of the thallus. A single water-conducting strand runs within the nerve. No specialized sieve element-like cells are present. x 115.
After Hébant (1964-5).

Fig. 56. *Symphyogyna sp.* (Metzgeriales). A multistrand form. General view of the thallus showing arrangement of strands. Dots in ovals show schematically the position of strands in midrib in cross section. x 3.5 approx.
Redrawn from Smith (1966).

Fig. 57. *Symphyogyna rubritincta* (Metzgeriales, Hepaticae). Transverse sections of "lamina" (fig. 57A) and rhizome (fig. 57B) showing multistrand condition. x 50.
Redrawn from Hässel de Menendez (1961).

Fig. 58. *Pallavicinia decipiens* (Metzgeriales, Hepaticae). Longisection of rhizome and lateral branch (LBR). The conducting strand (CS') of the branch does not merge with the central strand (CS) of the bearing axis.
Redrawn from Farmer (1894).

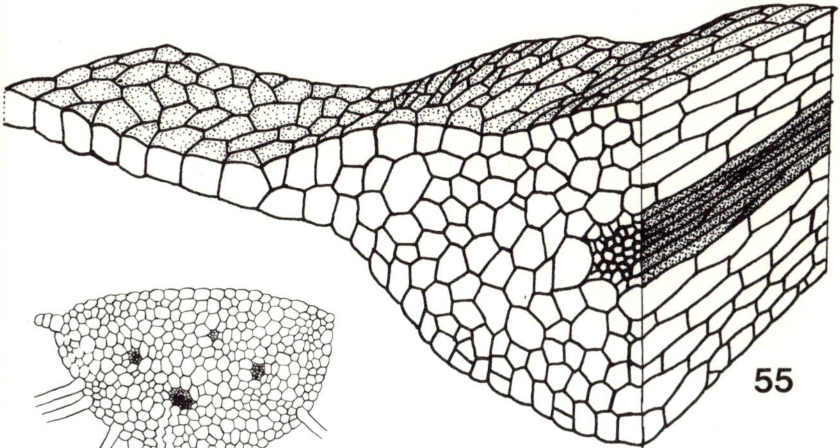

55

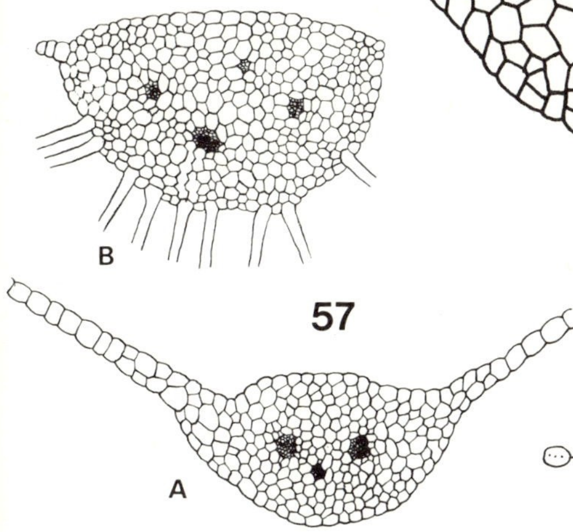

57

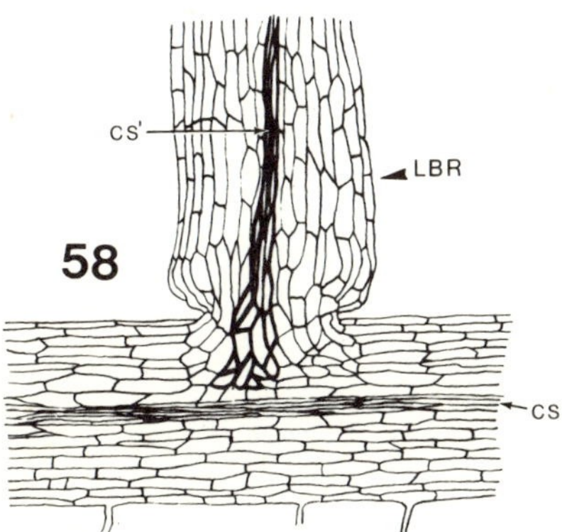

58

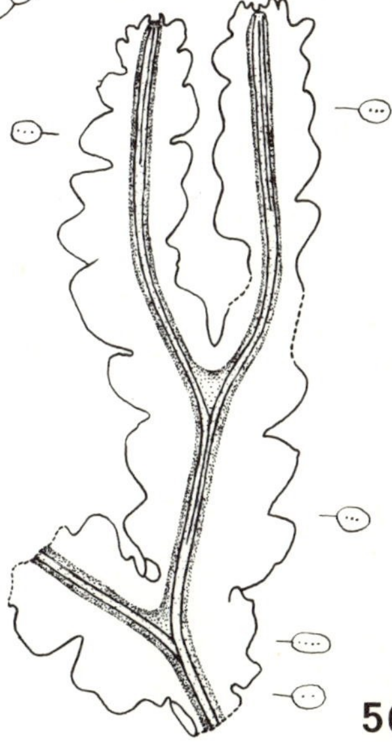

56

PLATE 14

Figs. 59 and 60. *Moerckia flotoviana* (Metzgeriales, Hepaticae).

Fig. 59. Portion of transverse section of the thallus showing one of the two strands (CS) which normally occur in the midrib of well-developed plants. x 55 approx.

Fig. 60. Portion of longisection of one of the conducting strands. The somewhat elongated strand-cells are dead and empty when mature, whereas the neighboring parenchyma cells retain a dense living protoplast in which large plastids are identifiable. x 140 approx.

Redrawn from Cavers (1903).

Figs. 61 to 63. *Takakia lepidozioides* (Takakiales, Hepaticae).

Fig. 61. Longisection of vegetative gametophyte showing outline of central strand. x 75.

Fig. 62. Detail of apical region showing origin of central strand within apical derivatives. x 310.

Fig. 63. Longisection of mature portion of stem. The central strand is sharply defined as it consists of dead cells, whereas cortical and epidermal cells retain a living protoplast. Note also that no relation exists between central strand and leaf. x 175.

After Hébant (1975b).

59

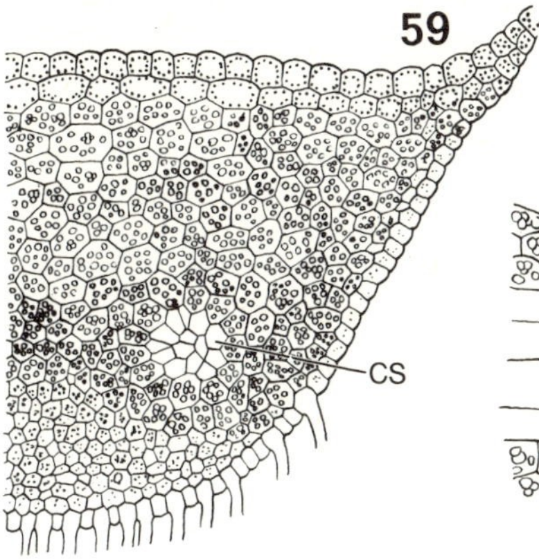
CS

60

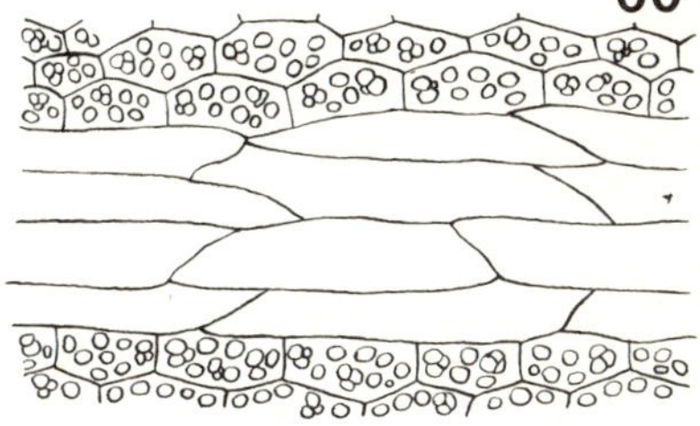

61

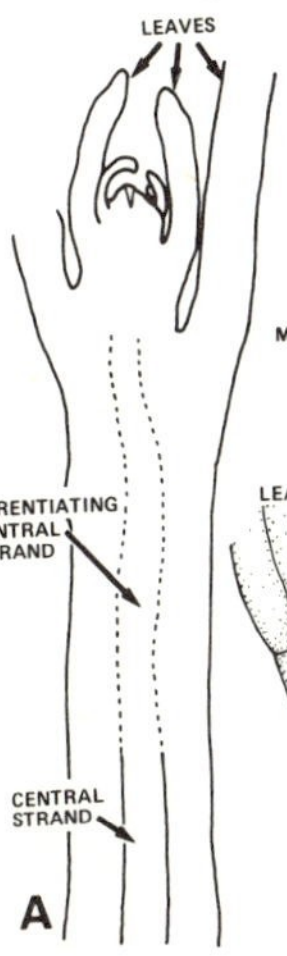
LEAVES
DIFFERENTIATING CENTRAL STRAND
CENTRAL STRAND
A

62

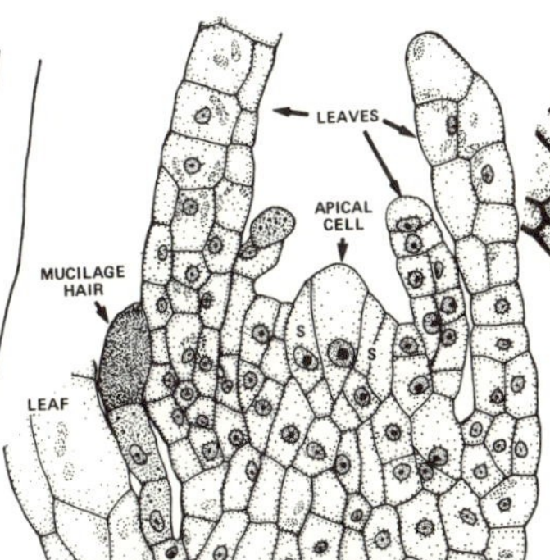
LEAVES
APICAL CELL
MUCILAGE HAIR
LEAF
S
S
B
INCIPIENT CENTRAL STRAND

63

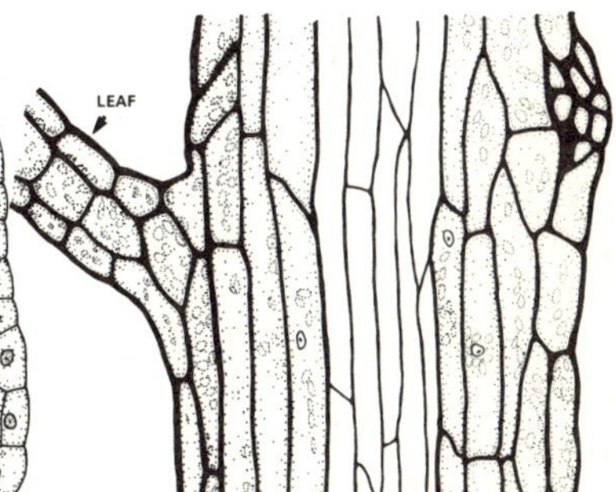
LEAF
C
CORTEX
CENTRAL STRAND

PLATE 15

Fig. 64. *Mnium hornum* (Eubryales, Musci). An early drawing by Hedwig (1784) depicting a longisection of a fertile male gametophyte, in which the central strand is clearly identifiable. Redrawn.

Fig. 65. *Mnium hornum.* An early representation of a transverse section of a gametophytic stem. Here again the conducting strand is readily identifiable.
Redrawn from Hedwig (1782).

Fig. 66. *Bryum latifolium* (Eubryales, Musci). General organization of gametophytic stem. The relationship between the transverse section of the stem and the various cell types as they appear in maceration is established by the letters a, b, c, d (d = hydroids !). Adapted from Unger (1861); redrawn : original presentation modified and simplified.

Fig. 67. Transverse section of the gametophytic stem of a moss (attributed to *Grimmia apocarpa*) without a conducting strand. Redrawn from Lorentz (1867-8).

64

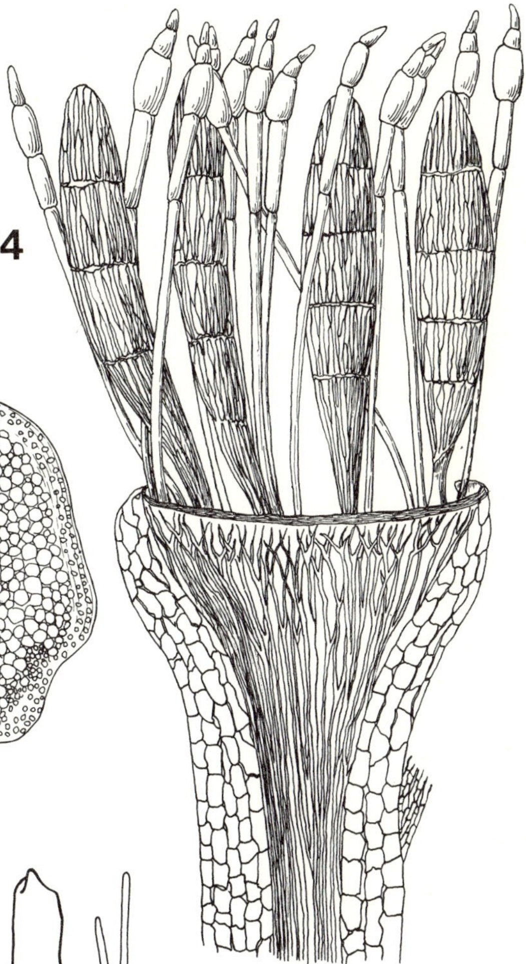

65

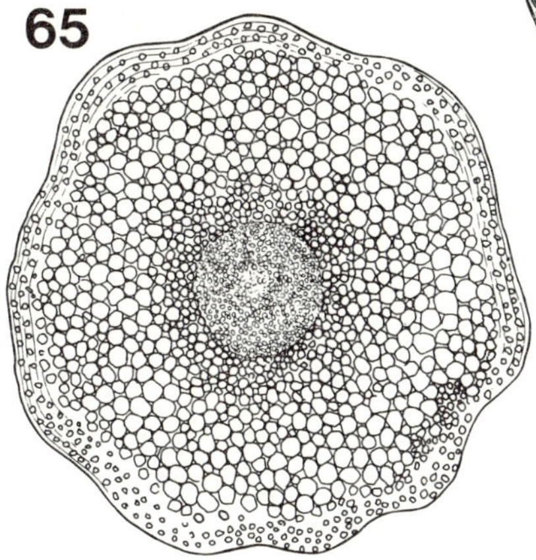

66

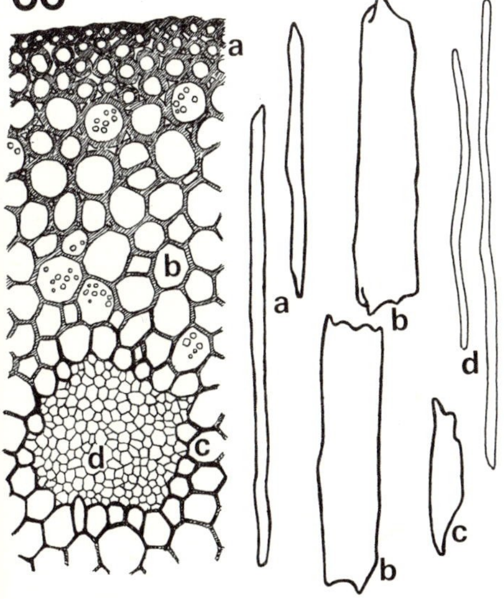

67

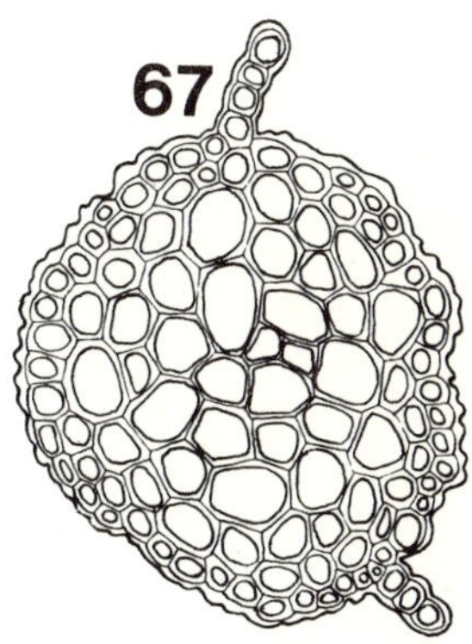

PLATE 16

Fig. 68. *Mnium punctatum* (Eubryales, Musci). Portion of longisection of erect green-leaved stem showing central strand of hydroids and surrounding parenchyma cells of inner cortex.
HYD, hydroids; M, mitochondrion; N, nucleus; P, plastid; PAR, parenchyma cell.
Camera-lucida drawing of a freezing microtome section; method of Baker (1946); original.

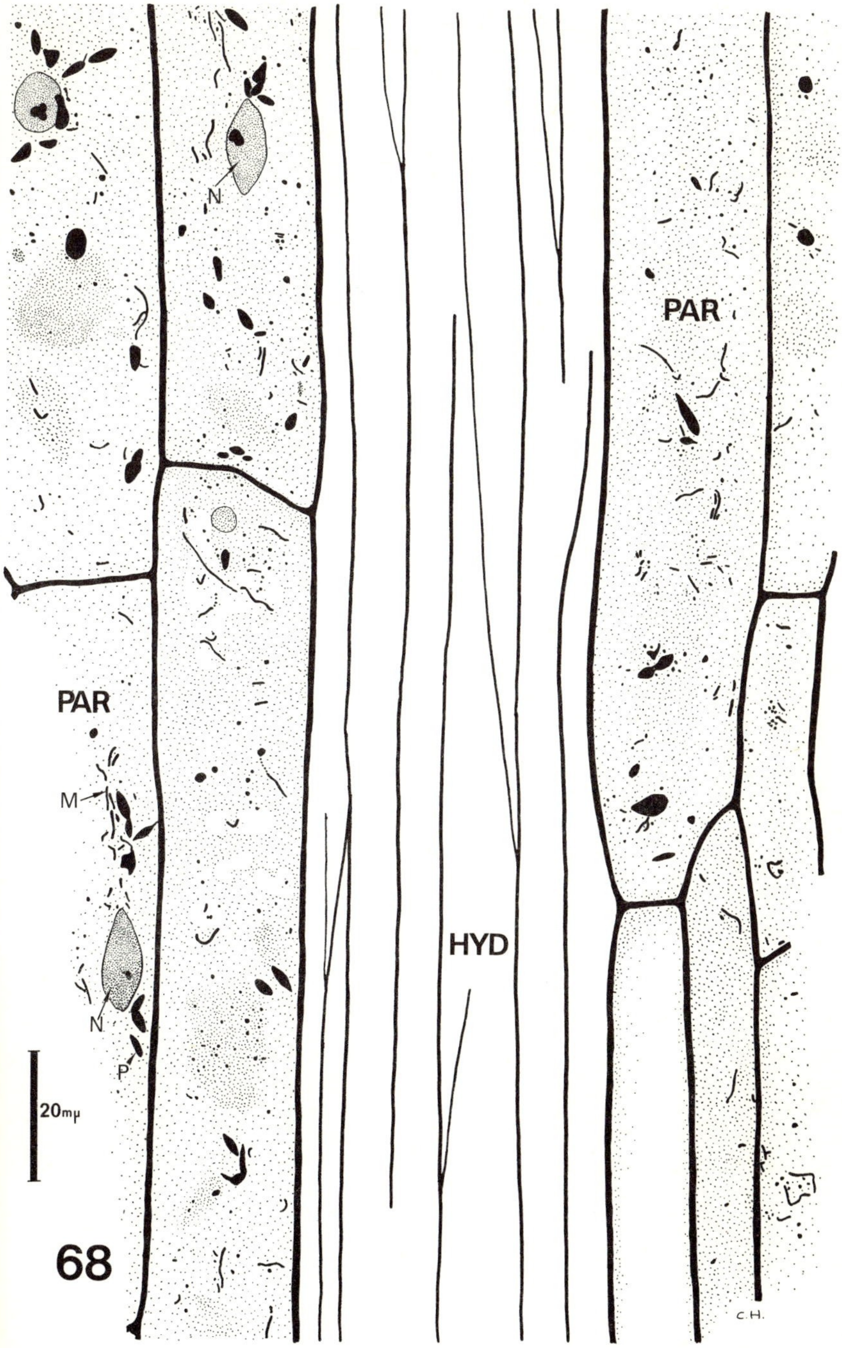

68

PLATE 17

Fig. 69. *Polytrichum commune* (Polytrichales, Musci). Median longisection of aerial leafy stem showing origin of conducting tissues, within apical derivatives. Note presence of large nuclei and nucleoli in developing leptoids, and in hydroids of median portion of central strand (developing hydroids of outer portion are shorter, and exhibit smaller nuclei and nucleoli; these hydroids are the first to reach maturity). Small plastids are seen in both hydroids and leptoids. In two of the segments cut off from the apical cell, wall "a" according to Leitgeb (1868) has been arrowed. A young presumptive inhibited lateral bud is marked with an asterisk (upper right).

A, apical cell; S, segment cut off from apical cell.

After Hébant (1973c).

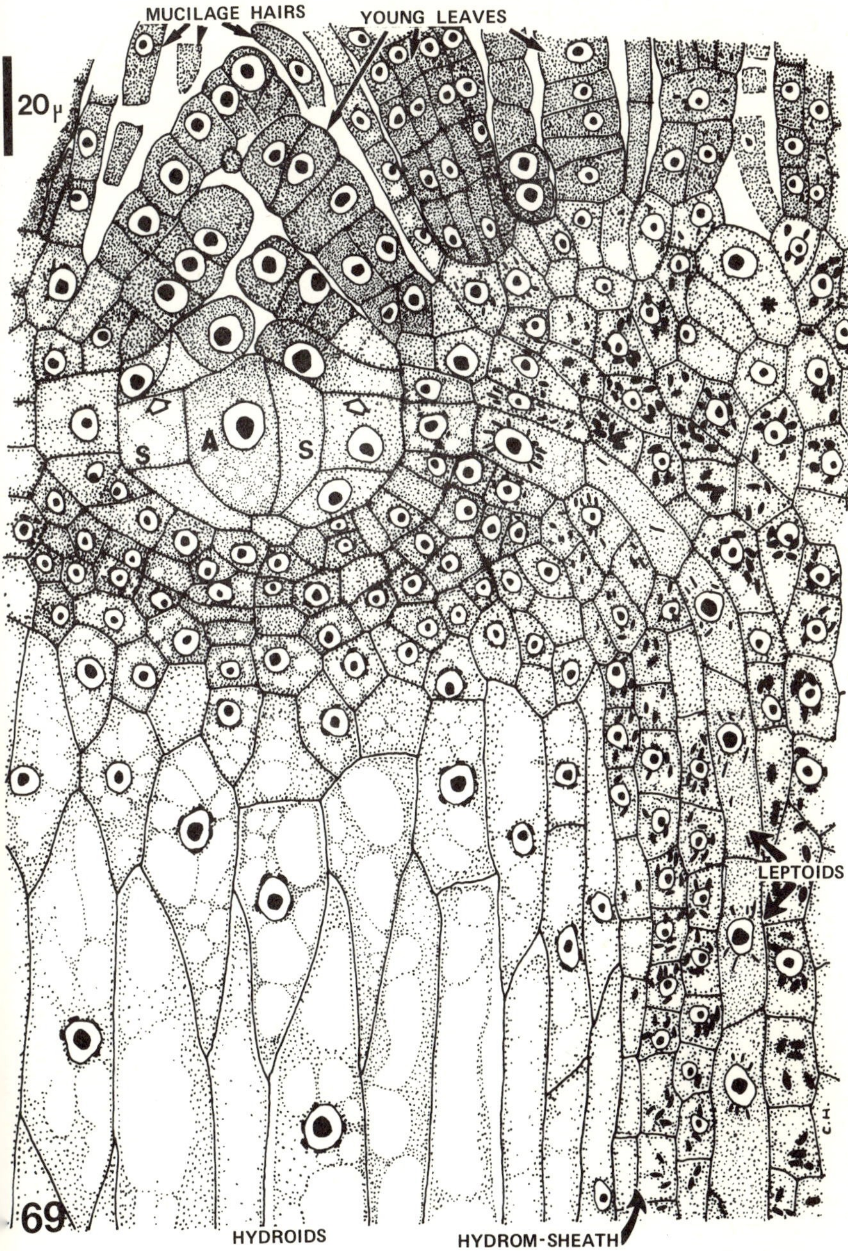
MUCILAGE HAIRS
YOUNG LEAVES
20μ
S
A
S
LEPTOIDS
HYDROIDS
HYDROM-SHEATH
C.H.

69

PLATE 18

Figs. 70 to 72. *Polytrichum commune* (Polytrichales, Musci). Relationships between apical segmentation and internal structure of stems at various levels of the gametophytes (fig. 70, “rhizome”; fig. 71, lower portion of erect leafy stem; fig. 72, upper green-leaved portion of aerial stem). Figs. 70A, 71A and 72A show the apical segmentation patterns, and figs. 70B, 71B and 72B the general organization of the stems at the corresponding levels.

All figures are of tranverse sections. A, apical cell; CS, central strand; LEP, leptoid; LT, leaf trace; S, segment cut off from apical cell.

Adapted from Wigglesworth (1956); fig. 71B of the present plate is original.

70

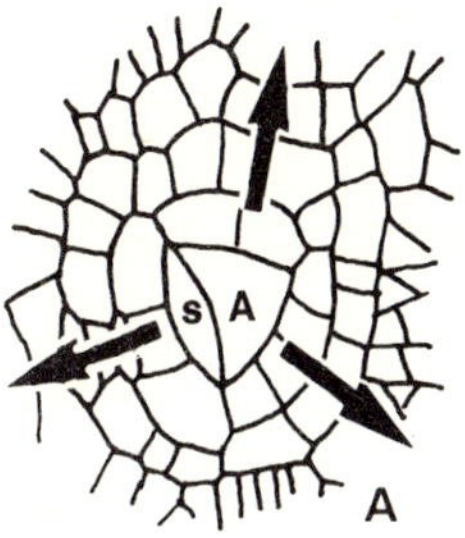

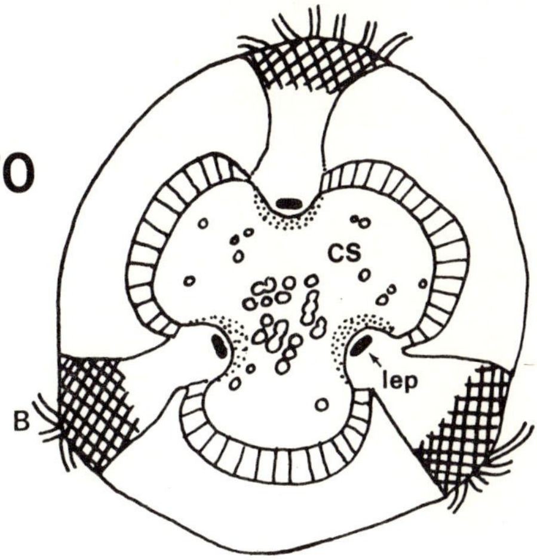

71

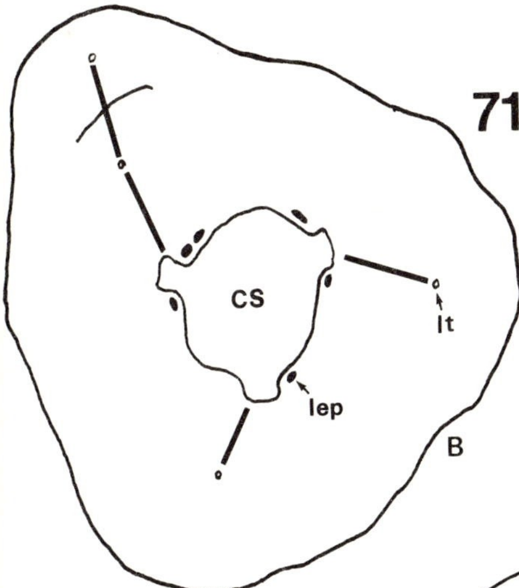

72

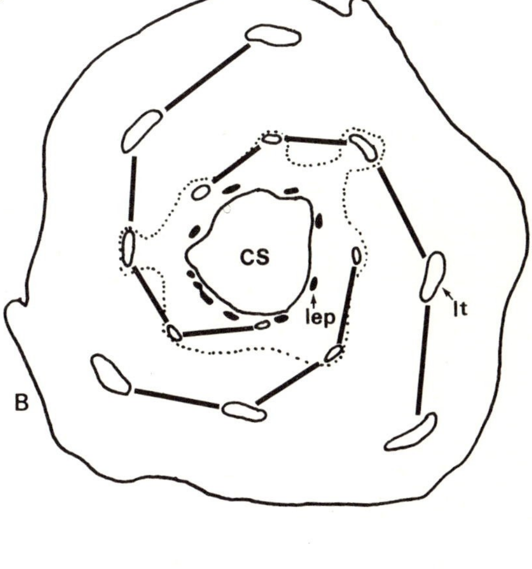

PLATE 19

Fig. 73 and 74. *Polytrichum alpinum* (Polytrichales, Musci). Relationships between apical segmentation and branching pattern.

Fig. 73. Transverse section of apex of leafy stem showing the apical cell (A) and the segments cut off from it (n, n-1, n-2, etc.). A secondary apical cell (A2) is seen in the segment n-10. Note its position on the abaxial side of the leaf (FR) belonging to the same segment.

Fig. 74. Transverse section in mature portion of stem at level where a lateral branch (R) and the leaf (FR) belonging to the same segment join the main stem. The correspondence with the apical segmentation pattern shown in fig. 73 is easily established.

CC, water-conducting strand of main stem; TR, water-conducting strand of lateral branch.

After Hébant & Berthier (1972).

Fig. 75. Diagram illustrating branching patterns in the gametophytes of *P. alpinum*. Small inhibited lateral buds (IB) coexist with well developed lateral branches (one of them shown : LBR). Only the latter possess a conducting strand which merges with the central strand of the bearing axis (for details, see Hébant & Berthier 1972).

Fig. 76. *Dendroligotrichum dendroides* (Polytrichales, Musci). Transverse section in erect aerial portion of gametophyte at the level where lateral branches are emitted. The water-conducting strand (TR4) of one lateral branch merges with the central strand (CC) of the main axis, whereas a second one (TR3) crosses the cortex of the latter. A third lateral branch is just merging with the main stem, and the leaf which belongs to the same segment is identifiable (FR).

After Hébant & Berthier (1972).

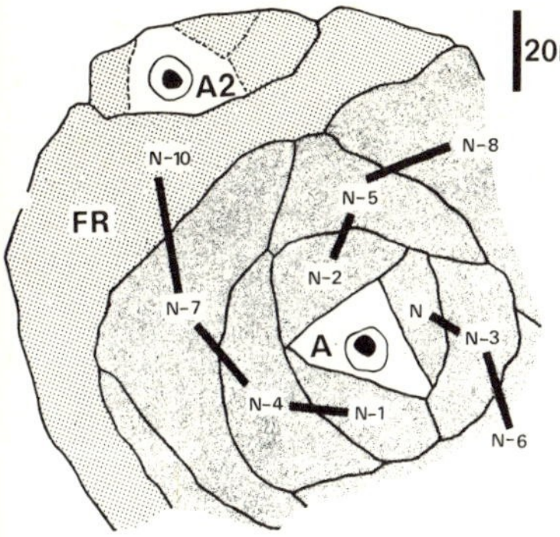
20μ
A2
FR
N-10
N-8
N-5
N-2
N-7
N
N-3
A
N-4
N-1
N-6

73

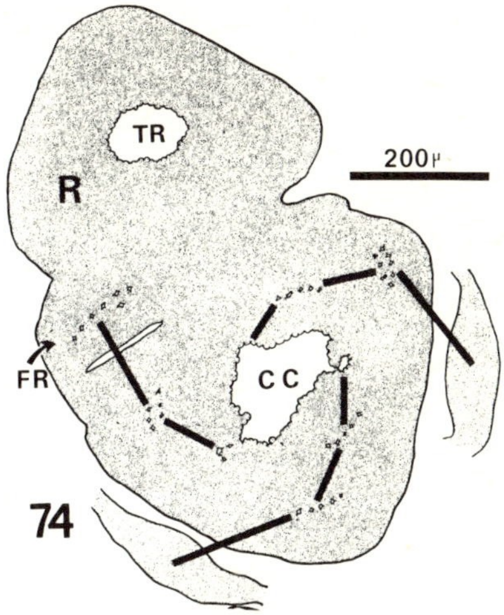
TR
R
200μ
FR
CC

74

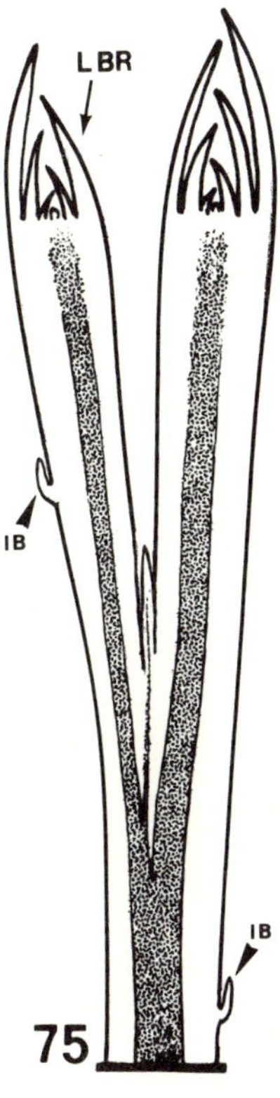
LBR
IB
IB

75

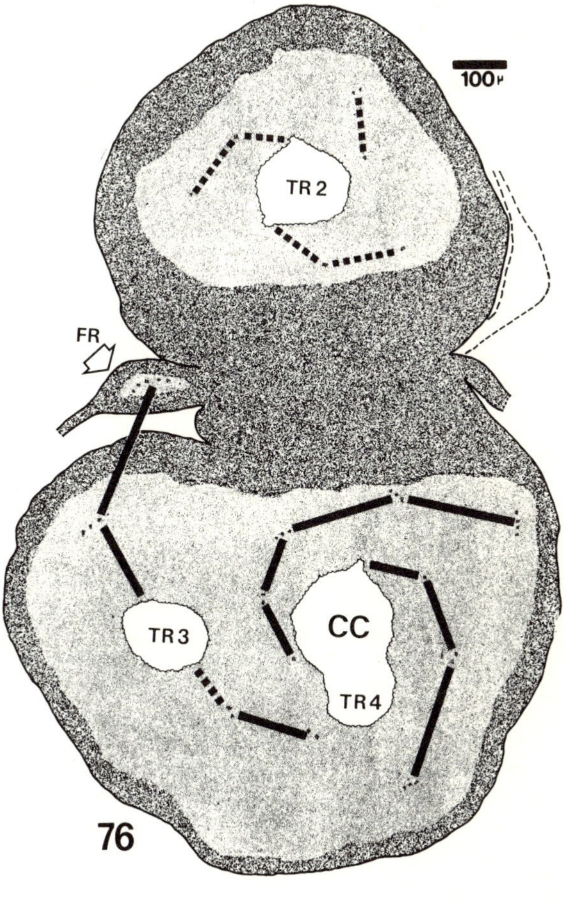
100μ
TR 2
FR
TR 3
CC
TR 4

76

PLATE 20

Fig. 77. *Polytrichum alpinum* (Polytrichales, Musci). Camera-lucida drawings of transverse sections at different levels of a gametophytic stem showing various stages in the penetration of the conducting strand (TR) of a lateral branch within cortical tissues and its linking up to the central strand (CC) of the main axis. Note that the trace (TF) of the leaf belonging to the same segment as the secondary branch merges with the strand (TR) of the latter before the ensemble thus constituted joins the central strand (CC) of the main axis. Fig. 77A, section at lower level; fig. 77D, section at higher level.

After Hébant & Berthier (1972).

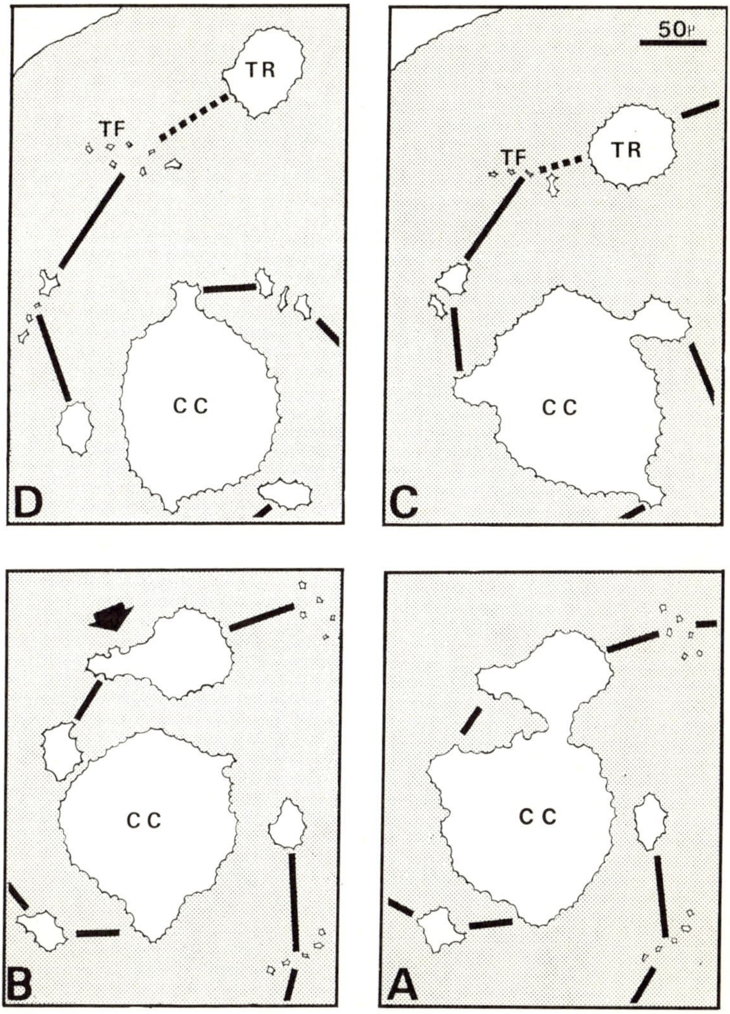

77

PLATE 21

Dawsonia superba (Polytrichales, Musci)

Figs. 78-81. Diagrams of transverse sections at various levels in gametophytic axes : "rhizome" (fig. 78), lower and "intermediate" regions of erect stems (figs. 79 and 80), green-leaved portion of aerial stem (fig. 81). Note changing patterns of the leaf-traces; those of lower regions of gametophytes irrigate only scale-like leaves.

Figs. 82-84. Outline drawings of transverse sections at three levels of a branched specimen. [One or more archegonia were presumably produced by the main stem. Two new branches of approximately the same size have originated from secondary apical cells].

CS, central strand; LT, leaf-trace.
All after Goebel (1906).

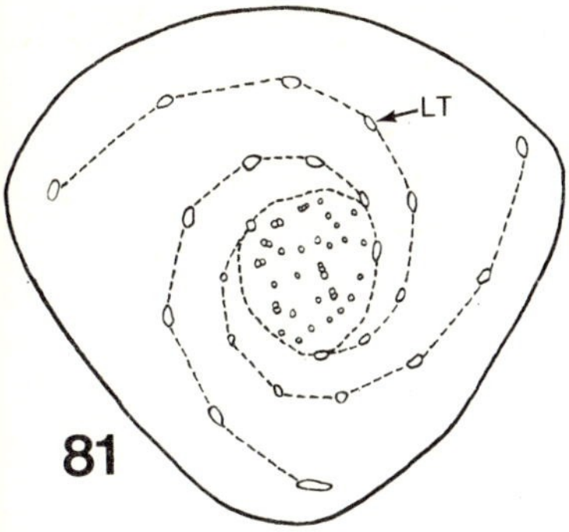

81

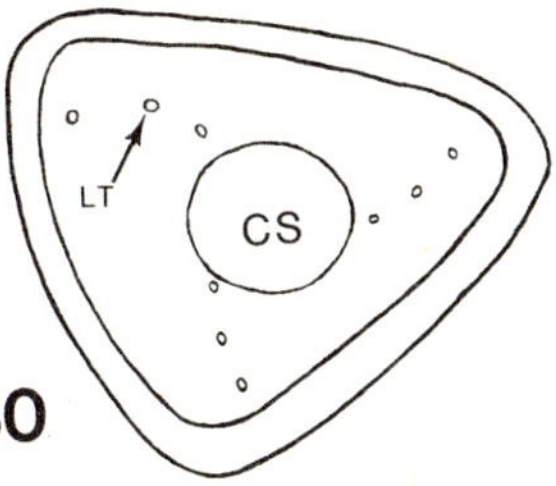

80

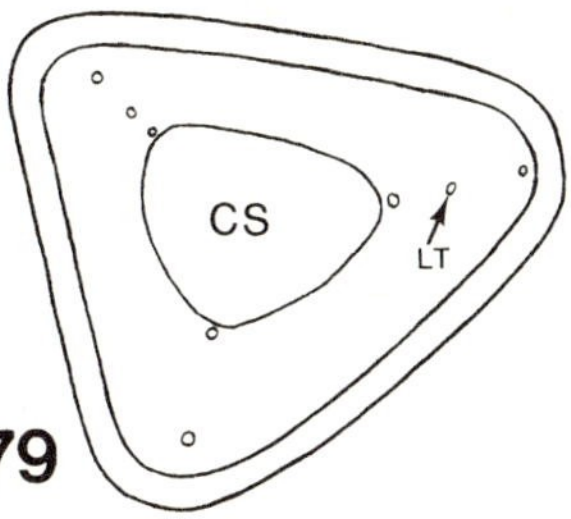

79

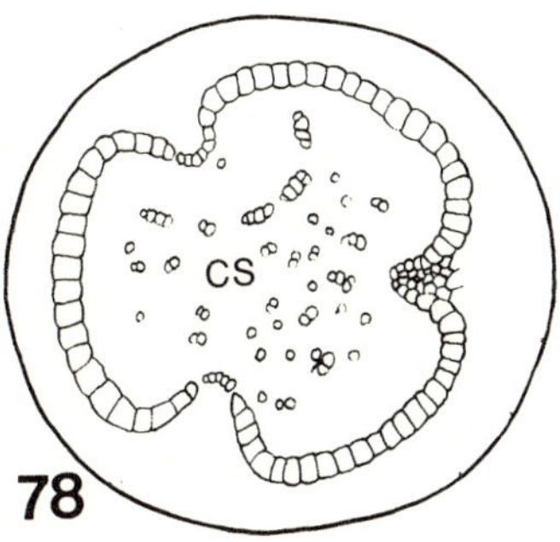

78

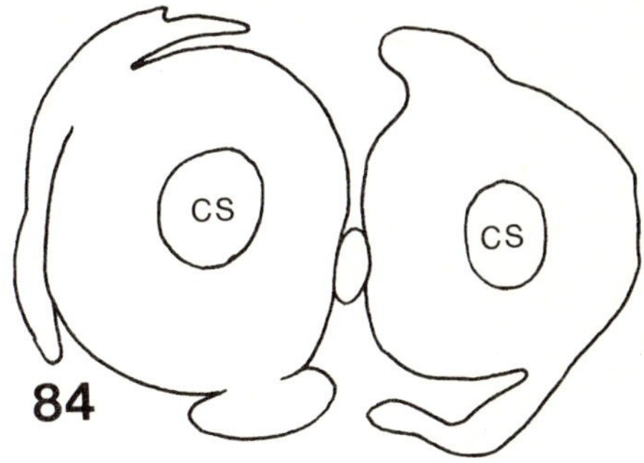

84

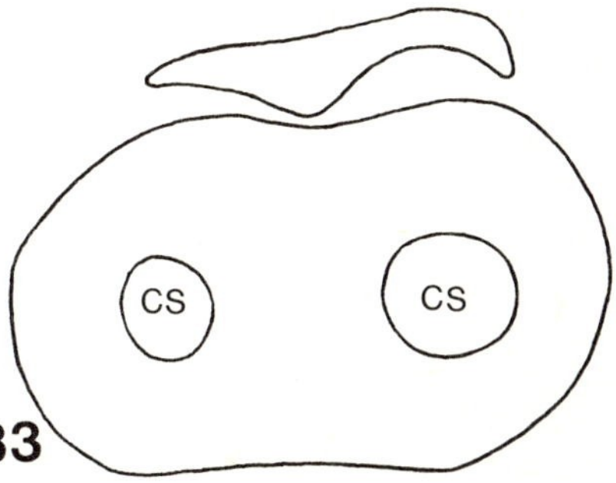

83

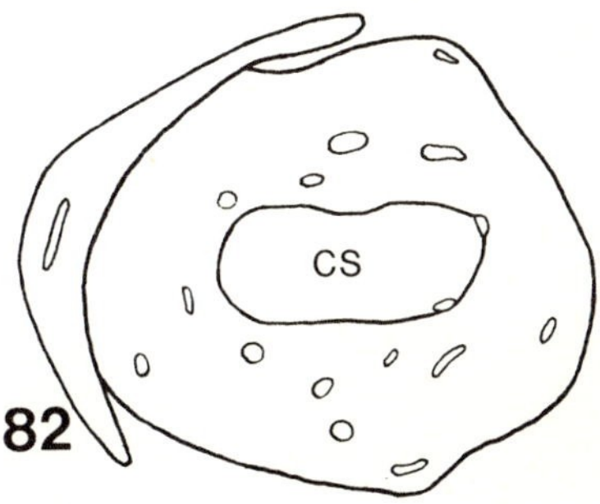

82

PLATE 22

Polytrichum juniperinum (Polytrichales, Musci). Histological structure of subterranean "rhizome-like" portions of gametophytes

Fig. 85. Diagram showing general organization of "rhizome".

Fig. 86. Cell of inner cortex (compared to an endodermis by certain authors; – see text p. 39).

Fig. 87. Hydroid and associated stereids (= parenchyma cells with thickened walls).

Fig. 88. Leptoids and associated parenchyma cell.

Fig. 89. Peripheral cells and rhizoids.

HYD, hydroid; LEP, leptoid; M, mitochondrion; N, nucleus; P, plastid; PAR, parenchyma cell; STER, stereid.
After Hébant (1967b).

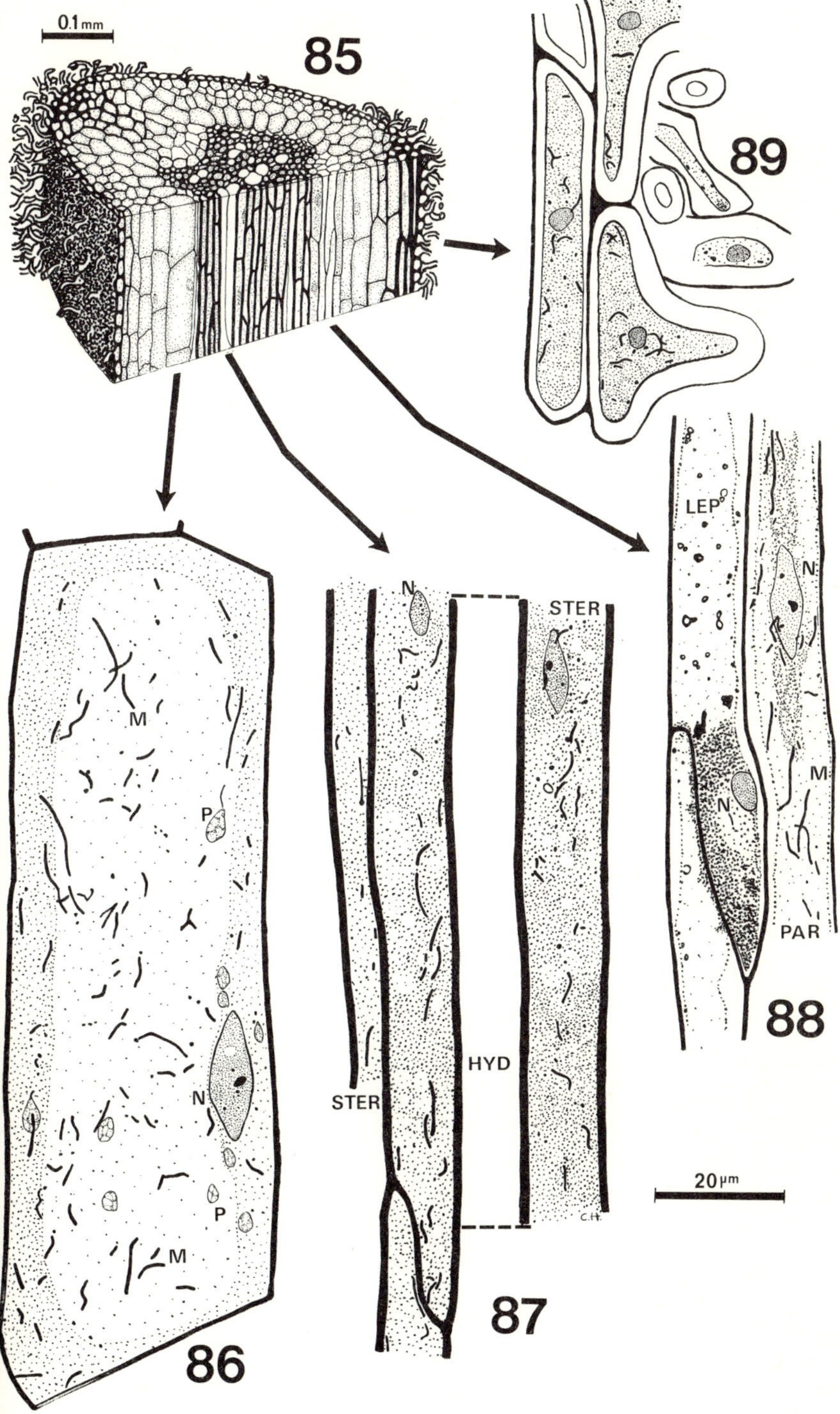
0.1 mm
85
89
LEP
N
M
PAR
88
N
M
P
N
P
M
86
N
STER
HYD
STER
87
20 µm
C.H.

PLATE 23

Leaf Traces

Fig. 90. An early representation of the leaf traces (LT) and their relationships with the central strand (CS) as seen in a transverse section of a gametophytic stem of *Dawsonia superba* (Polytrichales, Musci). x 37 approx.
Redrawn from Unger (1861).

Fig. 91. *Voitia nivalis* (Funariales, Musci). Transverse section of gametophytic stem showing central strand and "true" leaf traces (LT) : one of these actually merges with the central strand (CS). x 130 approx.
Redrawn from Lorentz (1867).
Note : In the study of complete series of transverse sections of vegetative portions of gametophytes of the same species, I could not observe "true" leaf traces but only "false" leaf traces – which do not join the central strand (– see text pp.41-42).

Fig. 92. *Bryum turbinatum* (Eubryales, Musci). Transverse section of gametophytic stem showing central strand (CS) and "false" leaf traces (LT) : the latter never join the central strand. x 250 approx.
Redrawn from Lorentz (1867-8).

Fig. 93. *Mnium rostratum* (Eubryales, Musci). Portion of transverse section of leafy stem. Detail of a "false" leaf trace within cortical tissues. Note the ribbon-like shape of the trace at this level. x 280. HYD, hydroids of the leaf trace.
Redrawn from Haberlandt (1886).

Fig. 94. *Polytrichum juniperinum* (Polytrichales, Musci). An early drawing by Haberlandt showing a portion of a longisection in the leptom zone. "Sieve element-like" cells are seen. (According to the present reviewer, this might possibly correspond to the lower portion of a leaf trace; the narrow elongated cell on the left would then be a hydroid, and the "sieve element-like" cells would be deuters). x 600.
Redrawn from Haberlandt (1886).

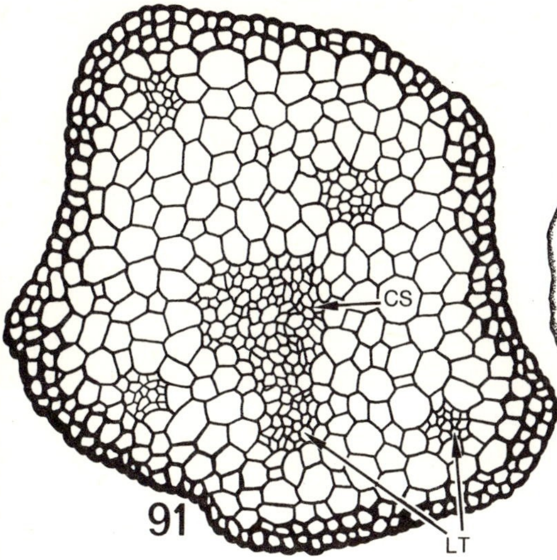

91

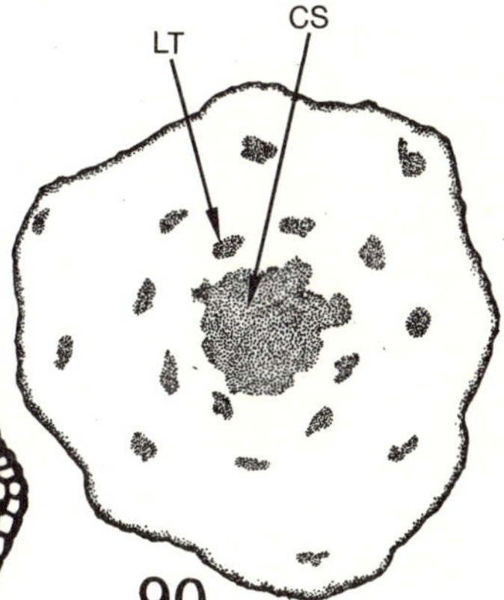

90

93

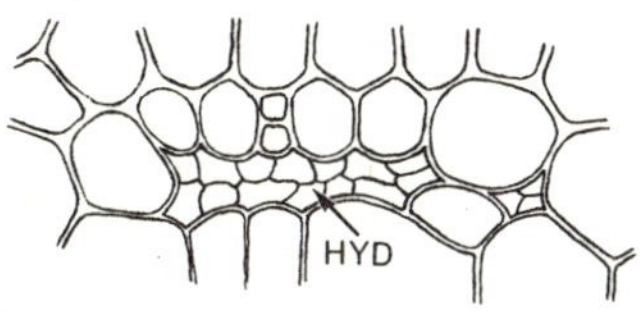

92

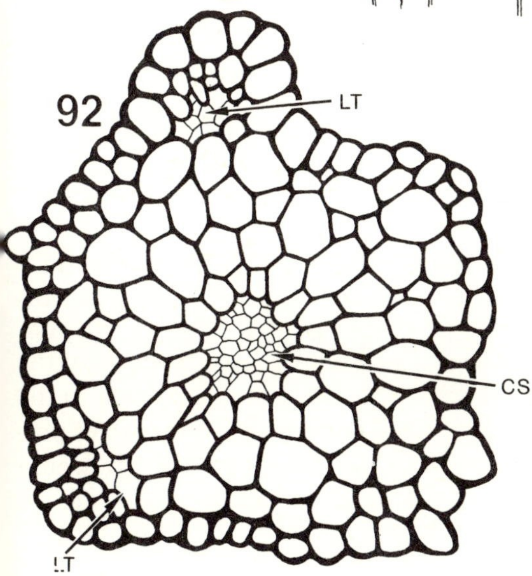

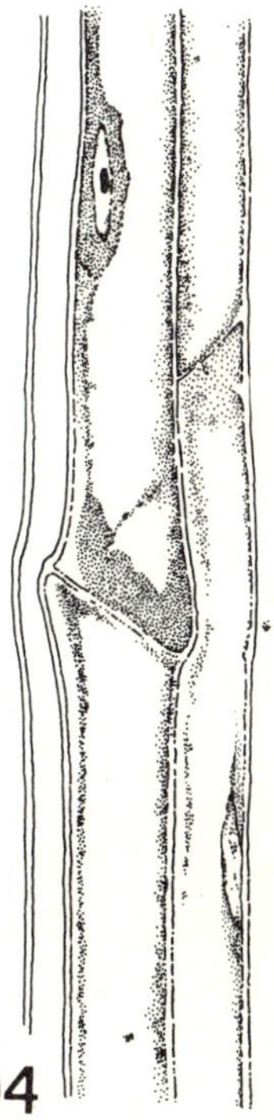

94

PLATE 24

Figs. 95-99. Some aspects of structural variability in the organization of the leaf nerves of mosses (transverse sections).

Fig. 95. *Webera cruda* (Eubryales, Musci). A complex internal organization is present, in which hydroids (HYD), cells of specialized conducting parenchyma (deuters, D) and supportive cells (stereids, STER) are identifiable. x 280.

Fig. 96. *Tetraphis pellucida* (Tetraphidales, Musci). Perichaetial leaf. A bundle of hydroids (HYD) runs within the nerve.

Fig. 97. *Dicranum undulatum* (Dicranales, Musci). Deuters (D) are seen but no hydroids are present in the nerve. x 280.

Fig. 98. *Encalypta ciliata* (Pottiales, Musci). Here again deuters (D) and stereids (STER) are present whereas hydroids are absent. x 280.

Fig. 99. *Hypnum commutatum* (Hypnobryales, Musci). Example of a type of leaf nerve with a simplified (=regressed ?) structure. x 280.

Figs. 95 and 97 to 99 redrawn from Lorentz (1867-8); fig. 96 redrawn from Lorentz (1869).

Fig. 100. A tentative generalized diagram illustrating the structure of hydroids (HYD) as seen in a transection of the central strand of aerial leafy stems of polytrichaceous mosses (with the exception of *Dawsonia spp.*; – see text pp. 36 and 55). EW, thin – usually hydrolysed – end walls; L, "lipid" droplet; LW, lateral walls; ML, middle lamella; PW, "primary" walls; SW, "secondary" walls. After Hébant (1974).

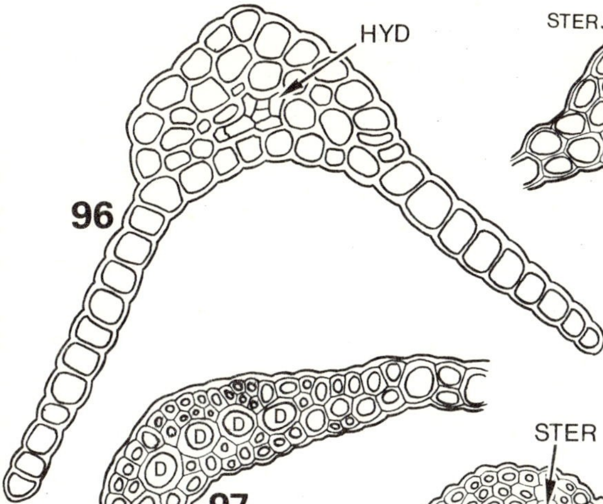

96

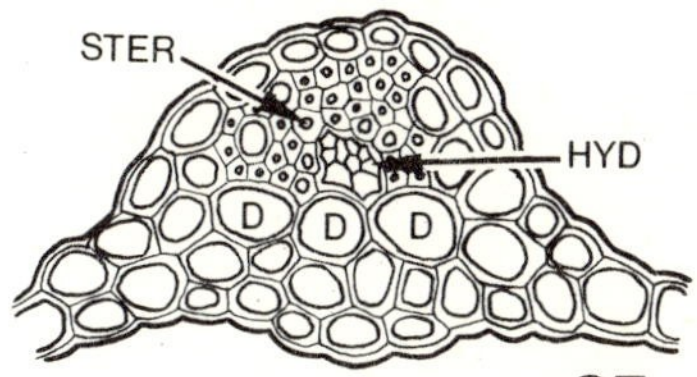

95

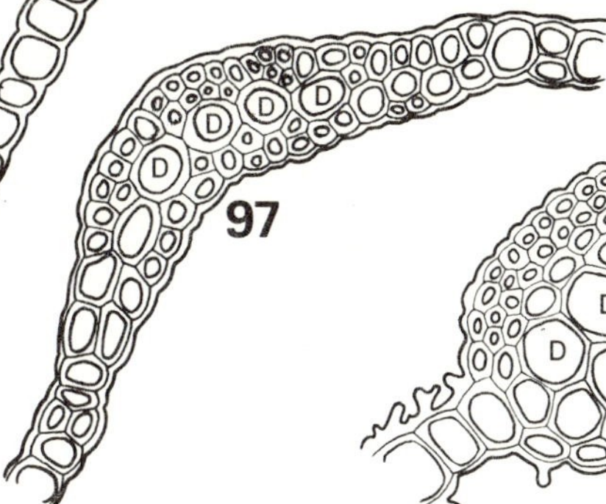

97

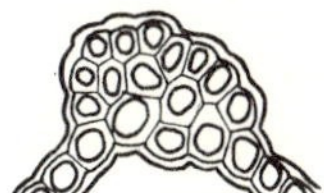

99

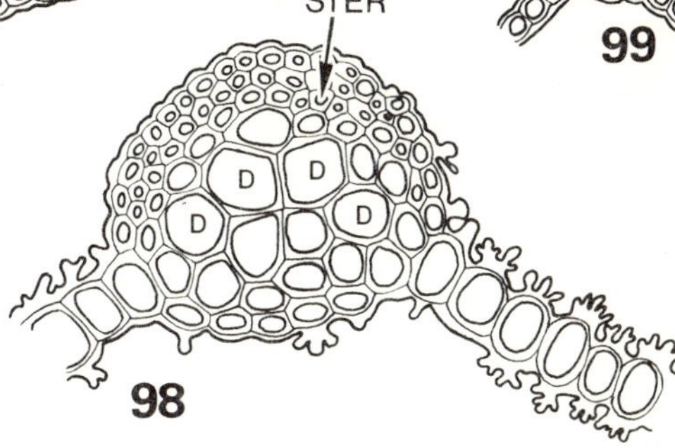

98

100

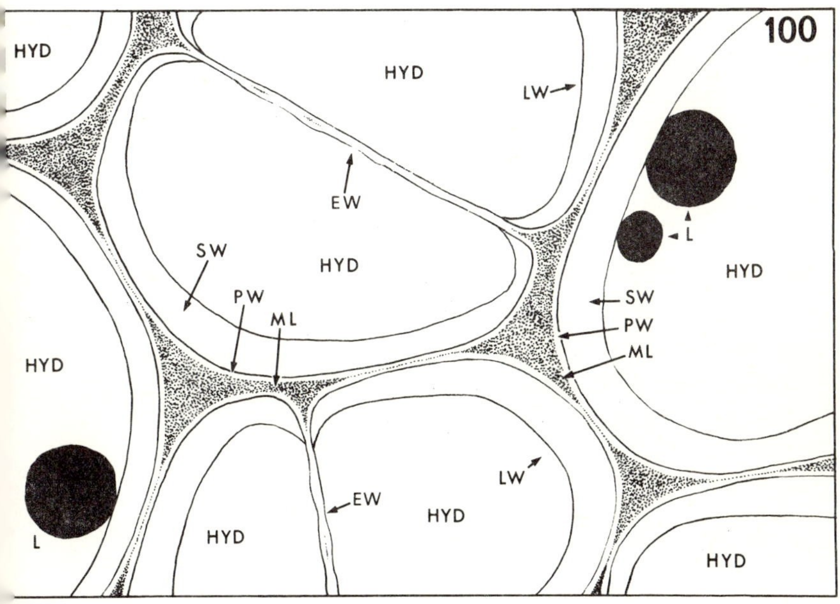

PLATE 25

Fig. 101. *Polytrichum commune* (Polytrichales, Musci) [specimen from pseudo-alpine grassland]. Diagram of transverse section in green-leaved region of gametophytic stem showing the relationships of the leaf traces with the axial conducting-system. The hydroids of the leaf traces and of the central strand are represented in black. The various cell types with "phloem" characteristics (viz. leptoids, deuters, transition elements) are dotted. Leaf traces are numbered n, n-1, n-2, etc. The trace n belongs to the leaf inserted at the uppermost level, whereas that numbered n-13 belongs to the leaf inserted at the lowest level.

Fig. 102. *P. commune.* Leafy stem. Some stages in the linking up of a leaf trace (TF) to the axial conducting-system. Drawn from serial transverse sections. Same conventions as in fig. 101. The higher level is in A and D is the lower, the vertical distance between A and D being ca. 500 μm.

Fig. 103. *P. commune.* Diagram showing, in a schematic and simplified form, the distribution of the elements with "phloem" characteristics at the periphery of the central strand of hydroids (compare with figs. 101 and 102). Hydroids have been left blank. Elements with "phloem" characteristics (leptoids, deuters, transition elements) are dotted. Hydroids of leaf traces not shown (except in the lowest region of each trace).
After Hébant (1972b).

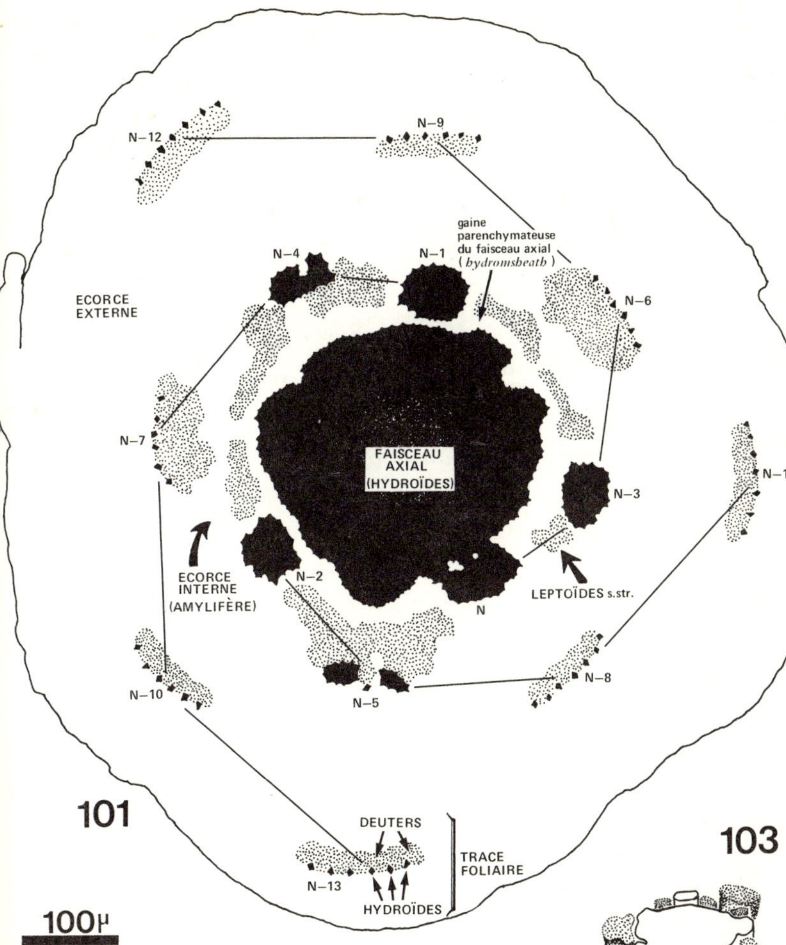
N–12
N–9
gaine
parenchymateuse
du faisceau axial
(*hydromsheath*)
N–4
N–1
ECORCE
EXTERNE
N–6
N–7
FAISCEAU
AXIAL
(HYDROÏDES)
N–11
N–3
N–2
ECORCE
INTERNE
(AMYLIFÈRE)
LEPTOÏDES s.str.
N
N–10
N–5
N–8
DEUTERS
TRACE
FOLIAIRE
N–13
HYDROÏDES
100µ

101

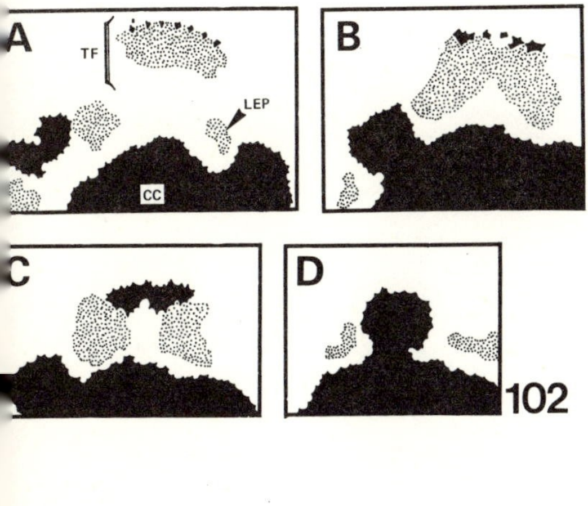
A
TF
LEP
CC
B
C
D

102

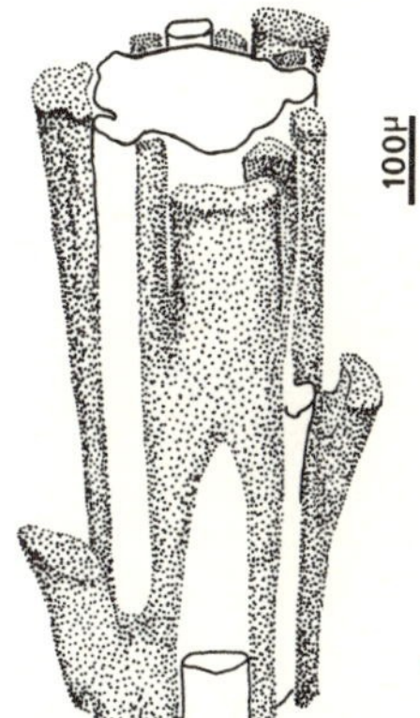
100µ

103

PLATE 26

Hydroids of leaf traces in *Polytrichum* (Polytrichales, Musci). I.

Fig. 104. *P. juniperinum.* Gametophyte. Transformations of the structure of a leaf trace during its penetration within cortical tissues. Drawn from serial transections. Only hydroids are shown : those which are in relation with the central strand are stippled; those which end blindly within cortical tissues have been left in blank (small crosses indicate the location of the hydroids which have just disappeared).

Fig. 105. Spatial reconstitution of the structure of the same leaf trace as in fig. 104. The various levels corresponding to the sections A, B, C, D, ... in fig. 104 are indicated by arrows.

After Hébant (1969b).

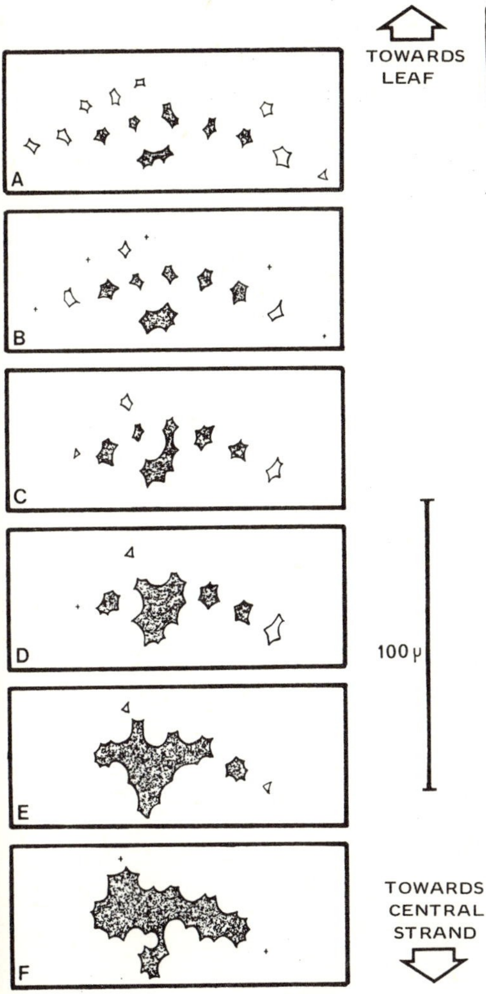

104

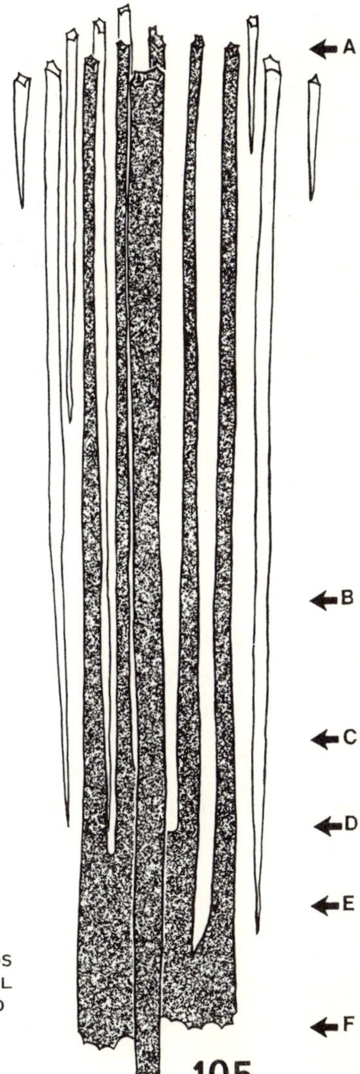

105

PLATE 27

Hydroids of leaf traces in *Polytrichum* (Polytrichales, Musci). II.

P. commune. Erect green-leaved portions of gametophytes. Transformations in the structure of the leaf traces during their penetration within cortical tissues. Same conventions as in fig. 104.

Figs. 106 and 107 illustrate the patterns observed in two specimens of the same species but from distinct sites : pseudo-alpine grassland (fig. 106) and peat-bog (fig. 107). In fig. 106, 8 hydroids of the leaf trace are in relation with the central strand whereas in fig. 107 (=specimen from peat-bog) 3 hydroids only are in relation with the central strand.
[The distance between the sections represented in 106A and 106G is 480 μm, and that between those represented in 107A and 107D is 610 μm].

After Hébant (1969b).

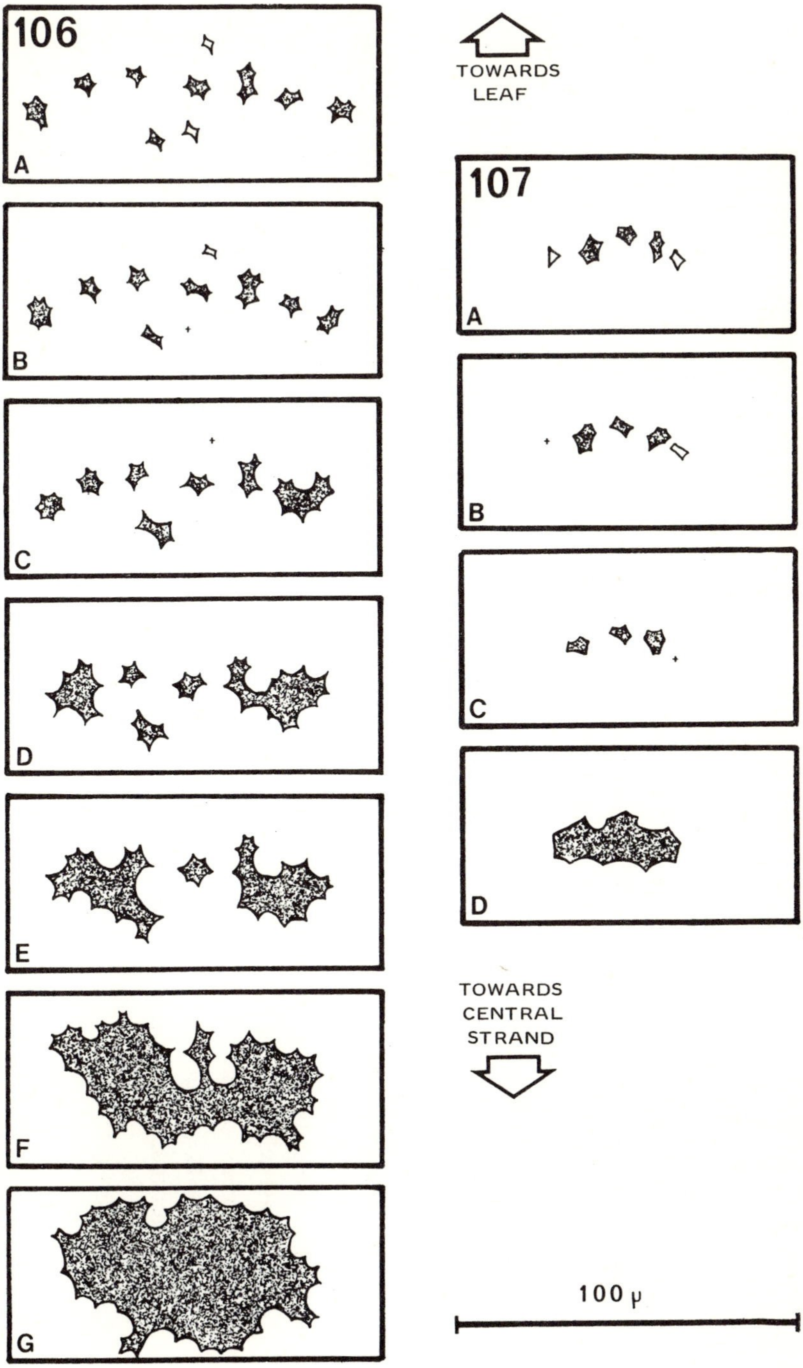

106
A
B
C
D
E
F
G
TOWARDS
LEAF
107
A
B
C
D
TOWARDS
CENTRAL
STRAND
100 µ

PLATE 28

Fig. 108. *Polytrichum commune* (Polytrichales, Musci). Leaf. Organization of the nerve in the region bearing photosynthetic lamellae (PL) [transverse section in lower third of expanded lamina].

Cells with "phloem" characteristics, the deuters (D, D'), form a continuous row. Those of the median region (D) are conspicuous because of the regression of their plastids (P) (cf. also fig. 252). Those on either side of the median region (D') possess better developed plastids and are closer to ordinary parenchyma cells in structure. "Passage cells" (PC) establish the connection between the deuters of the median region and the upper epidermis which bears the green lamellae (PL).

Several other cell types in the leaf nerve are worth noting, especially :

– the parenchymatous cells associated with the conducting elements (deuters, and hydroids). They correspond in part to the so-called *socii* (S) in Lorentz' terminology.

– the stereids (ST), elongated supportive cells with thickened walls. They are devoid of a living protoplast when mature.

– the cells of the photosynthetic lamellae (PL), in which associations between peroxisomes and chloroplasts similar those known in the leaves of vascular plants are found.

E, epidermis; H, hydroids; P, plastids.

After Hébant (1972b).

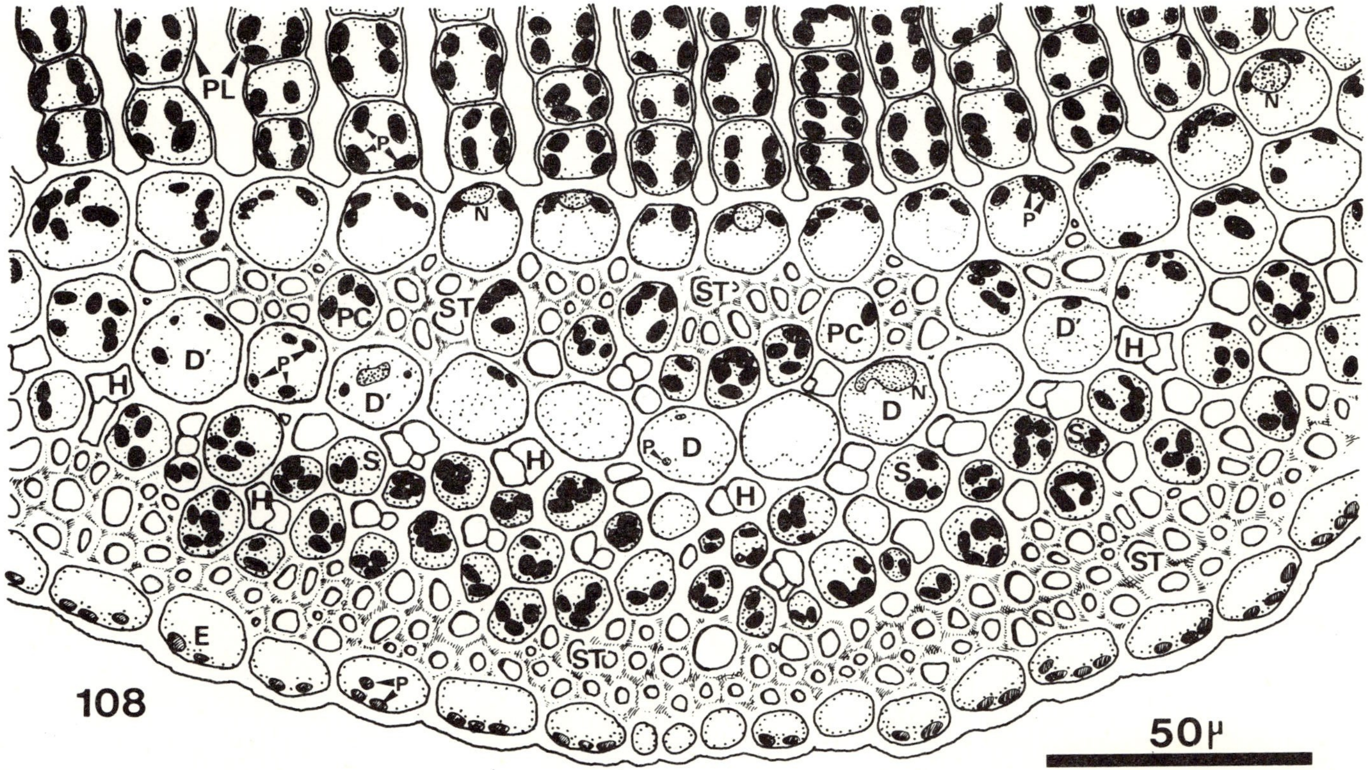

108

PLATE 29

Figs. 109-110. *Polytrichum commune* (Polytrichales, Musci). Portions of transections in green-leaved region of gametophyte showing the structure of the leaf traces at two successive stages of their penetration within cortical tissues. The food-conducting cells (deuters, D; leptoids, LEP; cells of specialized conducting parenchyma, CP) are easily identifiable through the clear appearance of their protoplast. In contrast, the neighboring cells of ordinary parenchyma possess large chloroplasts in which starch may frequently accumulate.

Figs. 109 and 110 correspond respectively to the leaf traces n-13 and n-7 in fig. 101. The white arrows indicate the center of the stem.

GT, ground tissue of the stem; H, HYD, hydroid; HS, hydrom-sheath; N, nucleus; P, plastid; S, socius.

After Hébant (1972b).

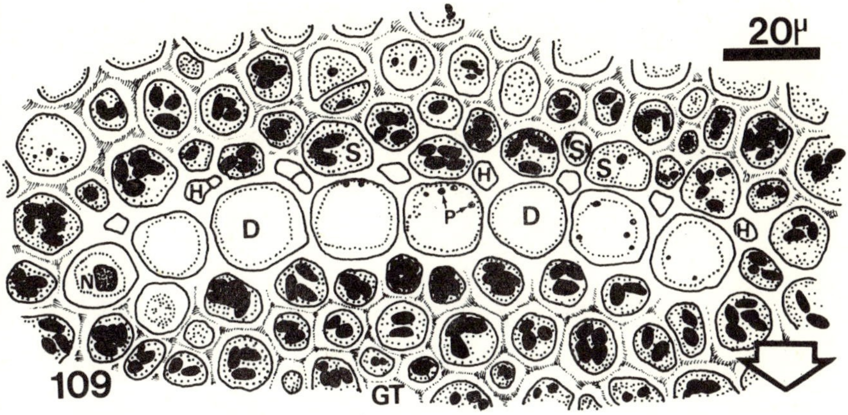

109

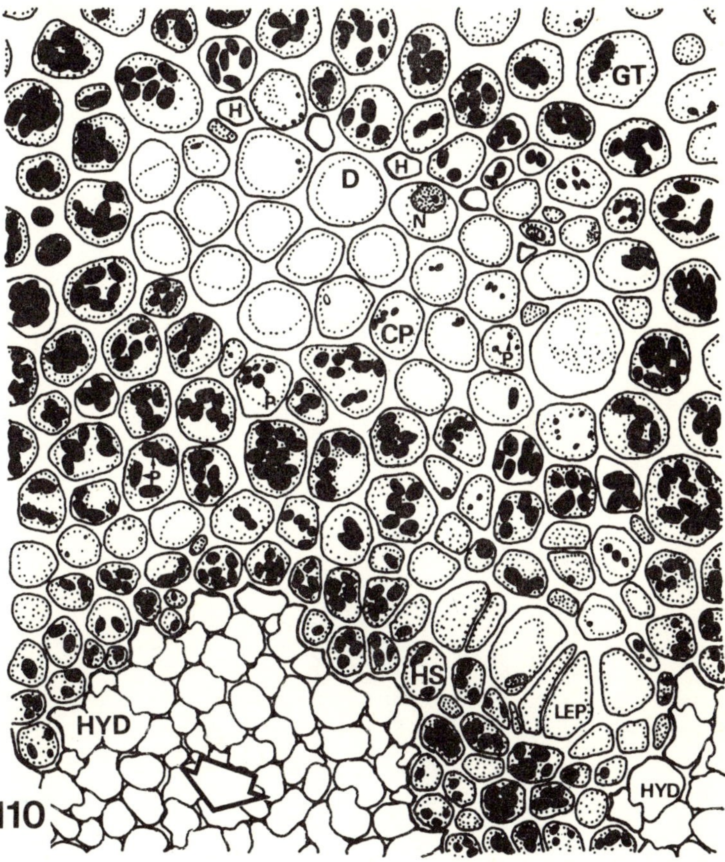

110

PLATE 30

Fig. 111. *Polytrichum commune* (Polytrichales, Musci). Leafy stem. Portion of transverse section at periphery of central strand showing relationships between the leaf traces and the axial conducting-system (compare with figs. 101-103). As in figs. 109-110, the various types of food-conducting elements are clearly distinguishable from the neighboring cells of the ground tissue because of the clear appearance of their protoplast. Here again, plastids are useful "markers" of the degree of differentiation of these elements : plastids are quite small and are not always easily seen with the light microscope in the most highly differentiated food-conducting cells, whereas ordinary parenchyma cells possess large starch-storing chloroplasts. Note that "intermediary" cells, in which plastids are smaller than in the ordinary ground tissue, also exist.

The white arrow points to the center of the section.

CP, conducting parenchyma; GT, ground tissue; HS, hydrom-sheath; HYD, hydroid; LEP, leptoid.

After Hébant (1972b).

111

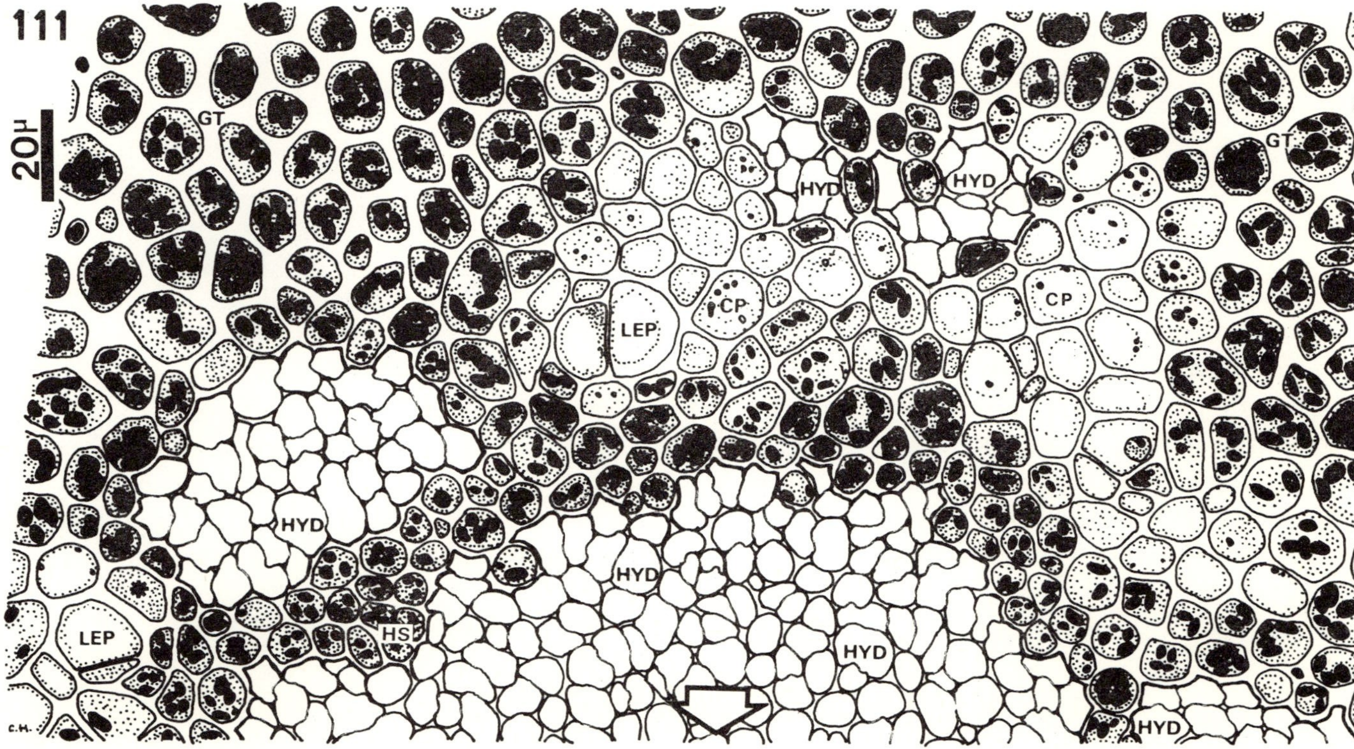

PLATE 31

Fig. 112. Diagrams illustrating the various types of organization which can be encountered in the leafy stems of mosses, with particular reference to the leaf traces (see text p. 42).

After Hébant (1969b).

112

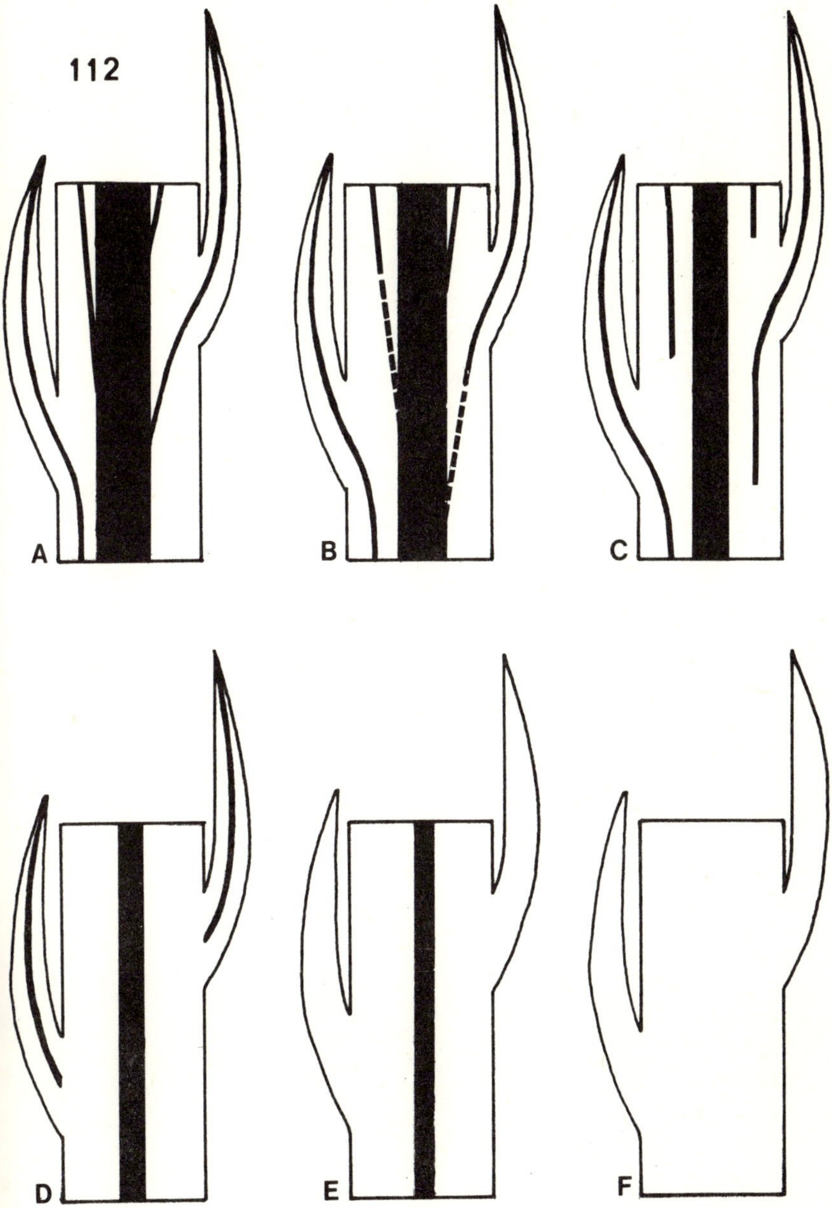

PLATE 32

Fig. 113. *Polytrichum juniperinum* (Polytrichales, Musci). Transverse section of mature seta. The central strand (CS) is anchored within an inner lacunar cortex (IC), quite distinct from the thick-walled outer cortex (OC). The conducting strand itself consists of a ring of leptoids (LEP; – "leptophloem" according to the terminology of Vaisey –) and of a solid central core of thin-walled hydroids (HYD; – "leptoxylem" according to the terminology of Vaisey –). x 365.

Fig. 114. *Atrichum undulatum* (Polytrichales). Longisection of the seta. The conducting strand (CS) runs within a lacunar inner cortex and presents the same basic organization as in *Polytrichum* (fig. 113). Note however that the ring of leptoids (LEP) is here two (sometimes three) cells thick only. x 365.

C, cortex; IS, intercellular spaces in the cortex.
Redrawn from Vaisey (1888).

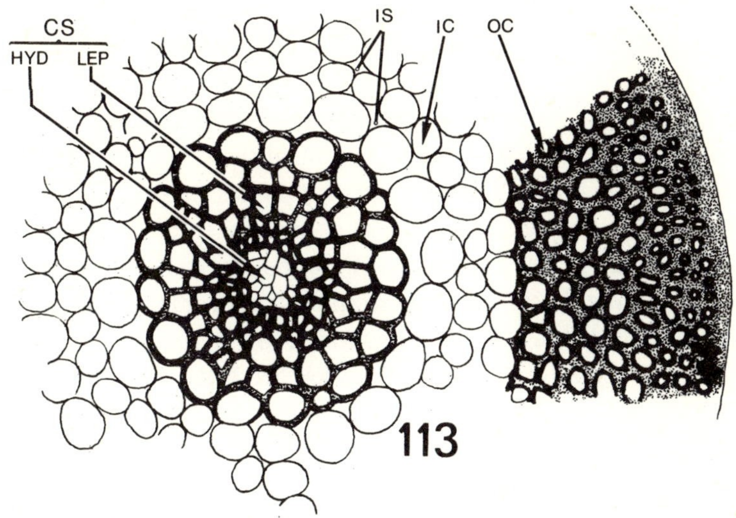

113

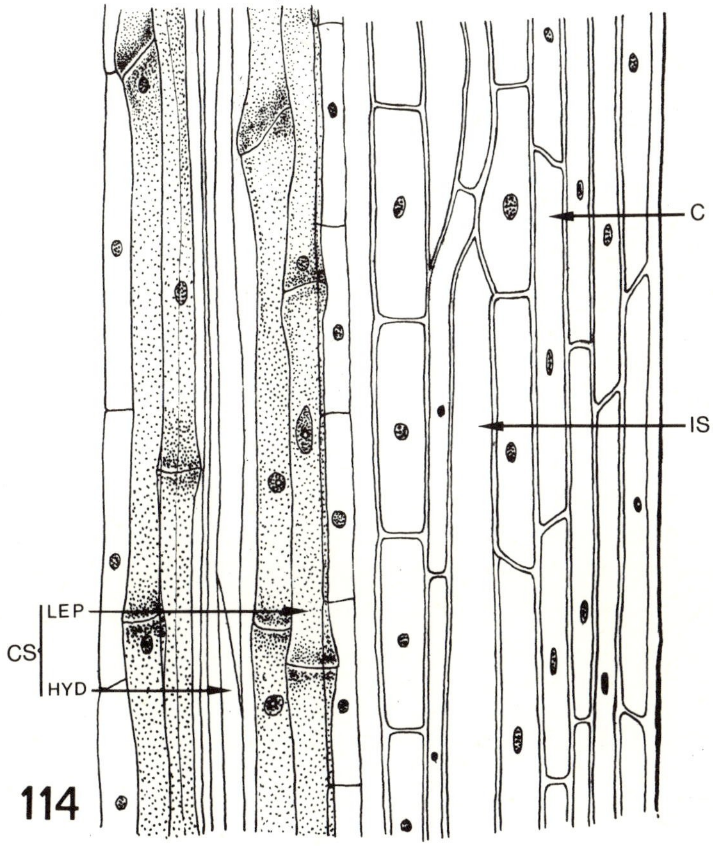

114

PLATE 33

Fig. 115. *Funaria hygrometrica* (Funariales, Musci). Transverse section of the seta. Its general organization is reminiscent of that found in such Polytrichales as *Polytrichum* or *Atrichum* (cf. figs. 113-114), although in a somewhat simpler form. In particular no intercellular spaces exist in the inner cortex and the central strand is much smaller. A discrete strand of hydroids (HYD; – "leptoxylem" of Vaisey –) is surrounded by a ring of cells ("leptophloem" according to Vaisey) which are interpreted here as residual leptoids (LEP) but in which many authors see a protective sheath (see text pp. 48-49).
Redrawn from Vaisey (1888).

Fig. 116. *Funaria hygrometrica.* Portion of longisection of seta showing from left to right :
. parenchyma cells of the inner cortex
. tibia-like cells (see text pp. 48-49 for details)
. hydroids
Redrawn from Haberlandt (1886). x 510.

Fig. 117. *Splachnum luteum* (Funariales, Musci). Portion of longisection (slightly tangential) of the central strand in the lower region (foot) of the sporophyte. The basic structure of this strand is quite similar to that of *Funaria* (compare with figs. 115 and 116).
Redrawn from Vaisey (1891).

Figs. 118 and 119. Transverse sections in the setae of *Dicranum undulatum* (Dicranales) (fig. 118) and *Webera polymorpha* (Eubryales) (fig. 119). A central strand of hydroids runs in the median region of each seta, but in none of them is a leptom clearly identifiable. x 320.
Redrawn from Lorentz (1867-8).

C, cortex; HYD, hydroid; LEP, leptoid.

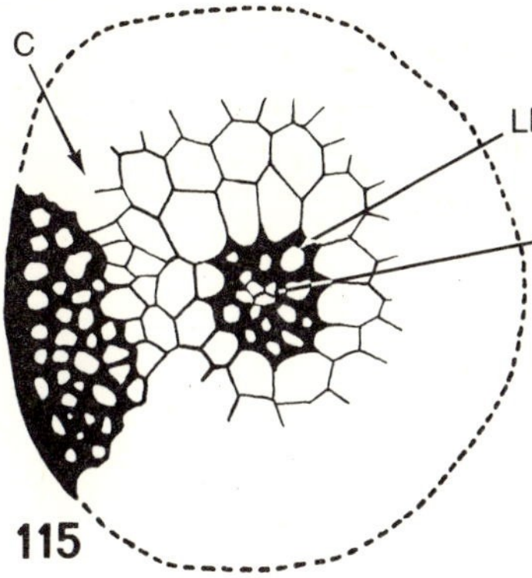

115

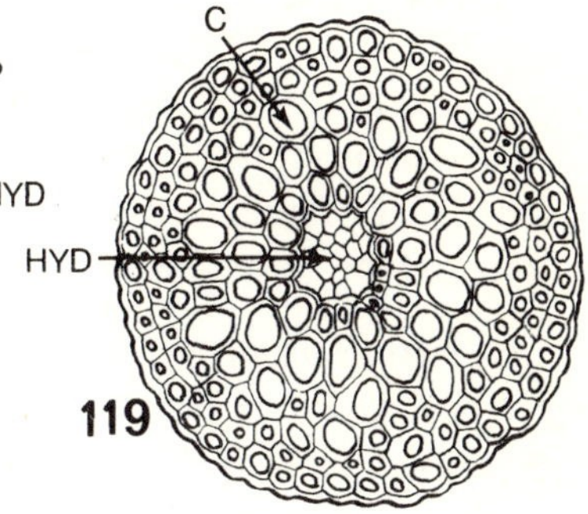

119

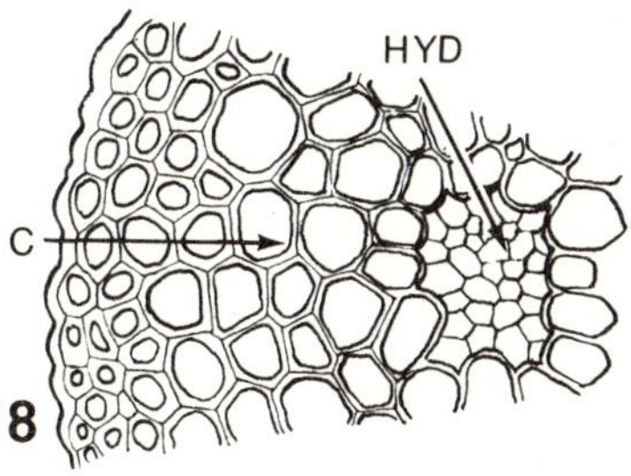

118

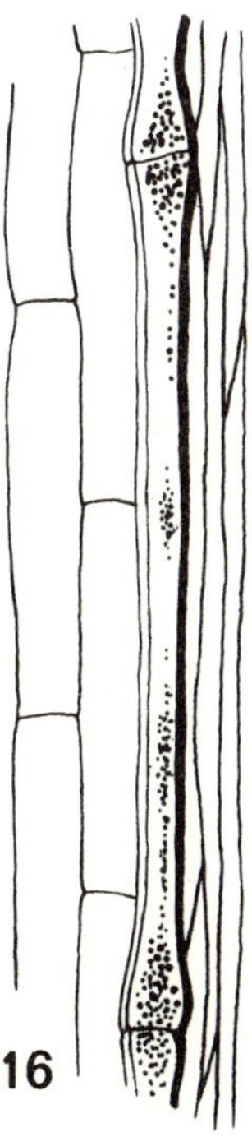
116

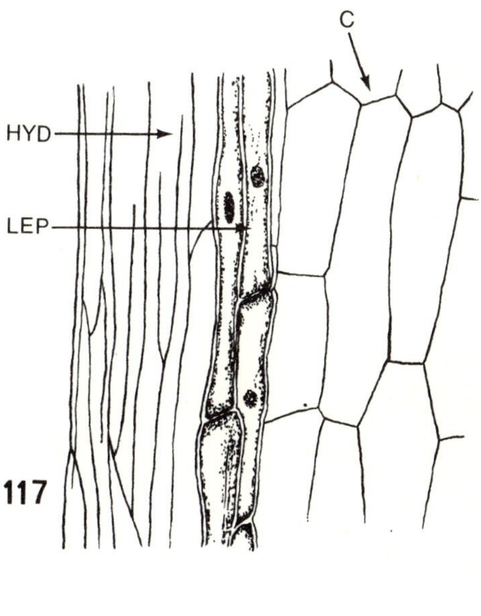

117

PLATE 34

Fig. 120. *Hedwigia ciliata* (Isobryales, Musci). An early representation by Rosander (1906) of the interphase between gametophyte (n) and sporophyte (2n) [longisection]. Presumable tranfer cells (TC) are identifiable in this drawing.
Redrawn from Rosander (1906).

Fig. 121. *Polytrichum formosum* (Polytrichales, Musci). Longitudinal section in lower extremity of foot, at base of sporophyte. This structure is deeply inserted within the central strand of the gametophyte. An "appendage" (APG) is identifiable at its extreme tip (this appendage corresponds to the lower part of the embryo which apparently plays an important role during its implantation within gametophytic tissues). Just above the appendage a "water-absorbing" organ (WO) (according to Vaisey's terminology) is seen, which represents the lower portion of the central strand of hydroids (HYD) of the sporogonium. Recent studies (Maier 1967; Maier & Maier 1972) have shown that the cells at the periphery of the foot are transfer cells (TC). x 420.
Redrawn from Vaisey (1888).

Fig. 122. A figure by Lorch (1931), remarkable for its accuracy, depicting the structure of a transfer cell in the foot of *Polytrichum commune*. The wall ingrowths are clearly identifiable.
Redrawn from Lorch (1931).

Fig. 123. *Bryum caespiticium* (Eubryales, Musci). Transverse section through vaginula (gametophytic tissue, (n)) and foot of sporophyte (2n). Note the important development of the outer cells of the foot. The central strand of hydroids (HYD) of the sporophyte is also seen.
Redrawn from Lorch (1925b).

Fig. 124. *Tayloria serrata* (Eubryales). Portion of transverse section in the interphase between gametophyte (n) and sporophyte (2n). Transfer-cells (TC) with well developed wall ingrowths are present on both sides. This figure is redrawn from part of another remarkable illustration by W. Lorch (1925a).

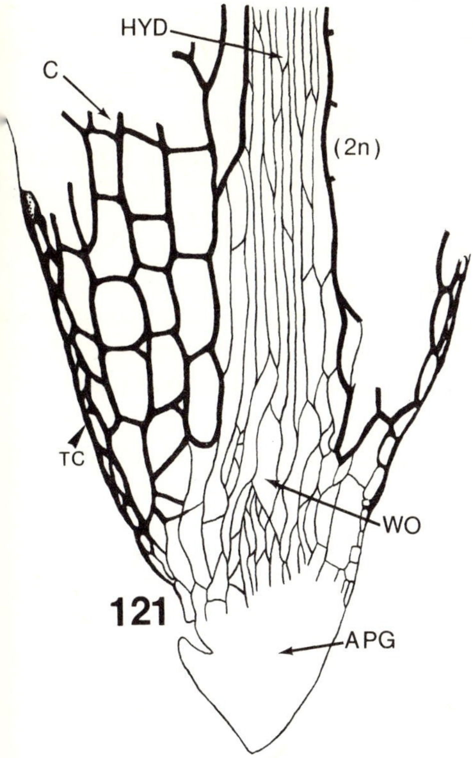

121

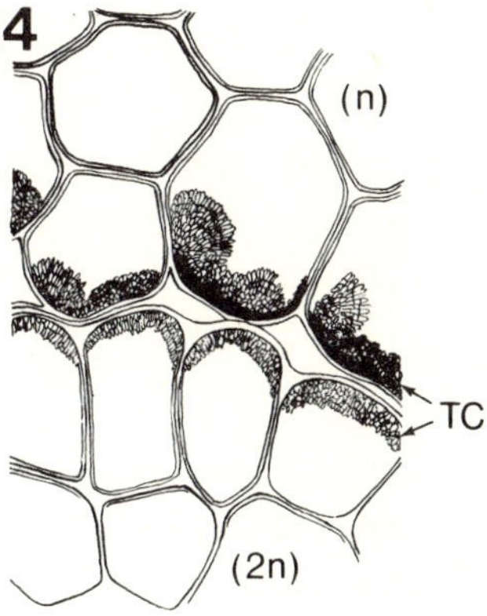

124

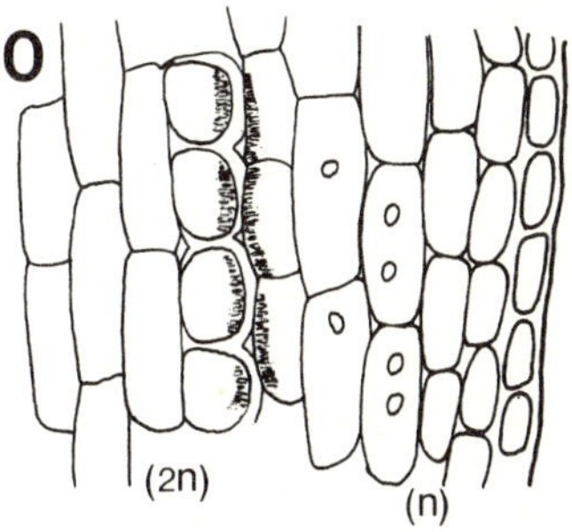

120

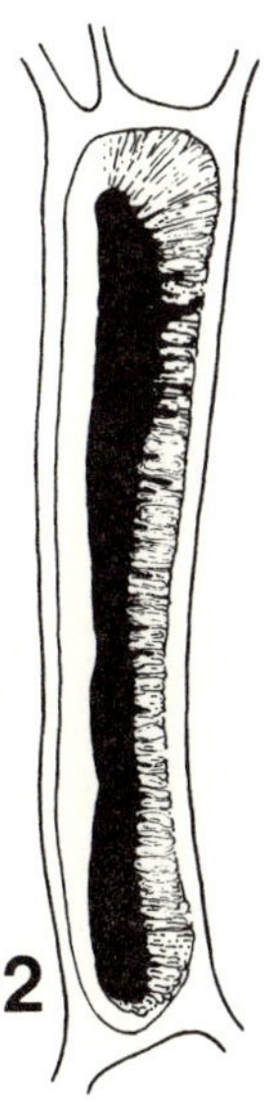

122

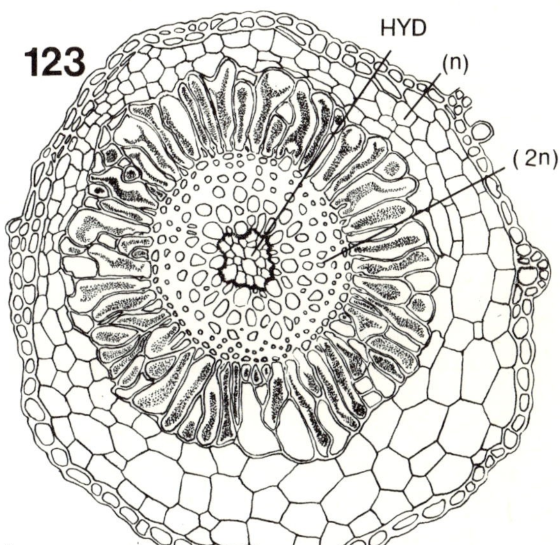

123

PLATE 35

Fig. 125. *Polytrichum commune* (Polytrichales, Musci). Diagrams illustrating the structure of the conducting tissues at the periphery of the central strand in leafy stem (fig. 125A) and seta (fig. 125B). Both generations possess hydroids and leptoids, and they both present a somewhat comparable organization pattern. Note however that no parenchyma cells are associated with the conducting elements in the seta, whereas such cells exist in the gametophyte (cf. hydrom-sheath; parenchyma cells of the leptom; ...). Differences also exist in the ultrastructure of the leptom in both generations, – the leptom in the short-lived seta apparently being less specialized than the corresponding tissue in the gametophyte (see text).
Redrawn from Hébant (1967a).

Fig. 126. *Funaria hygrometrica* (Funariales, Musci). Diagrams showing the internal structure of the leafy stem (fig. 126A) and the seta (fig. 126B). Both hydroids and "residual" leptoids [(1)] exist in the seta, whereas only a strand of hydroids is present in the gametophyte. x 220.
After Hébant (1964-5).

C, cortex; E, epidermis; HYD, hydroid; LEP, leptoid; LT, leaf trace; N, nucleus; PAR, parenchyma cell; ST, stereid.

(1) About the "residual" leptoids of the seta of *Funaria*, see also figs. 115-116 and text pp. 48-49).

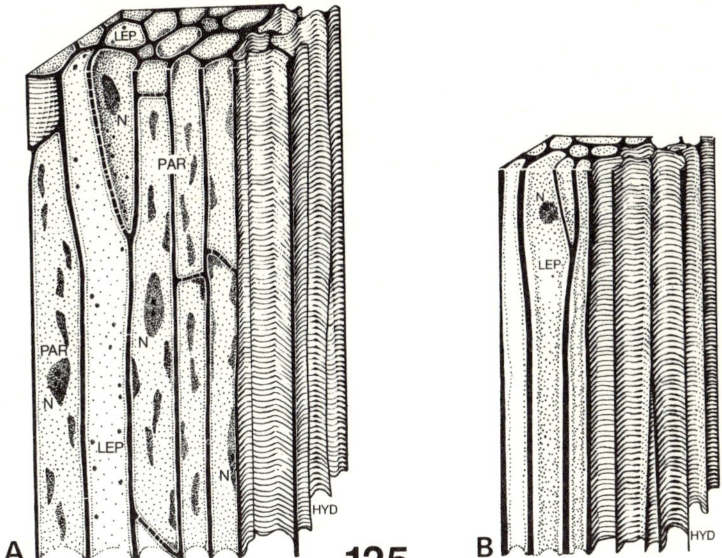

125

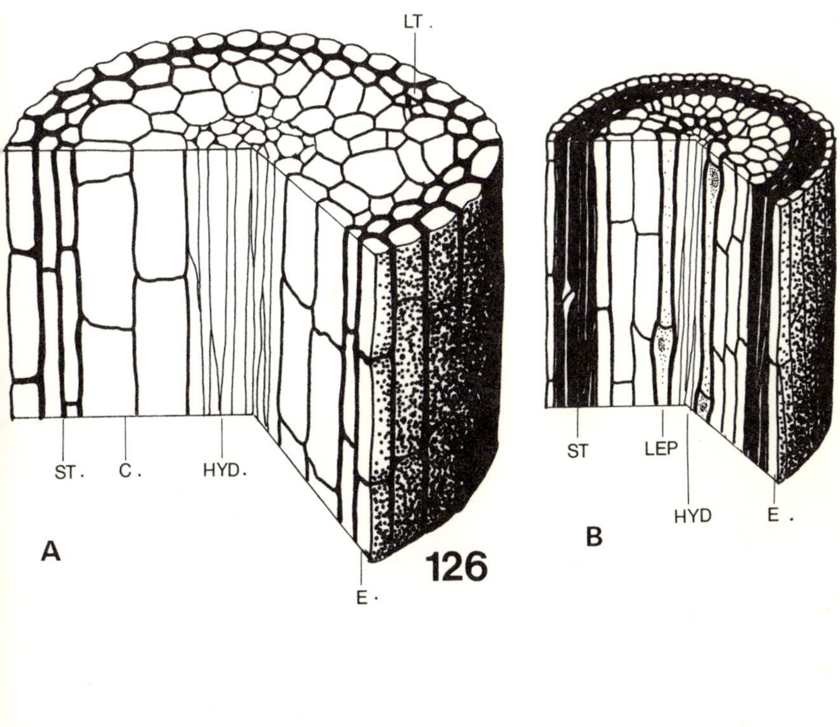

126

PLATE 36

Fig. 127. *Brachythecium sp.* (Hypnobryales, Musci). Internal structure of leafy stem (fig. 127A) and seta (fig. 127B). Both phases exhibit a strand of hydroids. They are devoid of leptoids.

Fig. 128. *Neckera crispa* (Isobryales, Musci). The gametophyte lacks a conducting strand (fig. 128A); a strand of hydroids exists (persists ?) in the seta (fig. 128B). There are no leptoids.

After Hébant (1970c).

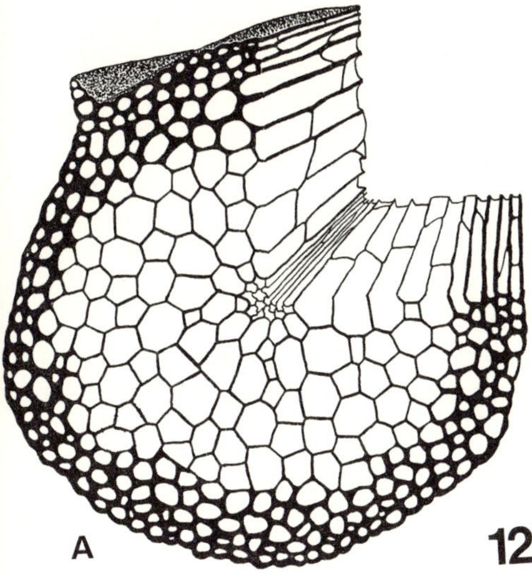

A

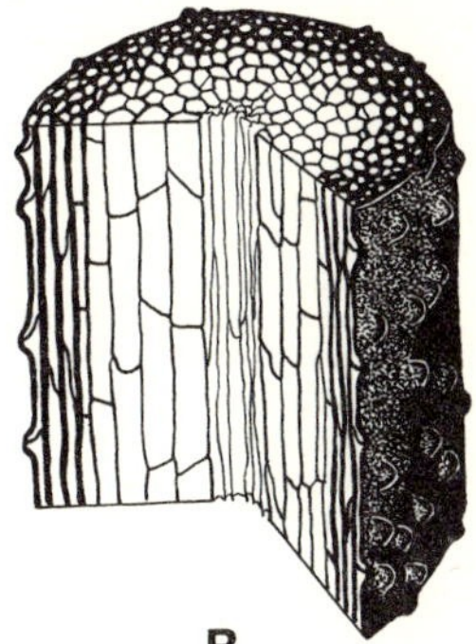

B

127

100 µm

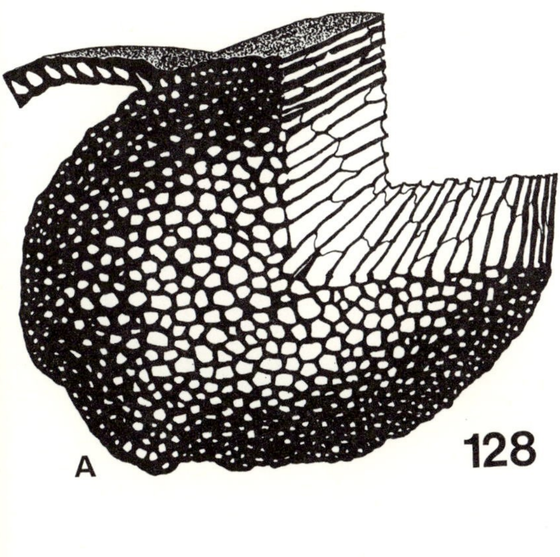

A

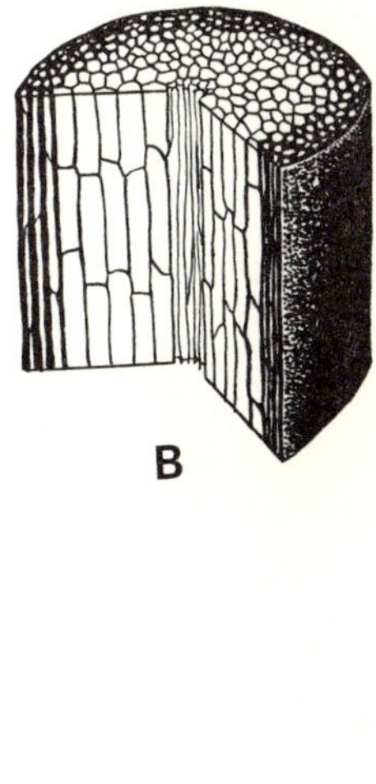

B

128

PLATE 37

Figs. 129A-G. *Aulacomnium palustre* (Eubryales, Musci). Leafy stem. Some stages in development of hydroids of central strand. Note increasing importance of vacuolar apparatus during elongation phase.

Fig. 130. *Aulacomnium palustre.* Portion of longisection in mature region of leafy stem. Hydroids of central strand and parenchyma cells of internal cortex are shown. Mature hydroids are elongated empty dead cells. Cortical cells retain a living protoplast in which a nucleus and various organelles are identifiable.

HYD, hydroid; M, mitochondrion; N, nucleus; P, plastid; PAR, parenchyma cell; V, vacuole.

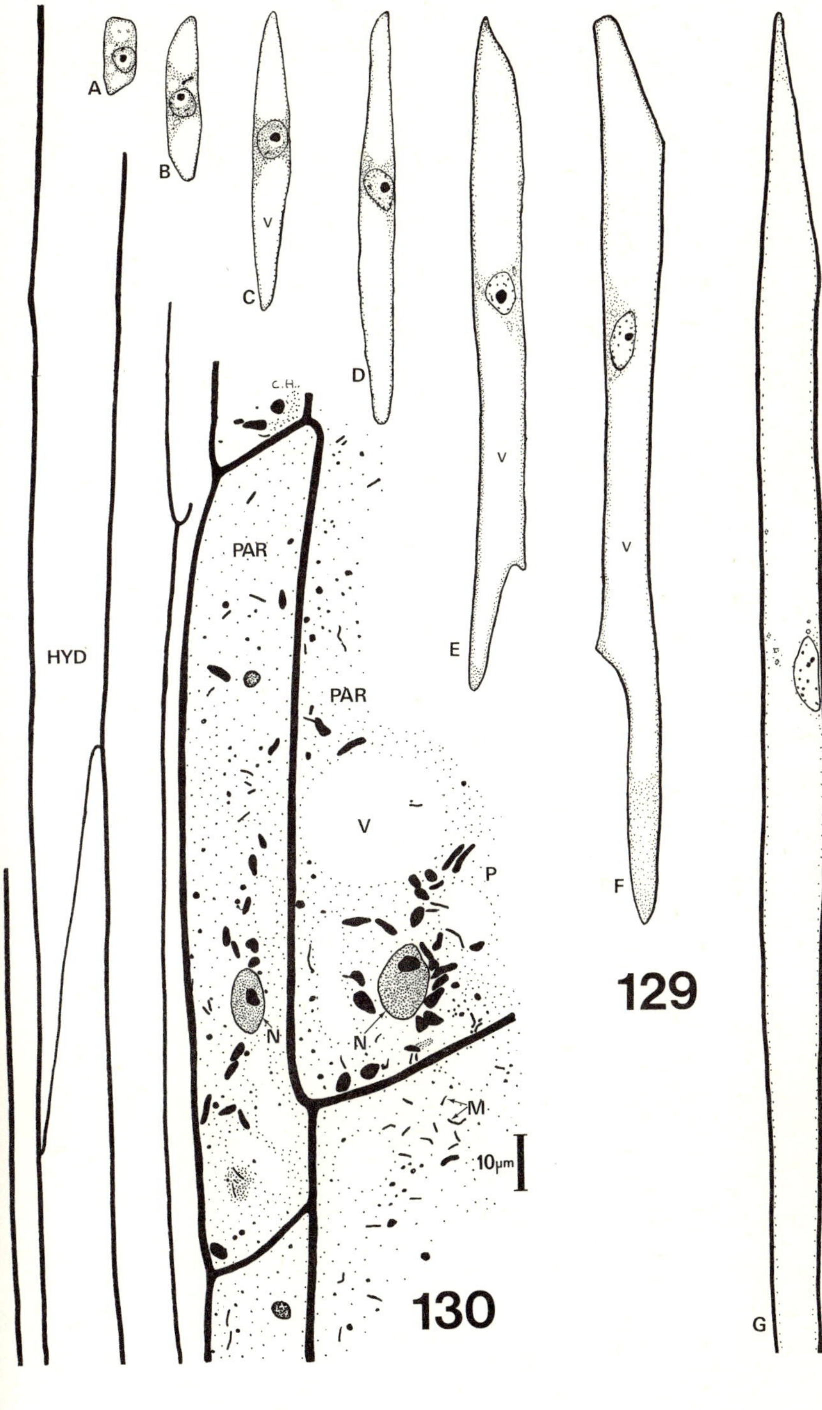
A
B
C
D
E
F
G
V
HYD
C.H.
PAR
PAR
V
P
N
N
M
10µm
129
130

PLATE 38

Figs. 131 to 135. *Polytrichum commune* (Polytrichales, Musci). Leafy stem. Some stages in development of hydroids of central strand.

Fig. 136. *Polytrichum commune.* Differentiating hydroid and associated elements in a young leaf trace.

FL, filamentous structure (mitochondrion or plastid); HYD, hydroid; M, mitochondrion; N, nucleus; P, plastid; PAR, parenchyma cell; V, vacuole.
After Hébant (1967b).

131

132

20 µm

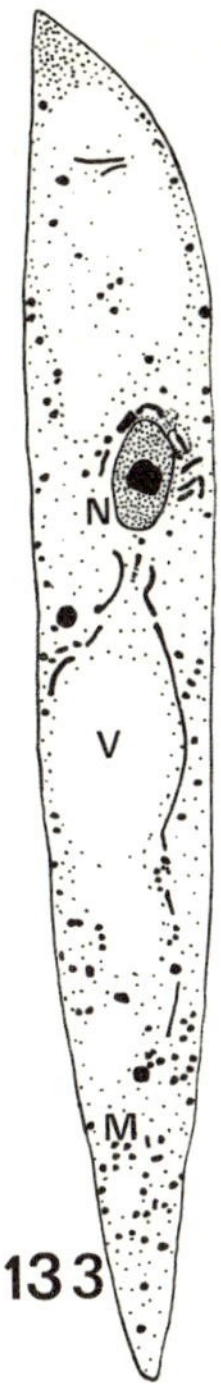

133

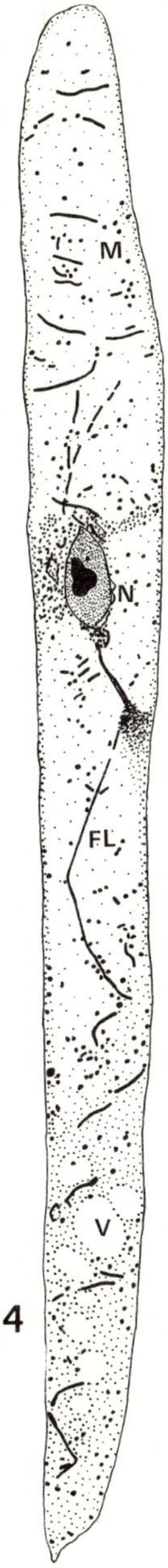

134

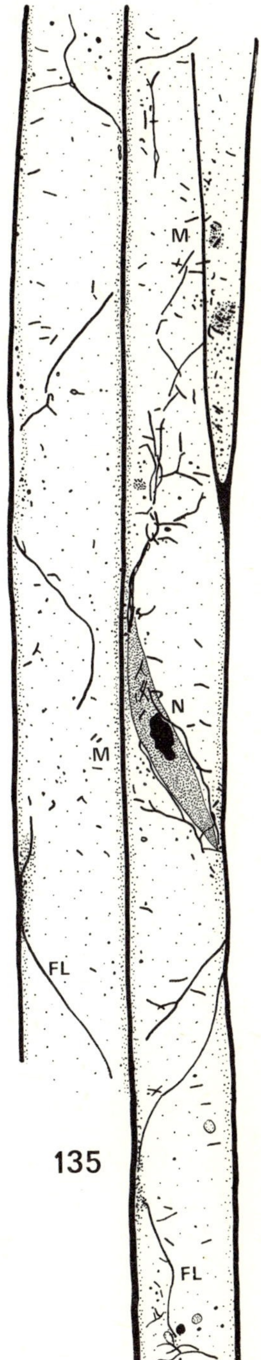

135

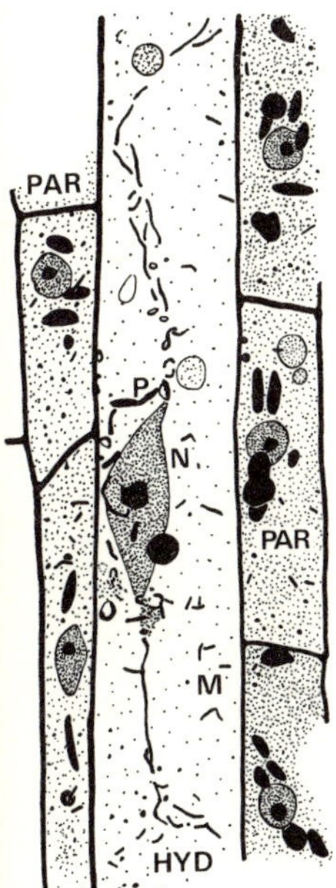

136

PLATE 39

Figs. 137 to 141. *Polytrichum commune.* Leafy stem. Some stages in the development of the leptoids *sensu-stricto*. Note difference in size between nuclei of differentiating leptoids and those of associated parenchyma cells (figs. 140 and 141).
After Hébant (1967b).

Fig. 142. *Polytrichum piliferum.* Leafy stem. Portion of longisection of leptom at maturity. Leptoids and associated parenchyma cells are shown. "Nacreous" walls are present in both cell types and therefore cannot be used for the identification of the leptoids themselves. Note that the nuclei of the leptoids are reduced in size and apparently degenerating.

LEP, leptoid; M, mitochondrion; N, nucleus; NW, nacreous wall; P, plastid; PAR, parenchyma cell; V, vacuole.

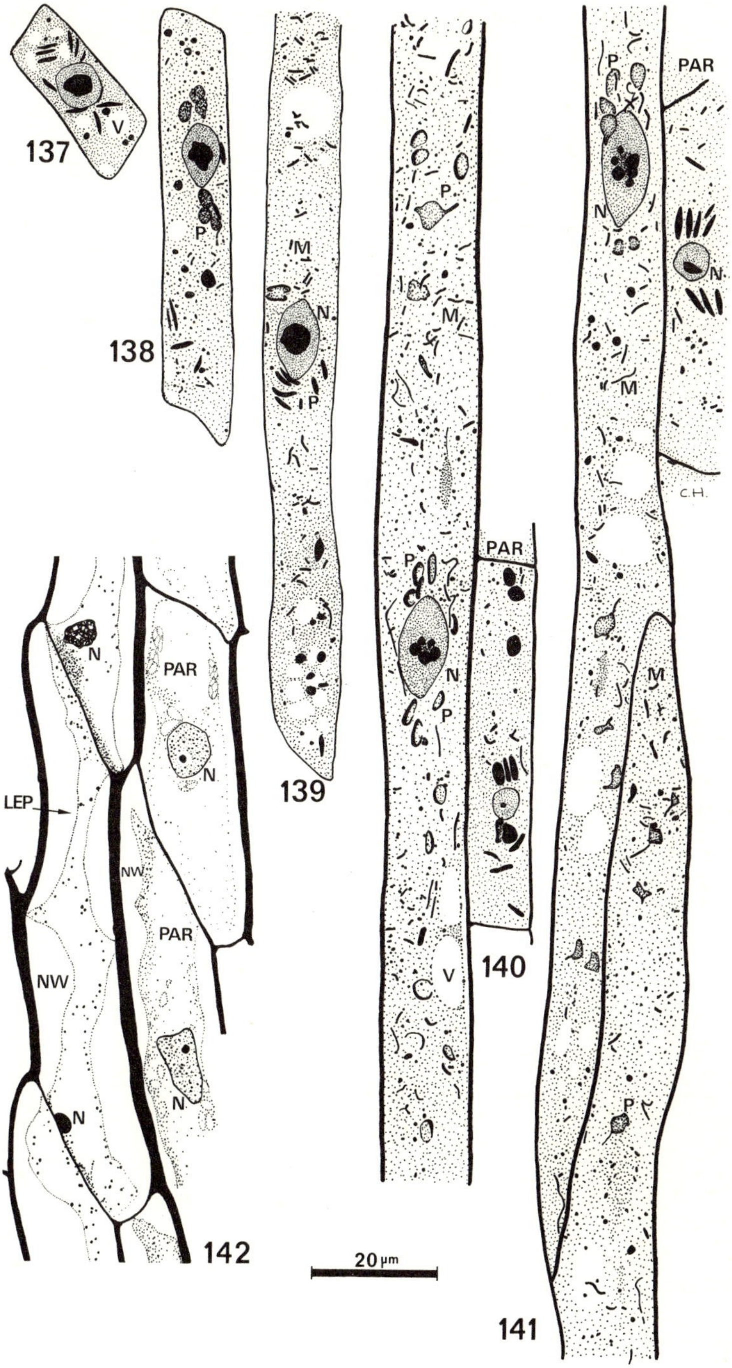
137
V
138
P
139
M
N
P
P
M
N
P
140
PAR
V
141
PAR
P
N
N
M
M
P
C.H.
142
N
PAR
N
LEP
NW
NW
PAR
N
N
20 µm

PLATE 40

Fig. 143. Diagram illustrating some of the most characteristic events in the development of the leafy gametophyte of *Polytrichum commune*, with particular reference to the conducting tissue-system (see text for details; data concerning gel electrophoresis reproduced from Kalhem & Hébant 1973).

143

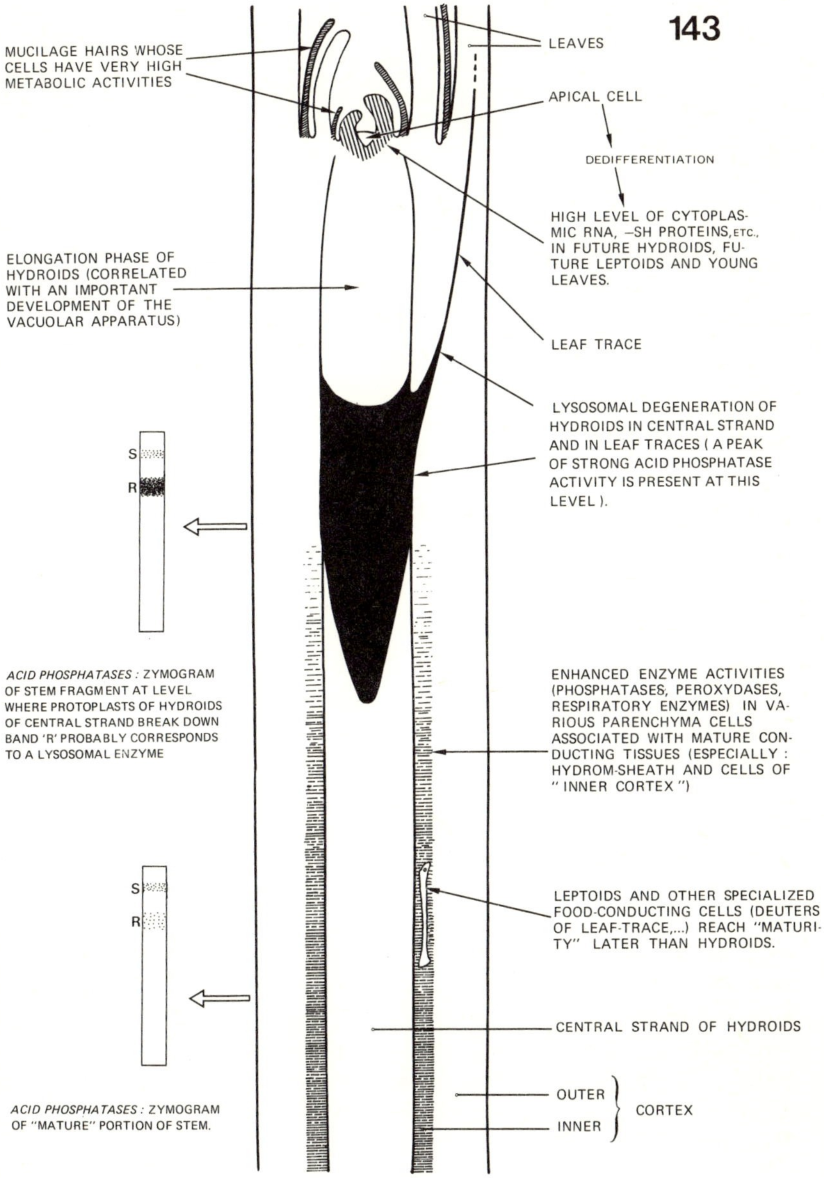

ACID PHOSPHATASES : ZYMOGRAM OF STEM FRAGMENT AT LEVEL WHERE PROTOPLASTS OF HYDROIDS OF CENTRAL STRAND BREAK DOWN BAND 'R' PROBABLY CORRESPONDS TO A LYSOSOMAL ENZYME

ACID PHOSPHATASES : ZYMOGRAM OF "MATURE" PORTION OF STEM.

PLATE 41

Physiology of conduction

Fig. 144. Synthetic diagrams showing the path taken by a solution of fluorochrome within the gametophytes and sporophytes of *Funaria hygrometrica* and *Polytrichum spp.* The sequence of events is illustrated from A to D. The solution absorbed at the base of the stem reaches the central strand of hydroids in which it rapidly circulates towards the upper regions of the gametophyte. A slower diffusion of the fluorochrome within cortical tissues also occurs (dotted). "True" leaf traces (as in *Polytrichum*) take their water supply directly from the central strand; in the case of "false" leaf traces (as in *Funaria*), the solution must first cross the parenchyma cells of the inner cortex before reaching the traces. Note also that the fluorochrome reaches very rapidly the vaginula where it is readily absorbed by the foot of the sporophyte, and then moves towards the capsule within the central strand of the seta. CS, central strand of hydroids; F, foot of sporogonium; FLT, "false" leaf trace; R, rhizoid; S, seta; TLT, "true" leaf trace; V, vaginula; (n), gametophyte; (2n), sporophyte.
Redrawn from Bopp & Weniger (1971).

Figs. 145 and 146. Two examples of experimental devices used in investigating water relations in bryophytes.
Fig. 145 shows the apparatuses used by Mägdefrau (1935-6) in his study of the relative importance of external versus internal conduction in mosses. The apparatus shown in fig. 145A is used in measurements of internal conduction alone : cut stems are immersed in a tube containing water; leaves are removed from a portion of each stem to eliminate external conduction and a thin layer of paraffin oil (dotted) is spread over the surface of the water to prevent evaporation. The device shown in fig. 145B is used in measurements of both internal and external conduction acting together. These apparatuses are kept in a chamber in which the relative humidity is controlled, and water losses are determined by weighing the tubes after specific intervals of time.
Fig. 146 shows the potometer and chamber used by Grubb (1970) for experiments on water balance in shoots of *Haplomitrium* and other bryophytes.
Fig. 145 redrawn from Mägdefrau (1935-6); fig. 146 reproduced from Grubb (1970).

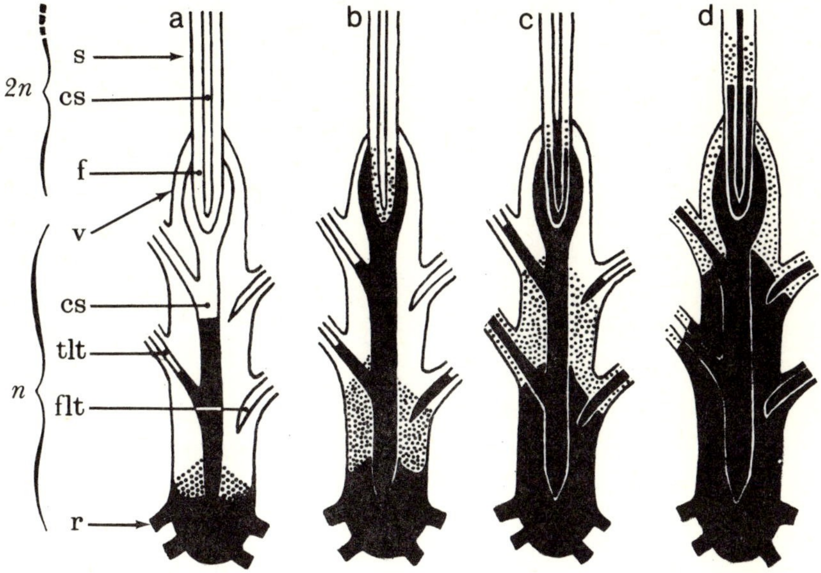
a
b
c
d
s
2n
cs
f
v
cs
tlt
n
flt
r

144

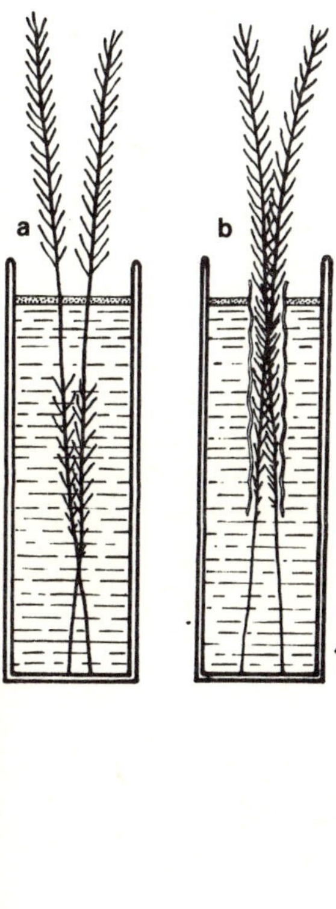
a
b

145

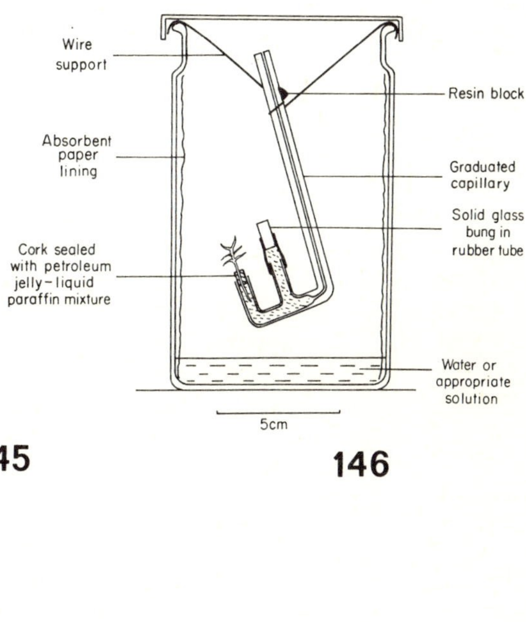
Wire support
Resin block
Absorbent paper lining
Graduated capillary
Solid glass bung in rubber tube
Cork sealed with petroleum jelly–liquid paraffin mixture
Water or appropriate solution
5cm

146

PLATE 42

The path of water in thalloid liverworts and anthocerotes as demonstrated with stain solutions (for details, see chapter 10)

Fig. 147. *Anthoceros laevis* (Anthocerotales, Anthocerotae). Section in gametophyte (N) and base of sporophyte (2N). Arrows indicate the path taken by the water.
C, columella; IN, involucre; MR, meristematic region; F, foot; NC, *Nostoc* colony; SO, sex organ; R, water reservoir; RH, rhizoids. Redrawn from Isaac (1941).

Fig. 148. *Pellia epiphylla* (Metzgeriales, Hepaticae). Longisection of thallus (N) and developing sporogonium (2N). Arrows indicate the path of water.
ANTH, antheridium; C, capsule; CAL, calyptra; F, foot; IN, involucre; RH, rhizoids; S, seta; TH, thallus. x 24.
Redrawn from Clee (1939).

Fig. 149. *Fegatella conica* (Marchantiales, Hepaticae). Transverse section of thallus. The arrows denote the path of entry of stains while the shaded area denotes the distribution of the so-called "mycorrhiza".
ASS.T., assimilating tissue; M.C., mucilage canal; M.R., mycorrhizal region; S.C., scale; S.W.R., smooth-walled rhizoid. x 28.
Redrawn from Clee (1943).

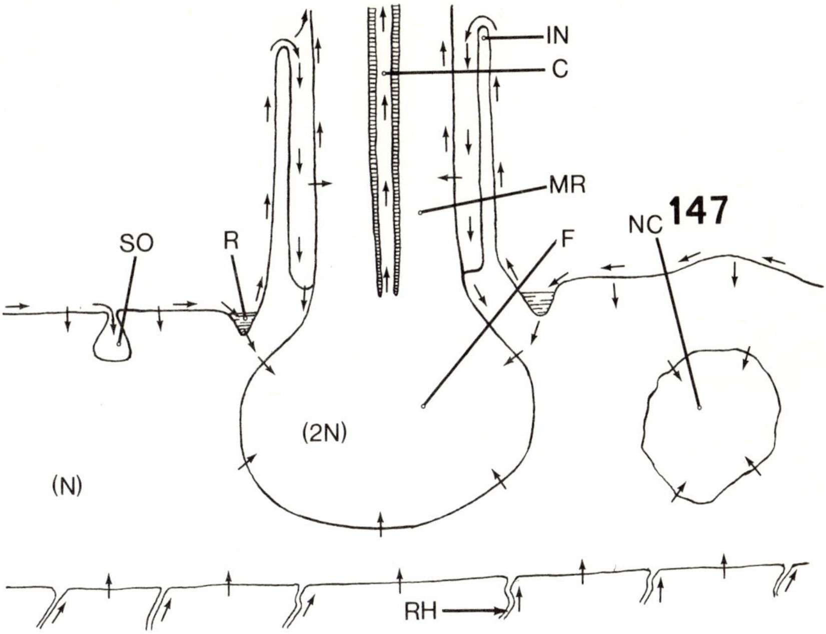

147

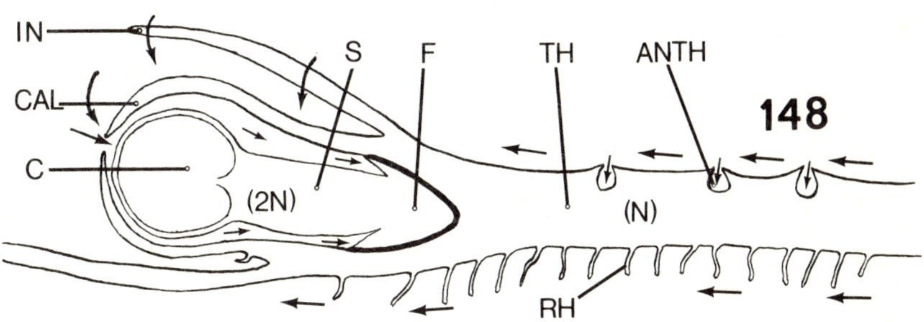

148

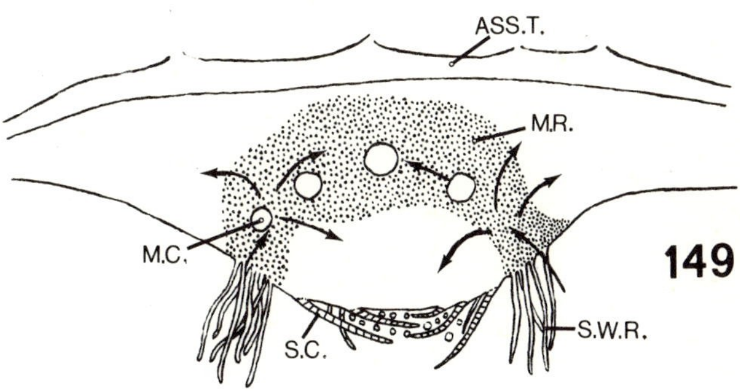

149

PLATE 43

Fig. 150. Diagram of the apparatus used by Clee (1937) in his experiments on external conduction in the leafy liverwort *Plagiochila asplenioides* var. *major* (Jungermanniales).

ES, eosin solution; GP, glass plate; PL, plant; T, tube.

The figures given are typical results of an experiment, and A & B the points reached by the stain in the succubous and incubous directions respectively in one minute.

Redrawn from Clee (1937).

Fig. 151. The "archegoniate pattern" as exemplified by bryophytes and lower vascular cryptogams. This diagram represents an attempt to summarize some of the most noteworthy characteristics shared by both bryophytes and the more primitive pteridophytes, characteristics which may also have occurred in their hypothetical common ancestors.

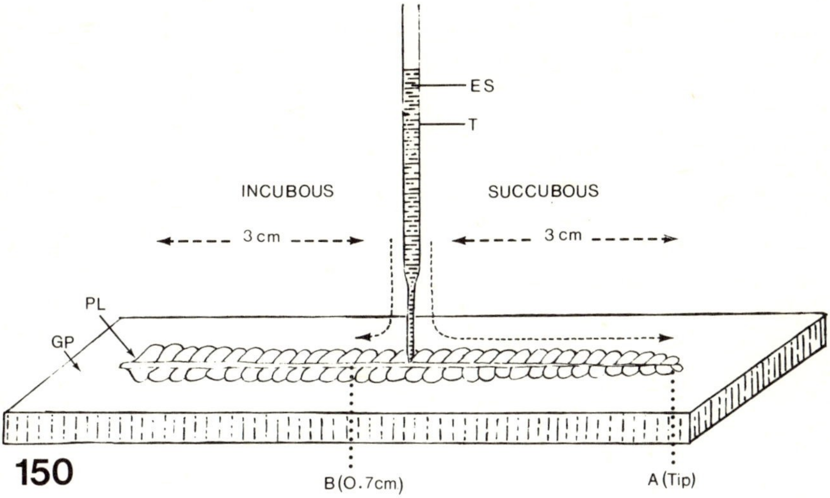

ES
T
INCUBOUS
SUCCUBOUS
3 cm
3 cm
PL
GP
B(O.7cm)
A (Tip)

150

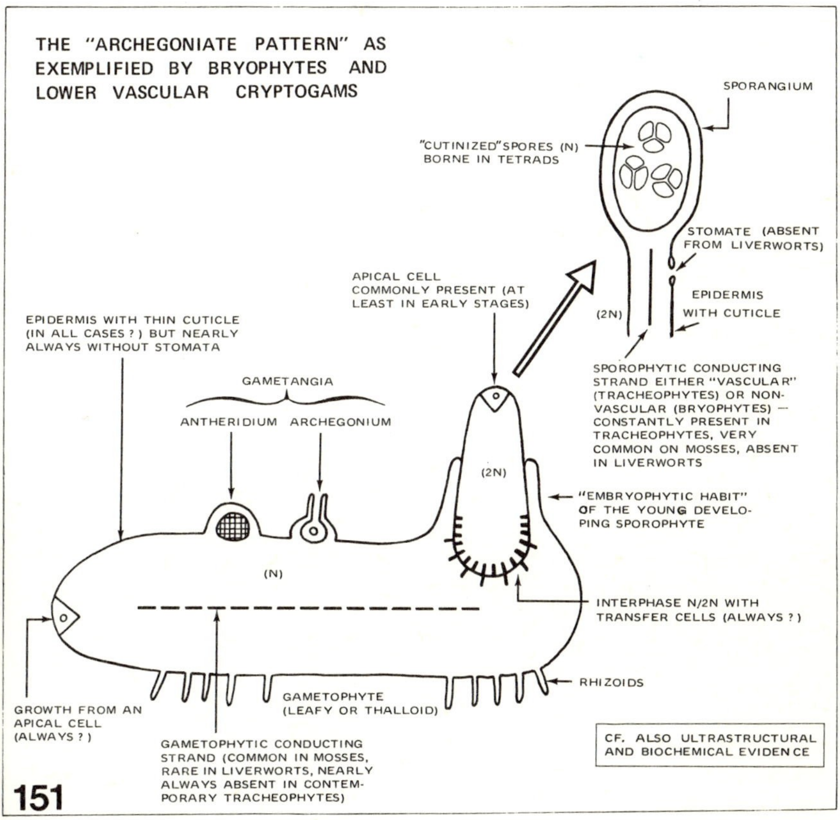

THE "ARCHEGONIATE PATTERN" AS EXEMPLIFIED BY BRYOPHYTES AND LOWER VASCULAR CRYPTOGAMS
SPORANGIUM
"CUTINIZED"SPORES (N) BORNE IN TETRADS
STOMATE (ABSENT FROM LIVERWORTS)
EPIDERMIS WITH CUTICLE
(2N)
APICAL CELL COMMONLY PRESENT (AT LEAST IN EARLY STAGES)
EPIDERMIS WITH THIN CUTICLE (IN ALL CASES ?) BUT NEARLY ALWAYS WITHOUT STOMATA
SPOROPHYTIC CONDUCTING STRAND EITHER "VASCULAR" (TRACHEOPHYTES) OR NON-VASCULAR (BRYOPHYTES) — CONSTANTLY PRESENT IN TRACHEOPHYTES, VERY COMMON ON MOSSES, ABSENT IN LIVERWORTS
GAMETANGIA
ANTHERIDIUM
ARCHEGONIUM
(2N)
"EMBRYOPHYTIC HABIT" OF THE YOUNG DEVELOPING SPOROPHYTE
(N)
INTERPHASE N/2N WITH TRANSFER CELLS (ALWAYS ?)
RHIZOIDS
GAMETOPHYTE (LEAFY OR THALLOID)
GROWTH FROM AN APICAL CELL (ALWAYS ?)
GAMETOPHYTIC CONDUCTING STRAND (COMMON IN MOSSES, RARE IN LIVERWORTS, NEARLY ALWAYS ABSENT IN CONTEMPORARY TRACHEOPHYTES)
CF. ALSO ULTRASTRUCTURAL AND BIOCHEMICAL EVIDENCE

151

PLATE 44

Fig. 152. Modified representation of Mehra's Phyletic Evolution in the Hepaticae (after Mehra & Sokhi 1973). The occurrence of "true" water-conducting strands [1] in the gametophytes at various levels of the phylogenetic tree has been indicated.

Redrawn from Mehra & Sokhi (1972); the information on the distribution of the conducting tissues has been added to the phylogenetic tree published by these authors.

(1) For the concept of water-conducting strands in bryophytes, see text p. 3.

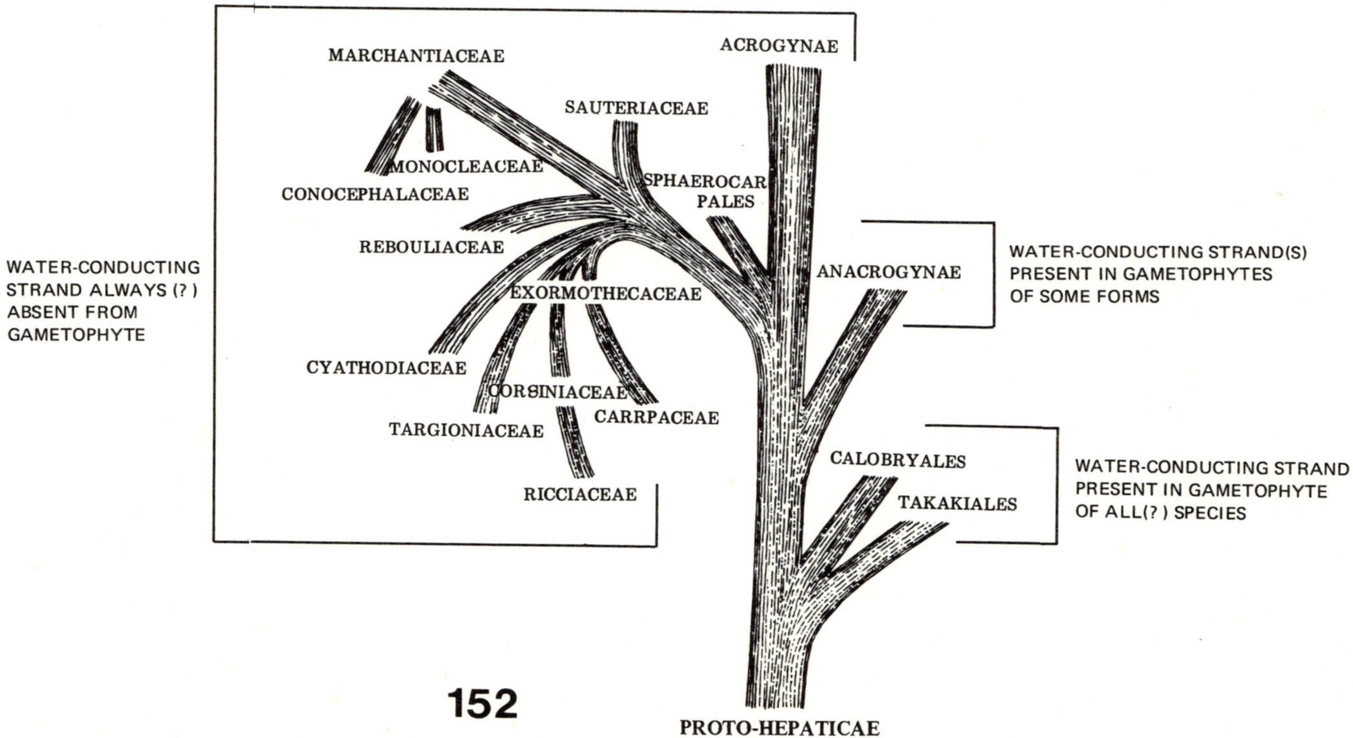
ACROGYNAE
MARCHANTIACEAE
SAUTERIACEAE
MONOCLEACEAE
CONOCEPHALACEAE
SPHAEROCAR PALES
REBOULIACEAE
EXORMOTHECACEAE
CYATHODIACEAE
CORSINIACEAE
TARGIONIACEAE
CARRPACEAE
RICCIACEAE
ANACROGYNAE
CALOBRYALES
TAKAKIALES
PROTO-HEPATICAE
WATER-CONDUCTING STRAND ALWAYS (?) ABSENT FROM GAMETOPHYTE
WATER-CONDUCTING STRAND(S) PRESENT IN GAMETOPHYTES OF SOME FORMS
WATER-CONDUCTING STRAND PRESENT IN GAMETOPHYTE OF ALL(?) SPECIES

152

PLATE 45

Organization of the thallus in Marchantiales (Liverworts)

Fig. 153. *Lunularia cruciata.* Vertical longitudinal section of apex. A complex organization pattern exists, but no "true" water-conducting strand differentiates within apical derivatives. x 90.

Fig. 154. *Conocephalum conicum.* Portion of transection through midrib of thallus. A zone of compact (conducting ?) parenchyma is seen (asterisk), but no "true" water-conducting strand is present. x 70.

Figs. 155-156. *Conocephalum conicum.* Gametophyte. Cells of the ventral parenchyma of the wings as they appear under polarized light. Large primary pit fields (dark in photographs) are separated by "bars" which belong to a reticular network of "secondary" wall material (white in photographs). 800 approx.

Figs. 157-158. *Marchantia polymorpha.* Macerated cells from wing (fig. 157) and midrib (fig. 158) of thallus. x 200 approx. (see also fig. 31 on plate 8).

Fig. 159. *Conocephalum conicum.* Bundle of tuberculate rhizoids from ventral surface of thallus. x 120.

A, apex; AS.T., assimilating tissue; MC, mucilage canal; MR, mycorrhizal region; PR, pore; RH, rhizoid; SC, scale.

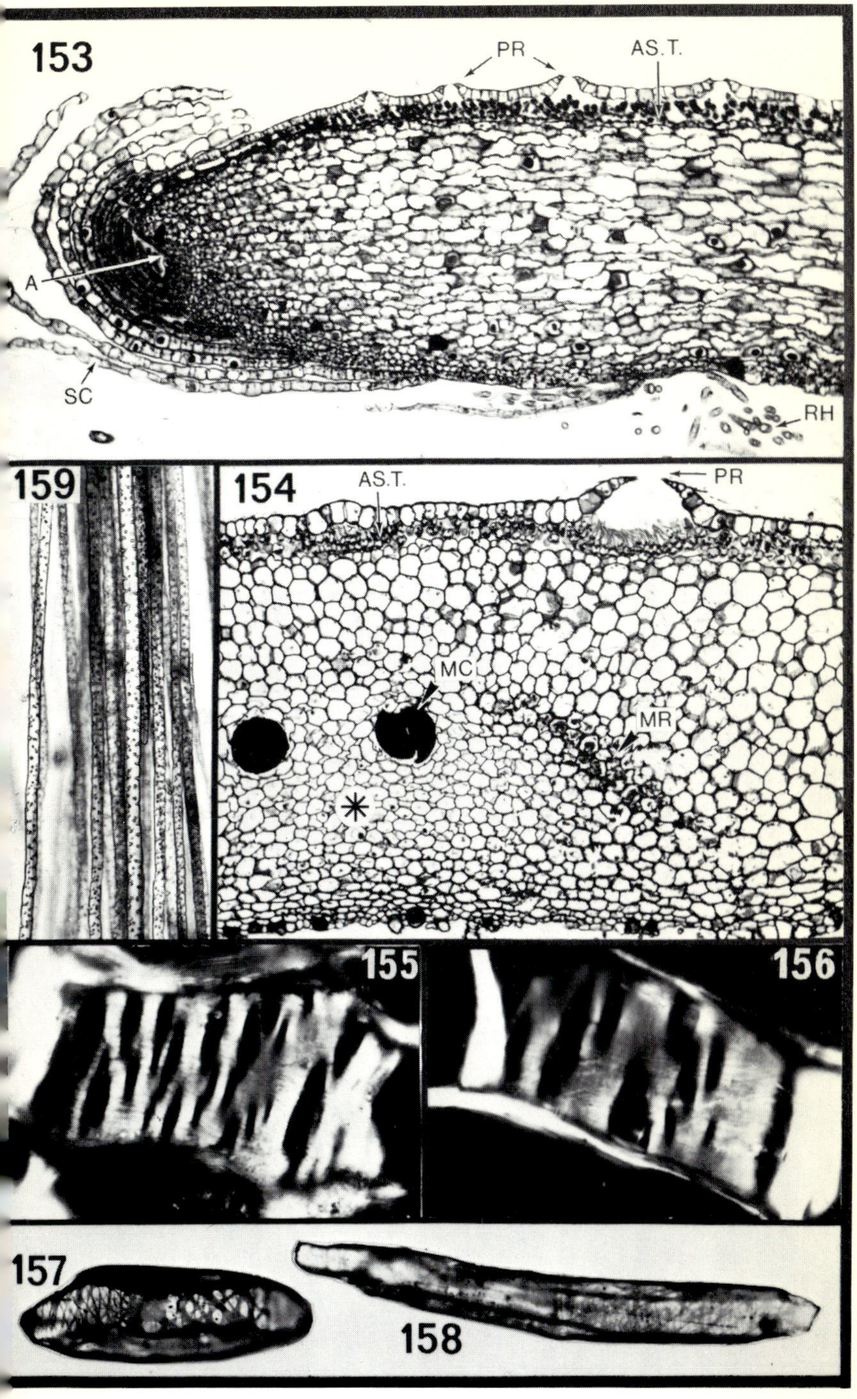
153
PR
AS.T.
A
SC
RH
159
154
AS.T.
PR
MC
MR
155
156
157
158

PLATE 46

Fig. 160. Cells isolated by maceration from mature leafy stems of the mosses *Climacium americanum* (a, b), *Thuidium delicatum* (c) and *Hylocomium splendens* (e); a, x 285; b, x 350; c, x 250; d and e, x 320.
After Finocchio (1967).

Fig. 161. *Hylocomium splendens* (Hypnobryales, Musci). Gametophyte. Numerous plasmodesmata are visible in the wall between two young cells of the sub-apical region. x 13,200.
After Bonnot (1967a).

Fig. 162. *Thuidium tamariscinum* (Hypnobryales, Musci). Portion of wall between two young cells from sub-apical region of gametophyte. Small "vacuoles" (VS) seem to literally "bud off" from the plasmodesmata. x 42,000.
By courtesy of E.J. Bonnot.

Fig. 163 to 165. Demonstration of pit callose by aniline blue fluorescence microscopy in end walls of cells of conducting parenchyma of mosses. Fig. 163, gametophytic stem of *Atrichum undulatum*; figs. 164 and 165, gametophytic stem and seta, respectively, of *Mnium affine*. x 260.
After Hébant (1968).

ER, endoplasmic reticulum; M, mitochondrion; PD, plasmodesmata; V, vacuole; VS, vesicles; W, wall.

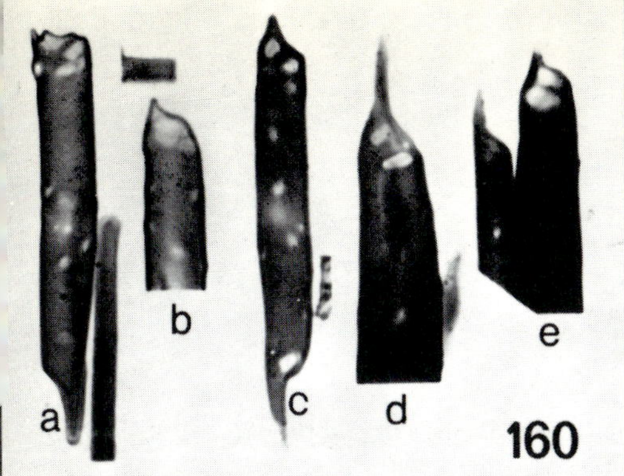
a
b
c
d
e
160

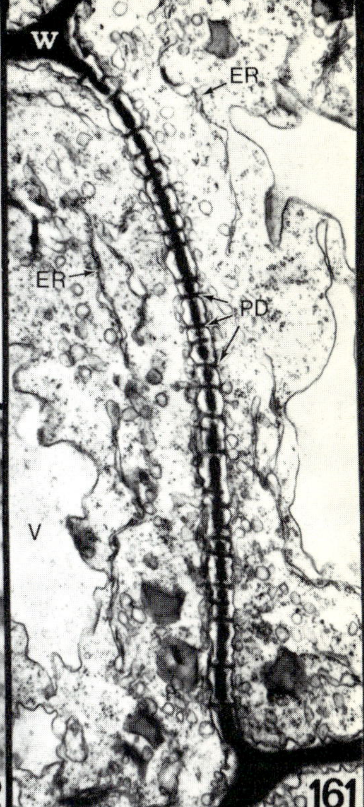
W
ER
ER
PD
V
161

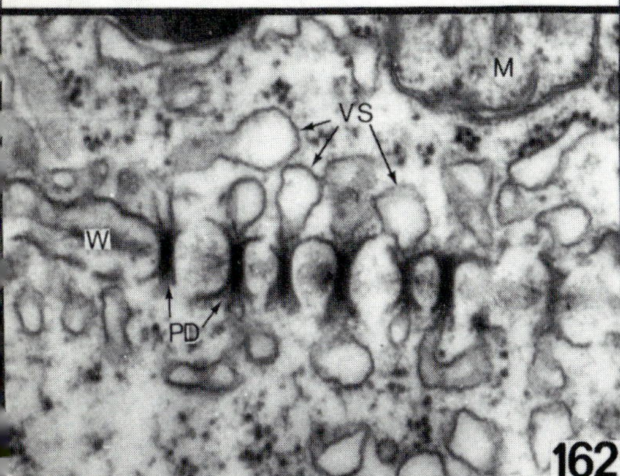
M
VS
W
PD
162

163

164

165

PLATE 47

Sphagnum palustre (Sphagnales)
(see also plate 10)

Fig. 166. Portion of a mature leaf in surface view; green cells (GC) and hyaline cells (HC) can be seen. x 220.

Fig. 167. Portion of a mature leaf from a lateral branch as seen under the scanning electron microscope. x 1 000.
After Fabre (1974a).

Fig. 168. Surface view of stem. "Pseudopores" with persisting membranes can be seen. S.E.M. x 1 000.
After Fabre (1974a).

Fig. 169. Mature leaf. Enlarged view of two adjacent pores. S.E.M. x 2,400.
After Fabre (1973).

Fig. 170. Portion of median longisection through sporogonium (2n) and pseudopodium [=part of gametophyte, (n)]. Cells with large primary pit fields exist in the sporophyte in the region (asterisk) located between the foot and the capsule. AR, archegonium; x 45.

Fig. 171. *idem.* Detail of the cells with large primary pit fields. x 450.

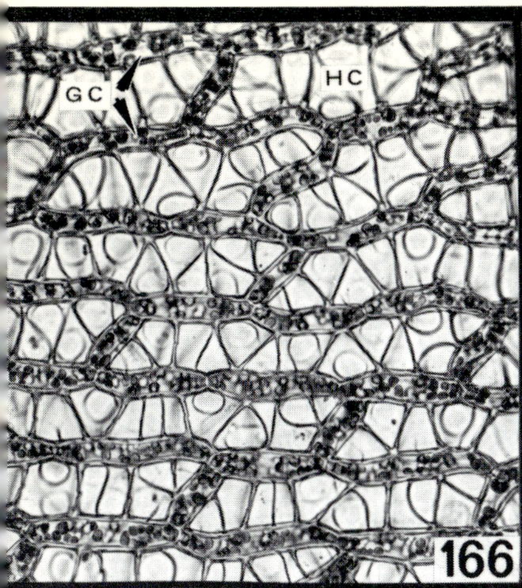
GC
HC
166

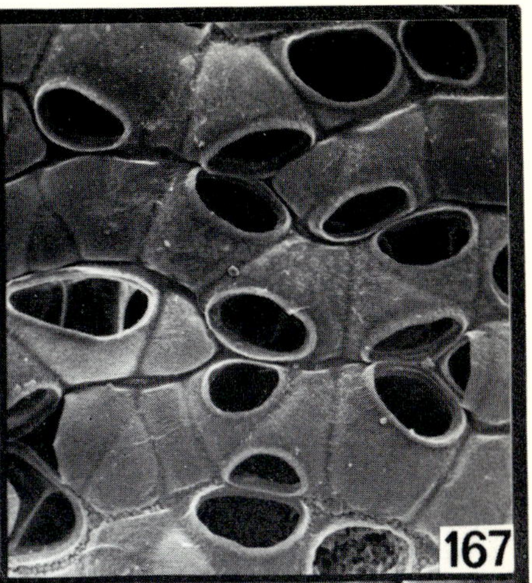
167

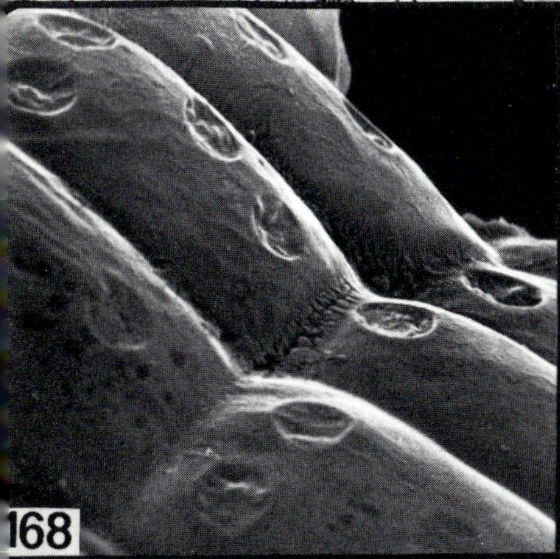
168

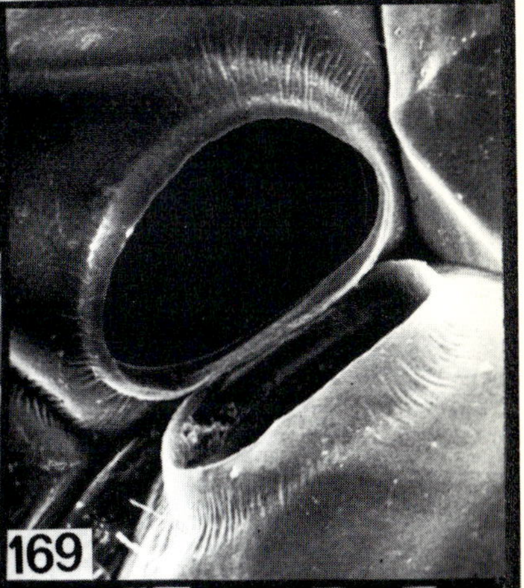
169

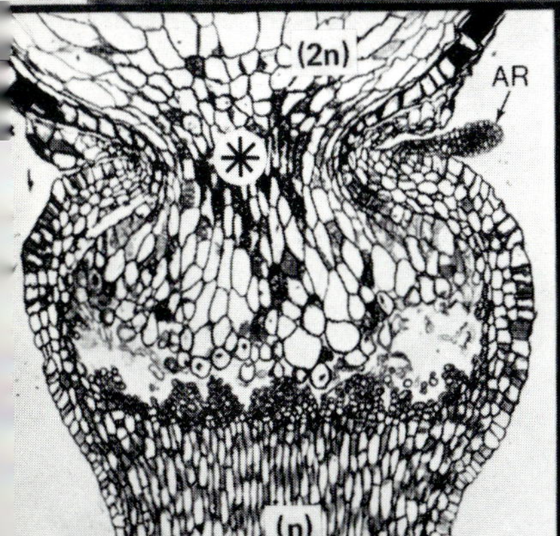
(2n)
AR
(n)
170

171

PLATE 48

Fig. 172. *Anthoceros sp.* (Anthocerotales; specimen from New Zealand). Longitudinal section through base of sporophyte (2n) embedded in gametophytic tissue (n). MR, meristematic region of sporophyte. x 100.

Fig. 173. *idem.* Detail of rhizoid-like cells (apparently not transfer cells !) at the periphery of the foot (=lower portion of the sporophyte). x 480.

Fig. 174. *Megaceros sp.* (Anthocerotales; specimen from New Zealand). Longisection of nearly mature portion of sporophyte. Note the columella (COL) at the center of the sporogonium. x 130.

Fig. 175. *idem.* Detail of columella. x 500.

Fig. 176 and 177. *Lophocolea bidentata.* Mature spores and elaters, as seen by light (fig. 176) and electron microscopy (fig. 177). Fig. 176, x 560; fig. 177, x 3,000.
After Lambert (1968).

COL, columella; EL, elater; SP, spore.

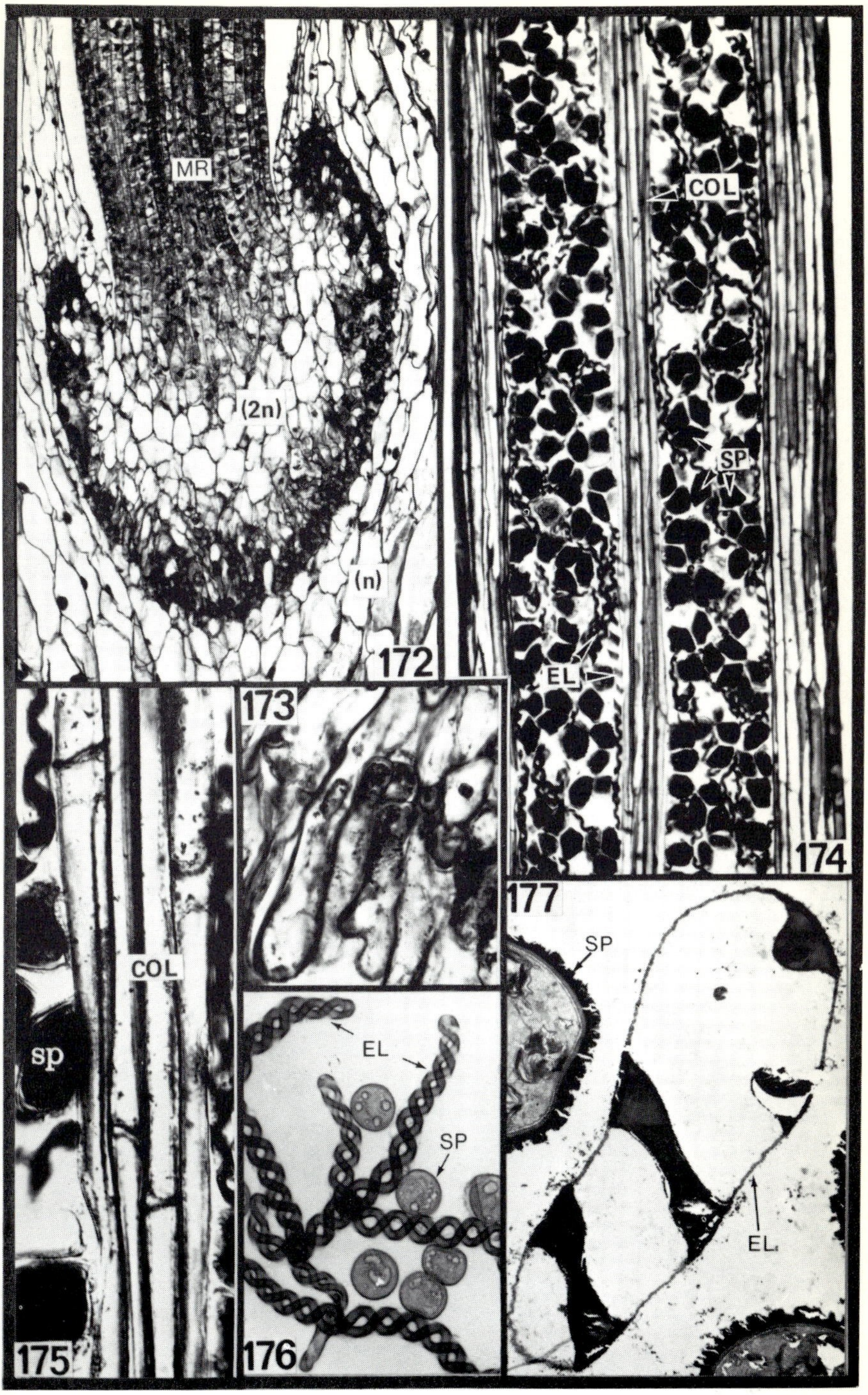
MR
(2n)
(n)
172
COL
SP
EL
174
173
175
COL
sp
176
EL
SP
177
SP
EL

PLATE 49

Takakia (Takakiales, Liverworts)
(see also figs. 61 to 63 on plate 14)

Fig. 178. *T. ceratophylla.* Transverse section of upright leafy stem. x 180.

Fig. 179. *T. lepidozioides.* Transverse section of gametophytic stem bearing a lateral branch. The water-conducting strand of the lateral branch joins the central strand of the main stem (this is not always the case; see text p. 25). x 330.

Fig. 180. *T. lepidozioides.* Transverse section of young region of leafy stem at level where the protoplast of water-conducting cells breaks down. A lysosomal acid phosphatase acitvity is revealed in the differentiating conducting cells. Burstone method. x 180 approx.

Fig. 181. *T. lepidozioides.* Transection of upright leafy stem showing portion of central strand and cortical tissues. The central strand is sharply defined as it consists of dead cells, whereas cortical cells retain a living protoplast. Low power electron micrograph; KM_nO_4 fixation; x 3,700.
After Hébant (1972a).

Fig. 182. *T. ceratophylla.* Ultrastructure of end wall of contact between elements of central strand. Plasmodesmata-derived pores (unlabelled arrows) can be seen. x 31,500.
After Hébant (1973d).

C, cortex; CS, central strand; L, leaf; LB, lateral branch.

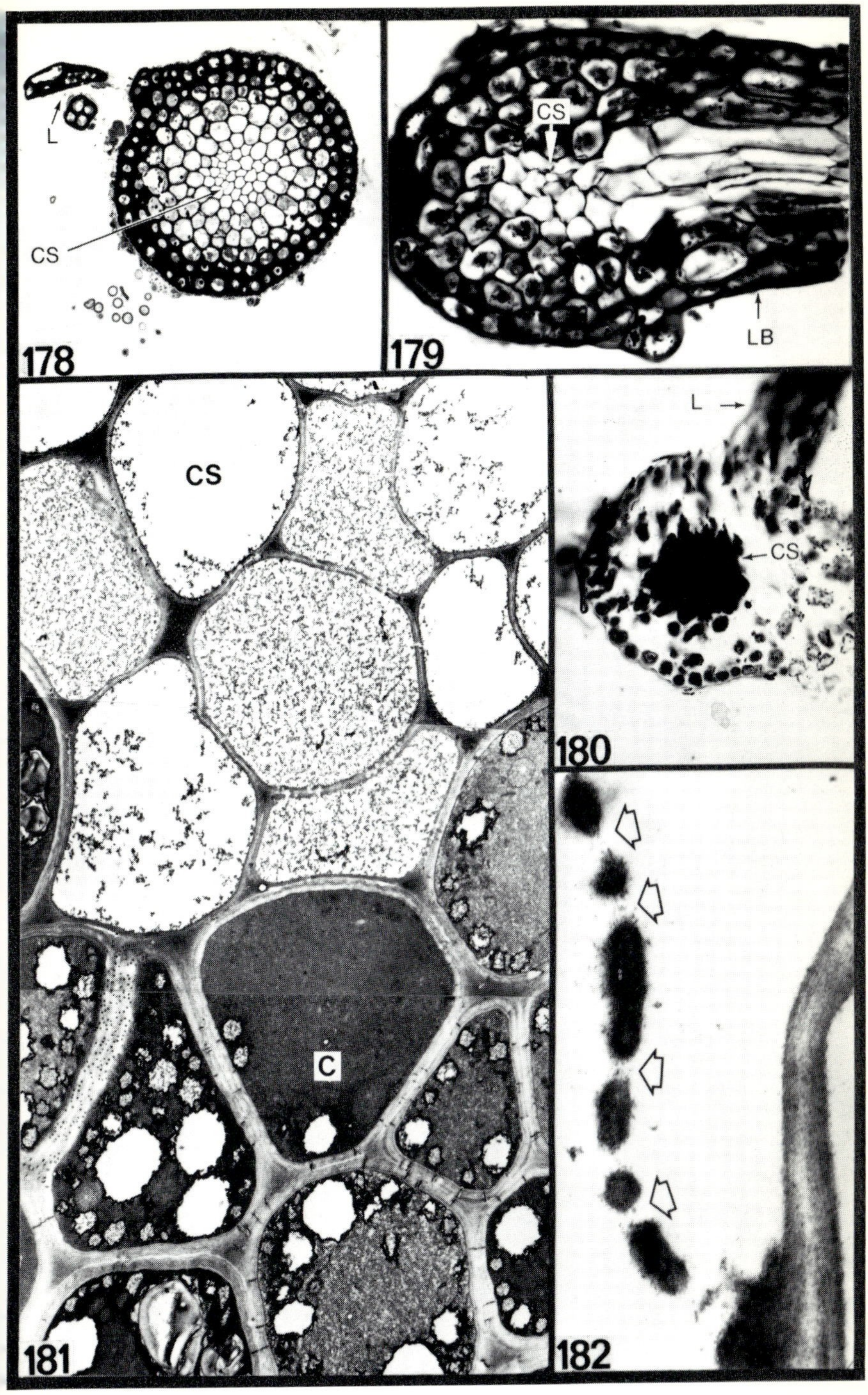

L
CS
178
CS
LB
179
CS
C
181
L
CS
180
182

PLATE 50

Haplomitrium gibbsiae (Calobryales, Liverworts)

Fig. 183. Longisection of vegetative gametophyte showing developing central strand. x 33.

Fig. 184. Strong lysosomal acid phosphatase activity (arrow) in the differentiating water-conducting cells of the central strand. Longisection of an erect female leafy branch. Burstone method. x 33.

Fig. 185. Longitudinal section of mature portion of stem. Water-conducting cells of central strand are dead and empty, whereas cortical and epidermal cells retain a living protoplast. x 82.

Fig. 186. Transverse section of gametophytic stem bearing a lateral branch. The water-conducting strand of the lateral branch is here in relation with the central strand of the main stem. x 54.

Fig. 187. As in fig. 186, but here cells of living parenchyma (arrow) are intercalated between the strand of the lateral branch and that of the main stem. No direct connection is established between the two strands. x 82.

Fig. 188. Scanning electron micrograph of end-wall of contact between two water-conducting cells of the central strand. Numerous plasmodesmata-derived pores can be seen. x 7,800.
After Burr et al. (1974).

Fig. 189. Transmission electron micrograph of a transverse section through an end wall between two water-conducting cells. Thiery method for demonstration of polysaccharides. x 12,000.

A, apex; CS, central strand; LB, lateral branch.

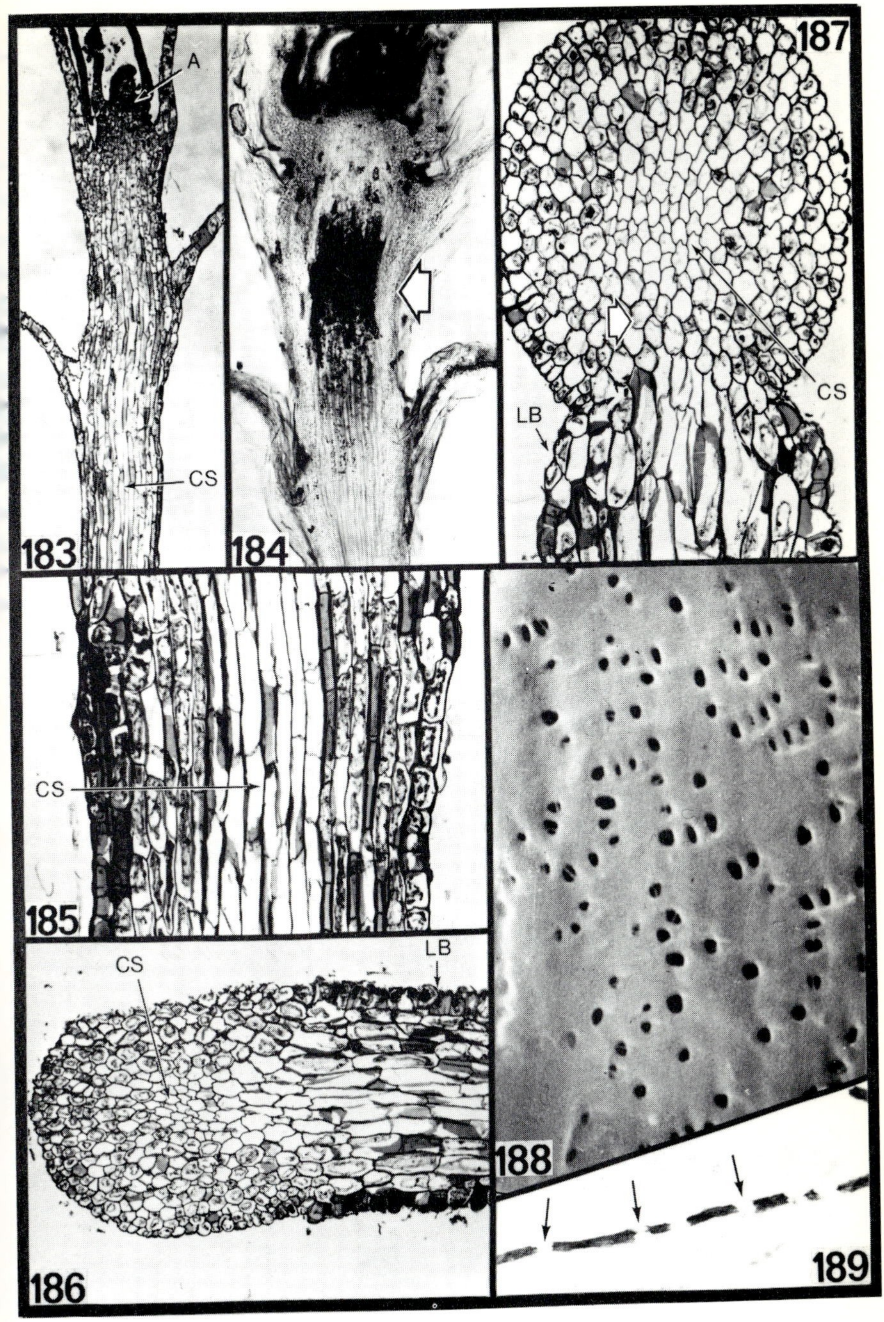
183
A
CS
184
187
CS
LB
185
CS
186
CS
LB
188
189

PLATE 51

Fig. 190. *Symphyogyna hymenophyllum* (Metzgeriales, Liverworts). Gametophyte. Longisection of apex of "rhizome" showing developing central strand. x 82.

Fig. 191. *Hymenophyton flabellatum* (Metzgeriales). Gametophyte. Transverse section in subapical region of young rhizome showing differentiating central strand and surrounding tissues. x 330.

Fig. 192. *idem.* Lysosomal acid phosphatase activity in the differentiating water-conducting strand. Burstone technique. x 54.

Fig. 193. *Pallavicinia lyellii* (Metzgeriales). Gametophyte. Lysosomal acid phosphatase activity in differentiating water-conducting cells of central strand. Freezing microtome longisection; Burstone method; x 330. After Hébant 1973b.

Figs. 194 and 195. *Symphyogyna undulata* (Metzgeriales). Transverse sections of mature portions of thalli showing conducting strands. In fig. 195, an unusual "dictyostelic" pattern (but without true leaf gaps !) is seen. x 54.

A, apex; CS, central strand.

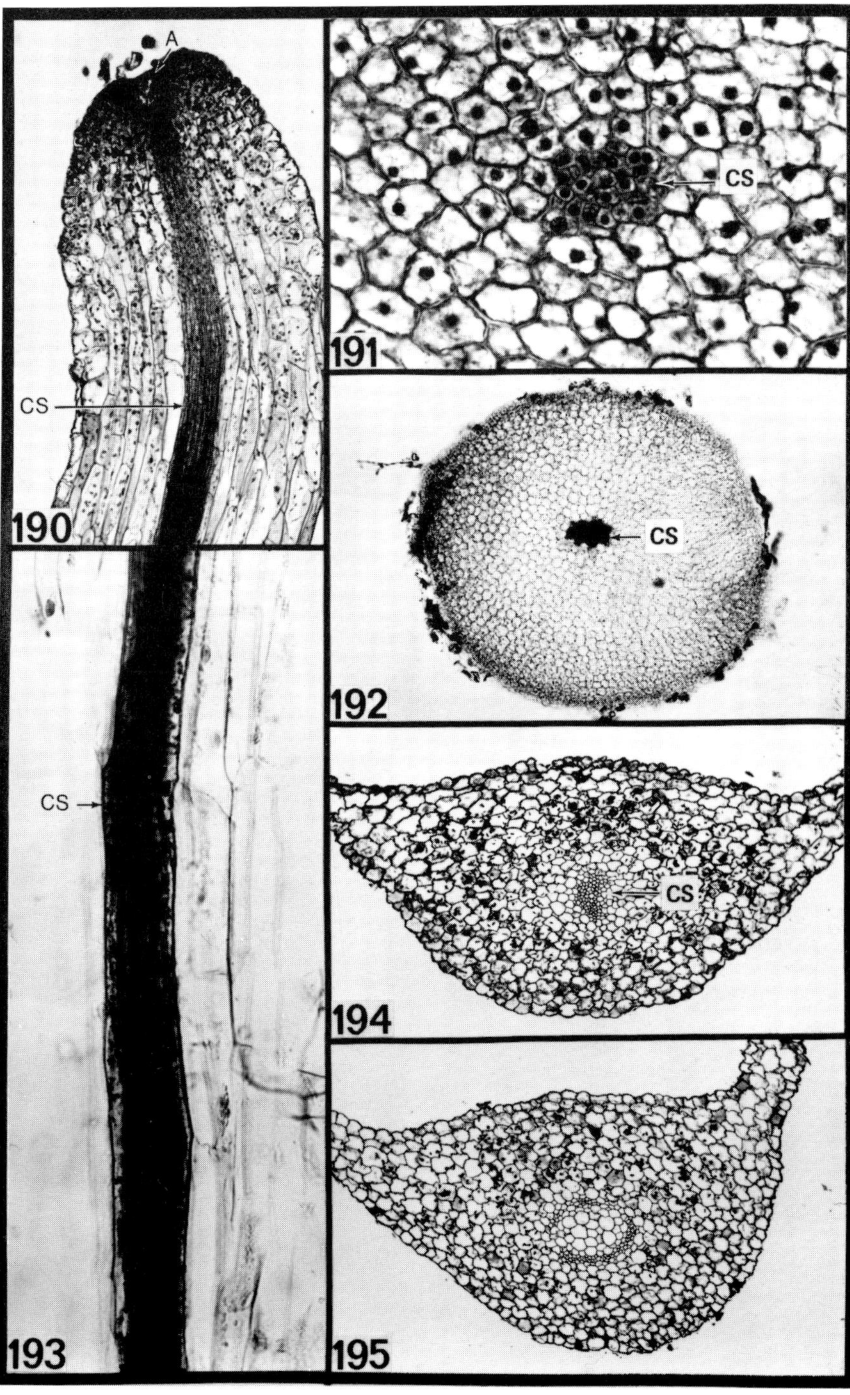
A
CS
190
CS
191
CS
192
CS
193
CS
194
195

PLATE 52

Fig. 196. *Symphyogyna hymenophyllum* (Metzgeriales, Liverworts). Gametophyte. Transverse section through "rhizome" bearing a lateral branch. No connection exists between conducting strand of lateral branch and that of bearing axis. x 72.

Fig. 197. *Hymenophyton leptopodon.* General morphology of a dendroid liverwort of the order Metzgeriales. x 1.

Fig. 198. *Hymenophyton flabellatum* (Metzgeriales). Gametophyte. Longisection of median portion of stalk showing single central strand. x 110.

Fig. 199 to 201. *idem.* Transverse sections at various levels in upper region of stalk showing "dichotomous" branching of conducting strand (each strand ultimately irrigates one frondlet). x 54.

Figs. 202-203. *Symphyogyna hymenophyllum* (another dendroid liverwort of the order Metzgeriales). Two steps in the branching of the central strand can be seen in these transverse sections. x 82.

CS, central strand; LB, lateral branch.

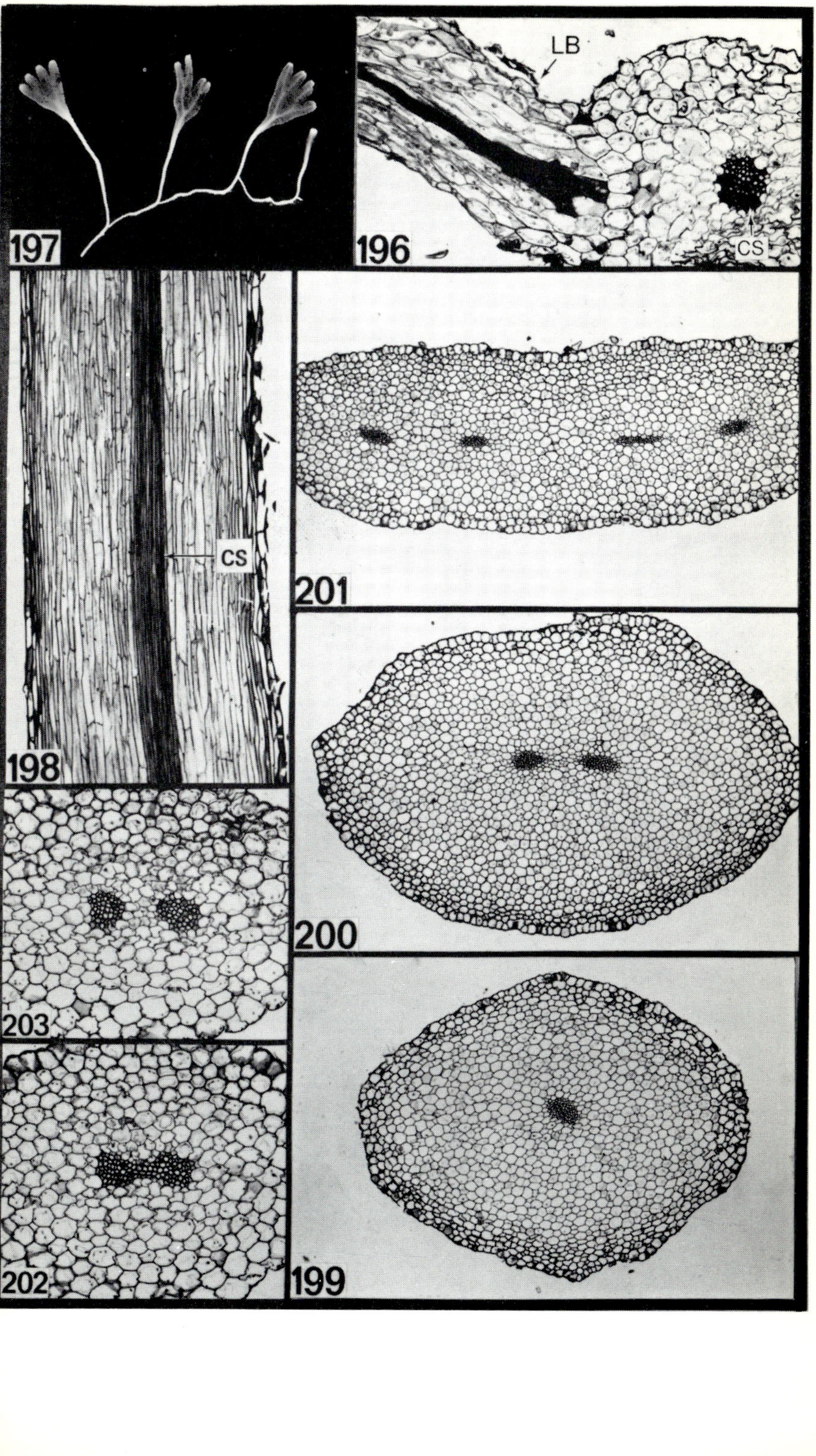
197
LB
196
CS
198
CS
201
200
203
202
199

PLATE 53

Fig. 204. *Pallavicinia lyellii* (Metzgeriales, Liverworts). Gametophyte. Conducting cells from central strand isolated by maceration. Phase contrast; x 250 approx.

Fig. 205. Transverse view of conducting strand in a representative of the Pallaviciniaceae. Scanning electron micrograph; x 1,900. After Hébant (1970c).

Figs. 206-207. *Hymenophyton flabellatum* (Metzgeriales). Transmission electron micrographs illustrating the wall structure of water-conducting cells after the Thiéry test for polysaccharides. The pores are indicated by darts. Fig. 206, x 21,000; fig. 207, x 19,300.

204

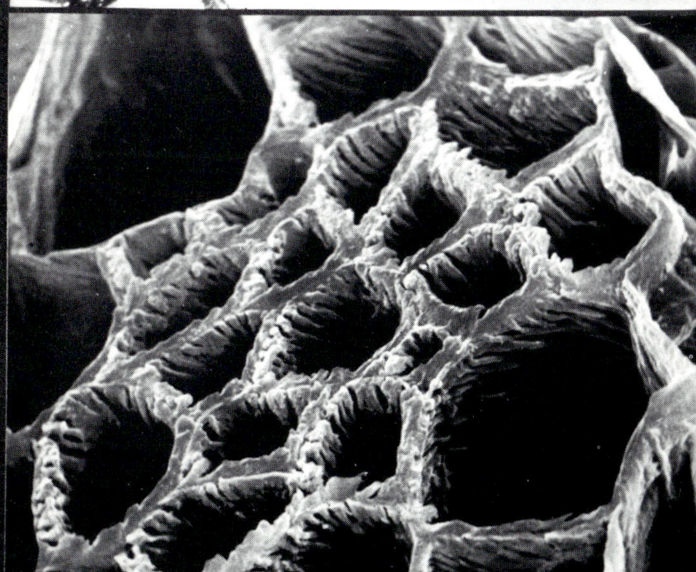
205

207

206

PLATE 54

Fig. 208. *Moerckia flotoviana* (Metzgeriales, Liverworts). Transverse section in mature region of thallus. Two discrete conducting strands (CS) can be seen. An endophytic fungus also occurs in the median region of the midrib. x 50.

Fig. 209. *idem.* Higher magnification of one of the conducting strands. x 310.

Figs. 210-211. *idem.* Wall structure of the water-conducting cells as seen under the electron microscope. Note that these walls are partially hydrolysed. Fig. 210, x 2,700; fig. 211, x 3,700.
After Hébant (1973d).

Fig. 212. *Treubia lacunosa* (Metzgeriales). Transverse section in mature region of thallus. No "true" water-conducting strand can be seen. A mycorrhizal region (MR) occurs in the median portion of the midrib. x 54.

Figs. 213-214. *Pellia epiphylla* (Metzgeriales). Transverse (fig. 213) and longitudinal (fig. 214) sections of the thallus. No conducting strand is present. Fig. 213, x 54; fig. 214, x 82.

Fig. 215. *Scapania undulata* (Jungermanniales, Liverworts). Longisection in apical region of leafy gametophyte. No conducting strand differentiates within apical derivatives. x 170.

A, apical cell; CS, conducting strand; HW, hydrolysed wall; MR, mycorrhizal region; RH, rhizoid.

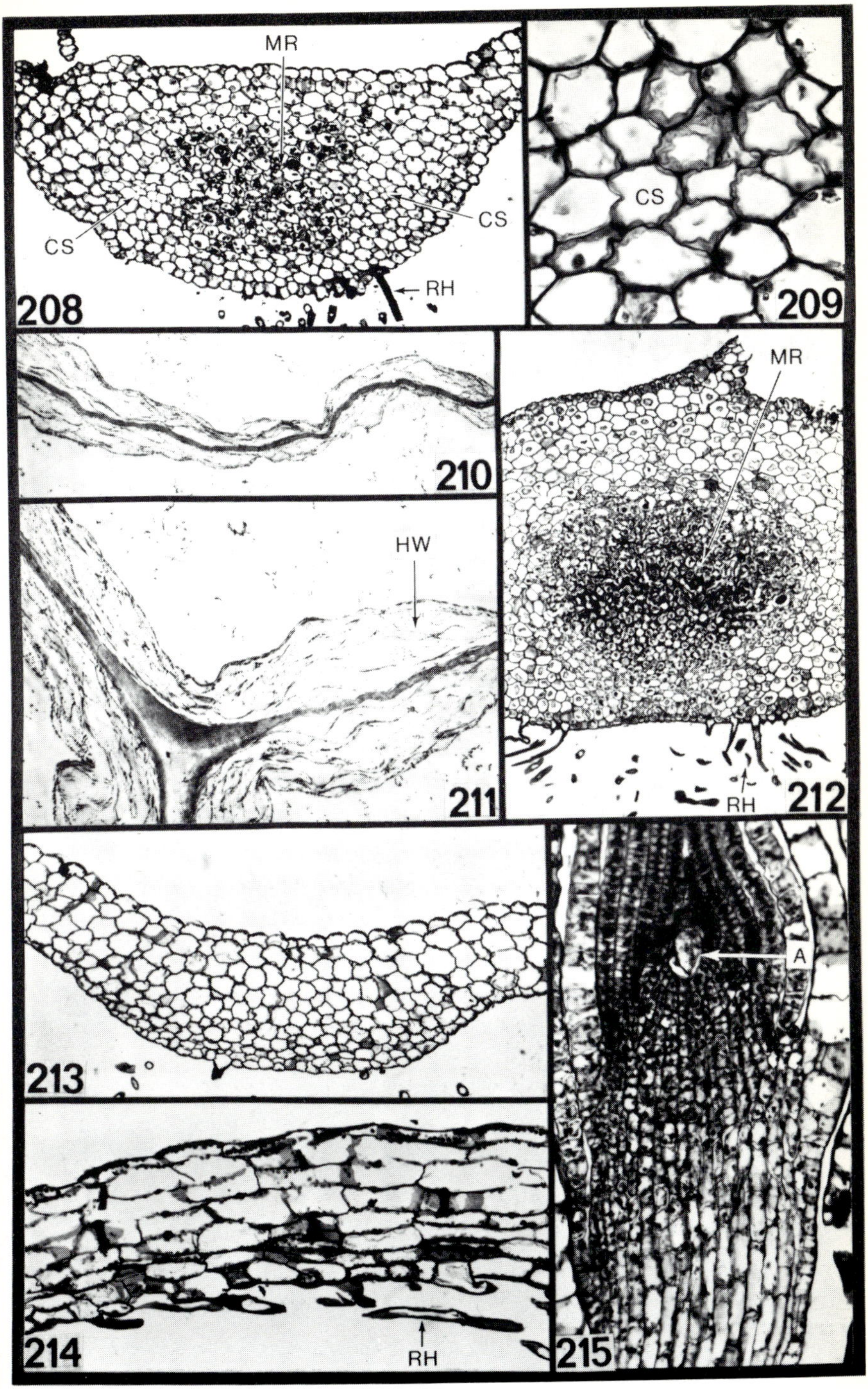
MR
CS
CS
RH
208
CS
209
210
HW
211
MR
RH
212
213
A
214
RH
215

PLATE 55

Fig. 216. *Fissidens polyphyllus* (Fissidentales, Musci). Longisection of apex of leafy stem showing central strand of hydroids. x 82.

Fig. 217. *Spiridens sp.* (Eubryales Pleurocarpi, Musci; specimen from New Caledonia). Longisection of apex of leafy gametophyte showing developement of central strand within apical derivatives. x 54.

Fig. 218. *Cyathophorum bulbosum* (Hookeriales, Musci). *idem.* x 82.

Fig. 219. *Hookeria (Pterygophyllum) lucens* (Hookeriales). Gametophyte. Only a small conducting strand is present here. x 82.

Fig. 220. *Dicranum majus* (Dicranales, Musci). Transection of gametophytic stem. A central strand of hydroids is present. x 90.

Fig. 221. *Thuidium tamariscinum* (Hypnobryales, Musci). Leafy gametophyte. Transection of main stem. Only a small residual strand of hydroids can be seen. x 90.

Fig. 222. *Hedwigia ciliata* (Isobryales, Musci). Transection of gametophytic stem. No central strand. x 90.

Fig. 223. *Orthotrichum affine* (Isobryales, Musci). Transection of gametophytic stem. No central strand. x 90.

Figs. 220 to 223 after Hébant (1966b).

A, apical cell; CS, central strand; L, leaf; PH, paraphyllia.

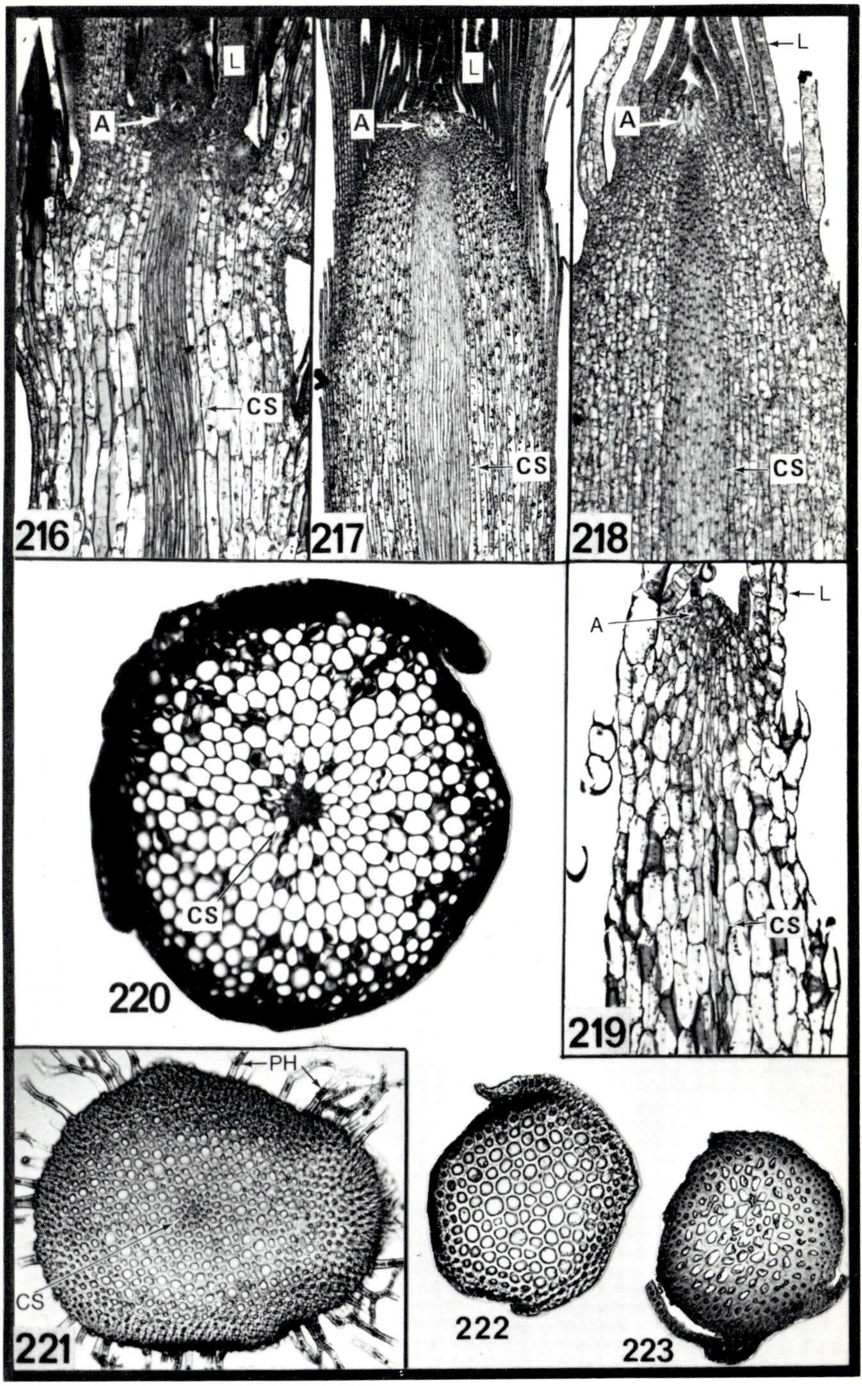

L
A
CS
216
217
218
219
220
221
222
223
PH

PLATE 56

Dawsonia superba var. *superba* (Polytrichales, Musci). Leafy stem

Fig. 224. Median longisection of apex showing apical cell and derivatives. Differentiating hydroids and stereids can be seen. Section of resin-embedded material; x 155.
After Hébant (1976a).

Fig. 225. Transection of leafy stem showing central strand and leaf traces. x 110.
After Hébant (1976a).

Fig. 226. Transverse view of central strand. Hydroids are intermixed with stereids. Scanning electron micrograph; x 1,000.
After Hébant (1974).

Fig. 227. *idem.* Low power transmission electron micrograph. Note that hydroids do not develop highly thickened lateral walls since the stereids contribute here to the strengthening of the stem (compare with fig. 275). x 3,900.

A, apical cell; HYD, hydroid; HW, hydrolysed wall; ST, stereid.

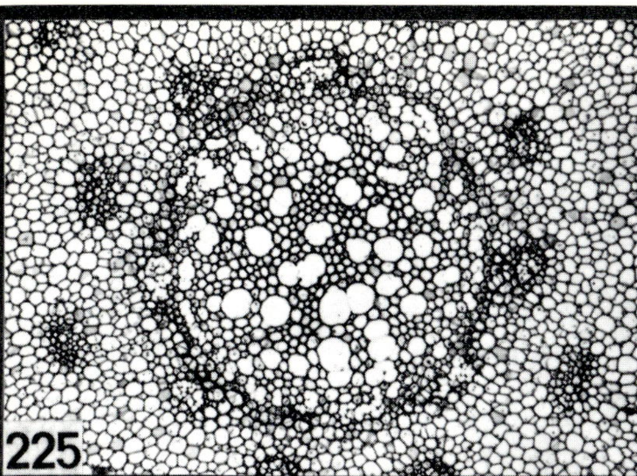
225

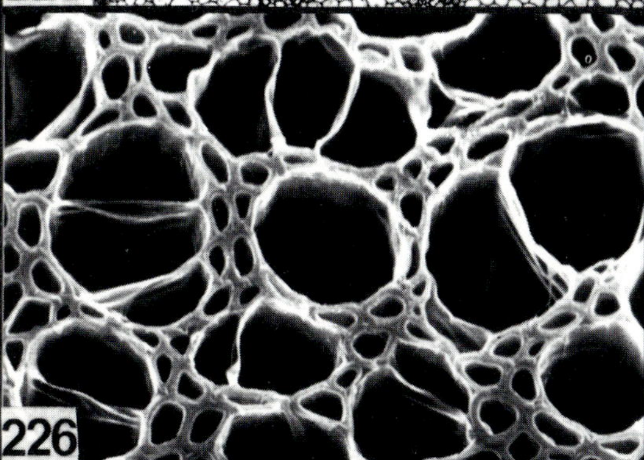
226

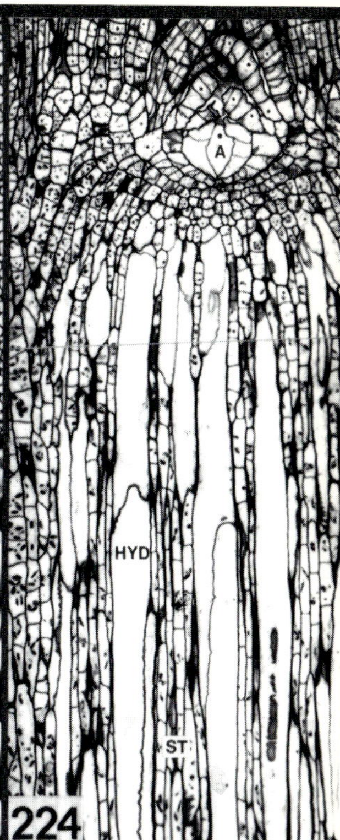
A
HYD
ST
224

HYD
ST
HYD
HW
HW
ST
HYD
227

PLATE 57

"Rhizomes" of Polytrichales (Musci)

Figs. 228 to 231. Transverse sections through "rhizome-like" portions of the gametophytes of *Polytrichum commune* (fig. 228), *Dendroligotrichum dendroides* (fig. 229), *Dawsonia polytrichoides* (fig. 230) and *Dawsonia longiseta* (fig. 231). (See also plate 22).
Figs. 228 and 231, x 100; figs. 229 and 230, x 68.
Figs. 230 and 231 after Hébant (1976a).

Fig. 232. *Dawsonia superba* var. *superba*. "Rhizome". Detail of conducting tissues at periphery of central strand. x 430.
After Hébant (1973a).

Fig. 233. *Polytrichum commune.* "Rhizome". Detail of leptoids and of an associated parenchyma cell in one of the three "furrows". Transmission electron micrograph; x 7,000.

HYD, hydroid; LB, lateral branch; LEP, leptoid; N, nucleus; P, plastid; PAR, parenchyma cell; PD, plasmodesmata; ST, stereid.

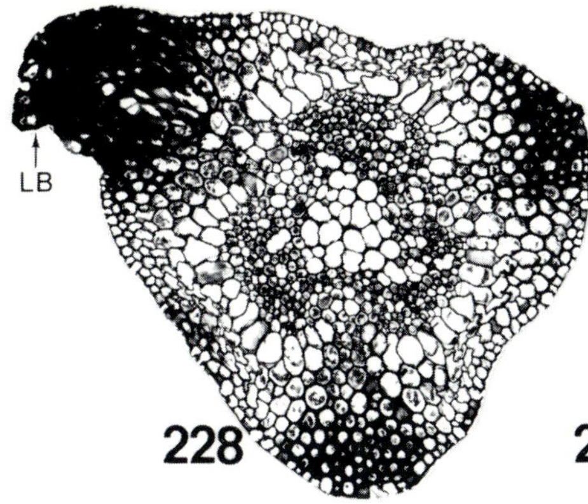

228

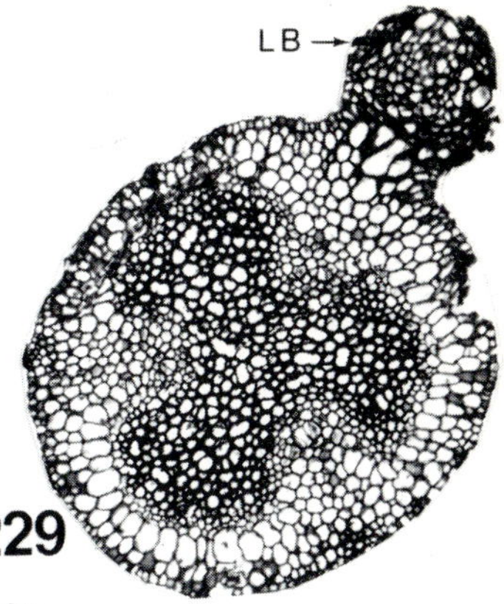

229

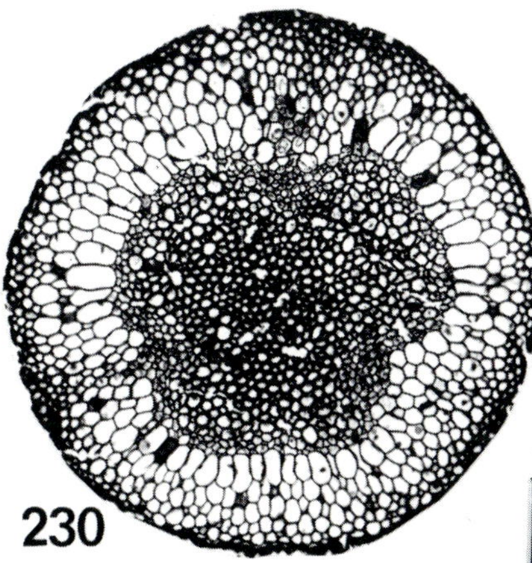
230

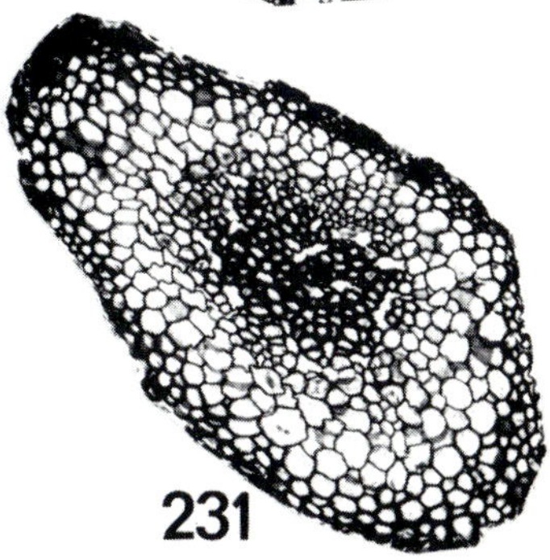
231

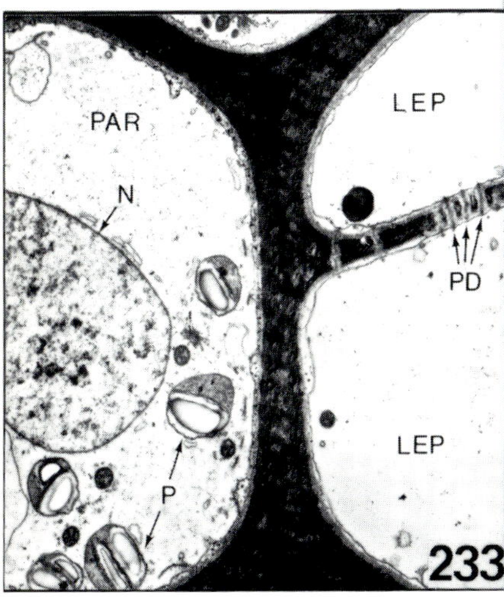

233

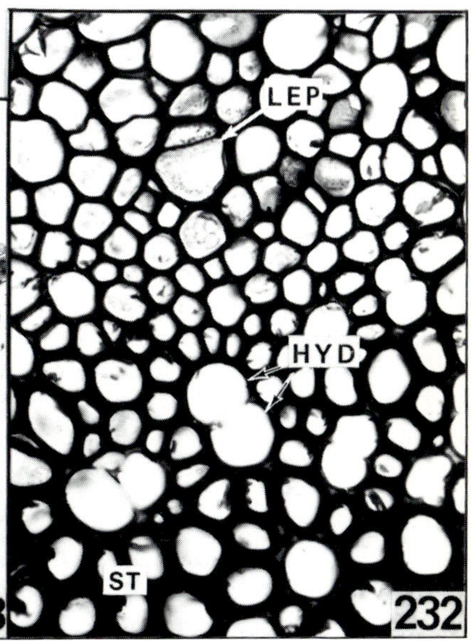

232

PLATE 58

Branching in moss gametophores

Fig. 234. *Polytrichadelphus magellanicus* (Polytrichales, Musci). Longisection of leafy stem bearing archegonium (AR). A secondary apical cell (A2) gives rise to a new vegetative branch. x 68.

Fig. 235. *Dendroligotrichum dendroides* (Polytrichales). Longisection (not median but distinctly tangential) of apical region of young upright stem in which branching has initiated. Two vegetative apices are here identifiable (arrows). A conducting strand differentiates below each of them. These strands merge with the central strand of the main stem. x 45.

Fig. 236. *Dawsonia papuana* (Polytrichales). Young inhibited lateral bud (IB). Note absence of connection with the conducting tissue-system of the main axis. x 145.
After Hébant (1976a).

Fig. 237. *Polytrichadelphus magellanicus* (Polytrichales). Transverse section of a branched specimen showing conducting strands of main stem and of two lateral branches. x 35.

Fig. 238. *idem.* Portion of transection of another branched specimen showing conducting strands of main stem (CS) and of a lateral branch (CS'). x 82.

Figs. 239-240. *Hypnodendron dendroides* (Eubryales Pleurocarpi).
Fig. 239. Longisection of upright gametophytic stem bearing a lateral branch (LB). x 33.
Fig. 240. Detail of the connection between the conducting strand (CS') of the lateral branch and the central strand (CS) of the main stem. x 67.
After Frey (1974ab).

Figs. 241-242. *Funaria hygrometrica.* Transverse sections at two levels (fig. 241, lower level; fig. 242, higher level) of gametophytic stem bearing a lateral branch (LB). The conducting strand (CS') of the lateral branch ends blindly within ground tissue without linking up with the central strand (CS) of the main stem. x 82.

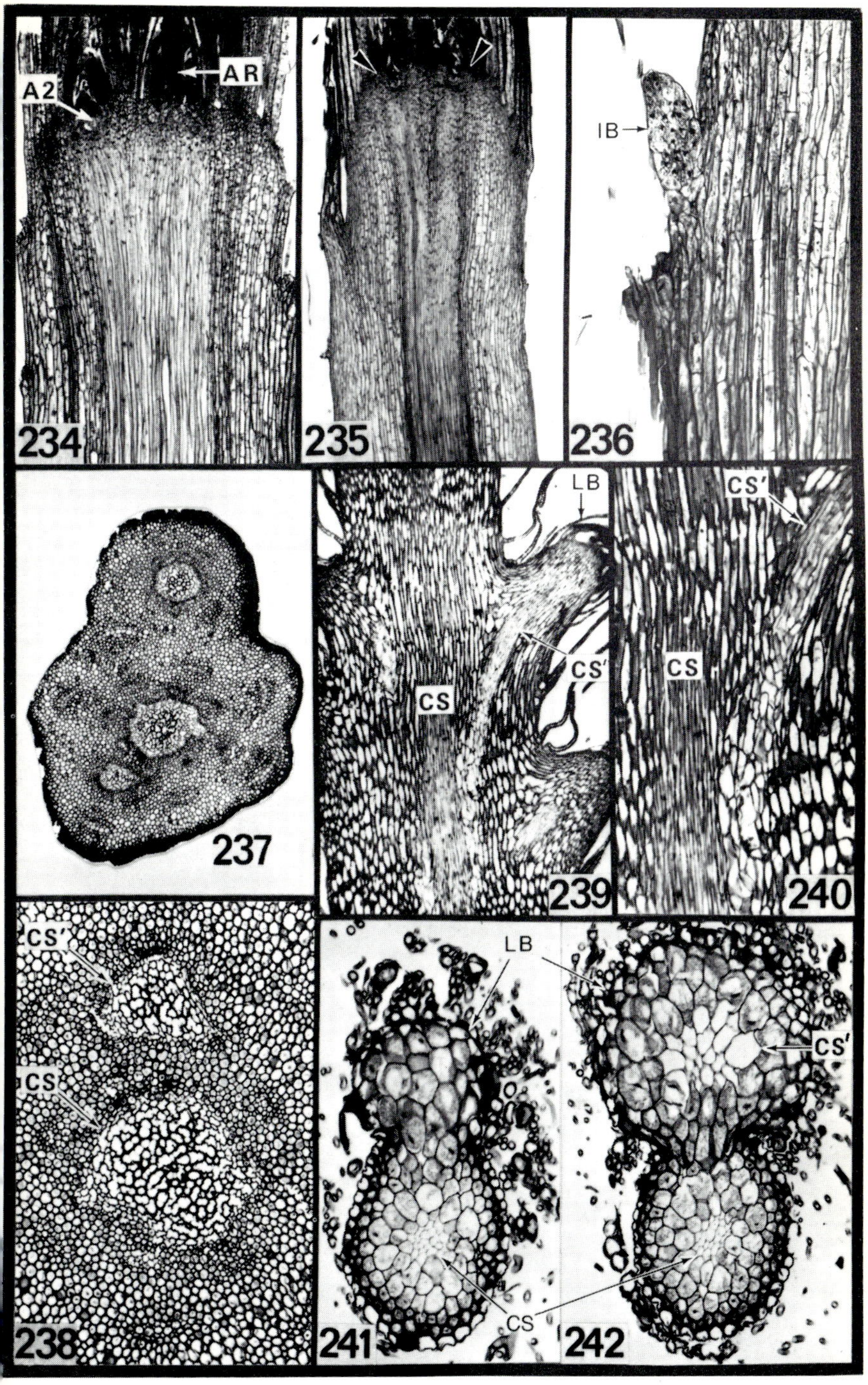

AR
A2
IB
LB
CS'
CS
234
235
236
237
238
239
240
241
242

PLATE 59

Dendroligotrichum dendroides (Polytrichales, Musci)
Leafy gametophyte

Fig. 243. Transverse section in apical region of a young upright stem in which branching has initiated. Apices of main stem (Al) and of several lateral branches (arrows) can be seen (the section is just above the level of the apical cells). x 105.

Fig. 244. Transverse section of the same specimen, just below the apex, to show the differentiating conducting strands. x 105.

Fig. 245. Portion of transverse section of mature upright stem at the level where branching occurs. Central strand (CS) of main stem and conducting strands (CS'1, CS'2, CS'3) of lateral branches can be seen. x 165.

(For details, see Hébant & Berthier 1972).

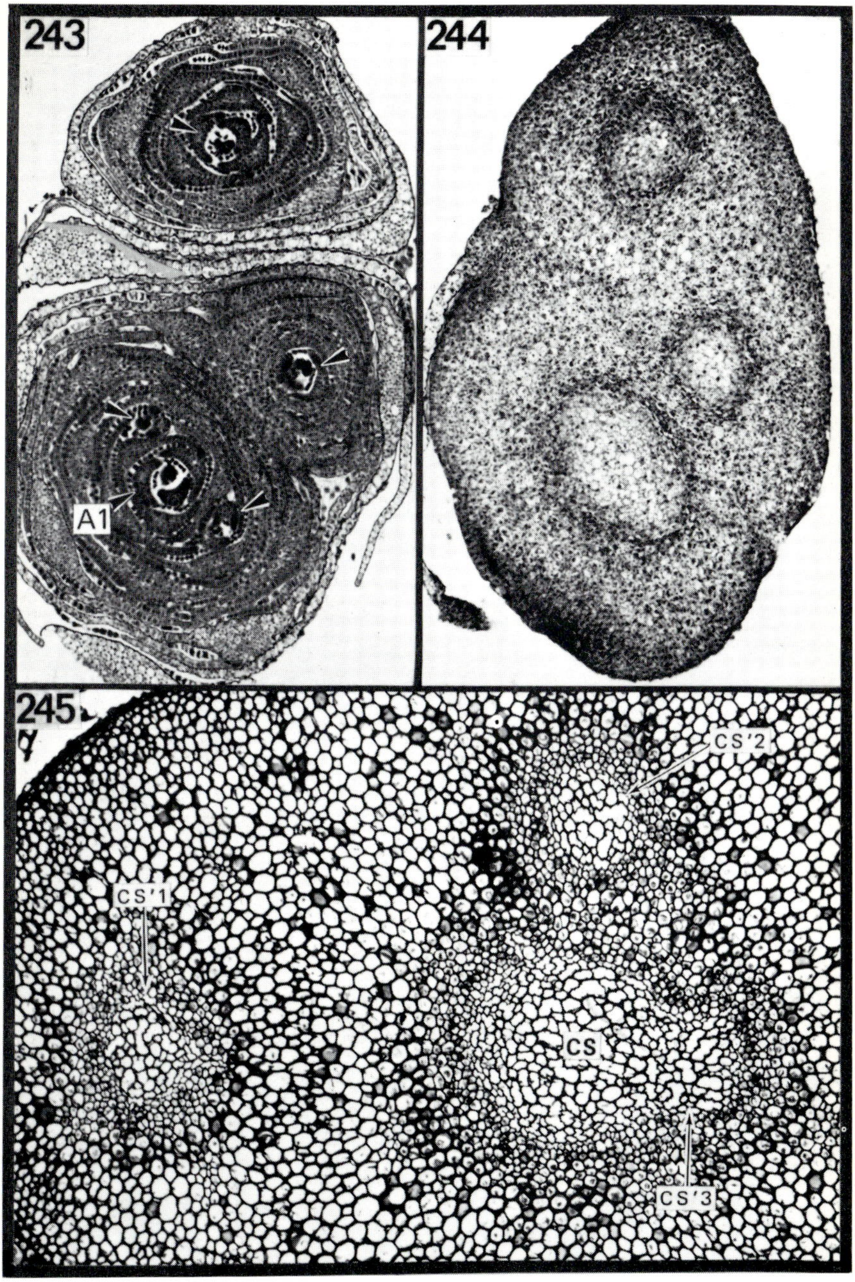
243
A1
244
245
CS'2
CS'1
CS
CS'3

PLATE 60

Fig. 246. *Polytrichum commune* (Polytrichales, Musci); specimen from pseudo-alpine grassland. Portion of transverse section of aerial leafy stem showing part of the central strand of hydroids (CS), two leaf traces (LT; one of them merging with the central strand), and leptoids (LEP). GT, ground tissue. x 800.

Fig. 247. *Dawsonia superba* var. *superba* (Polytrichales). Leaf. Transverse view of the nerve. Hydroids (H), deuters (D) and stereids (STER) can be seen. Scanning electron micrograph, x 2,000 approx.

Fig. 248. *Dawsonia superba* var. *superba.* Portion of a leaf trace in the outer cortex. A hydroid (H), and a deuter (D) with a transverse end wall, can be seen. Scanning electron micrograph; x 2,500 approx.

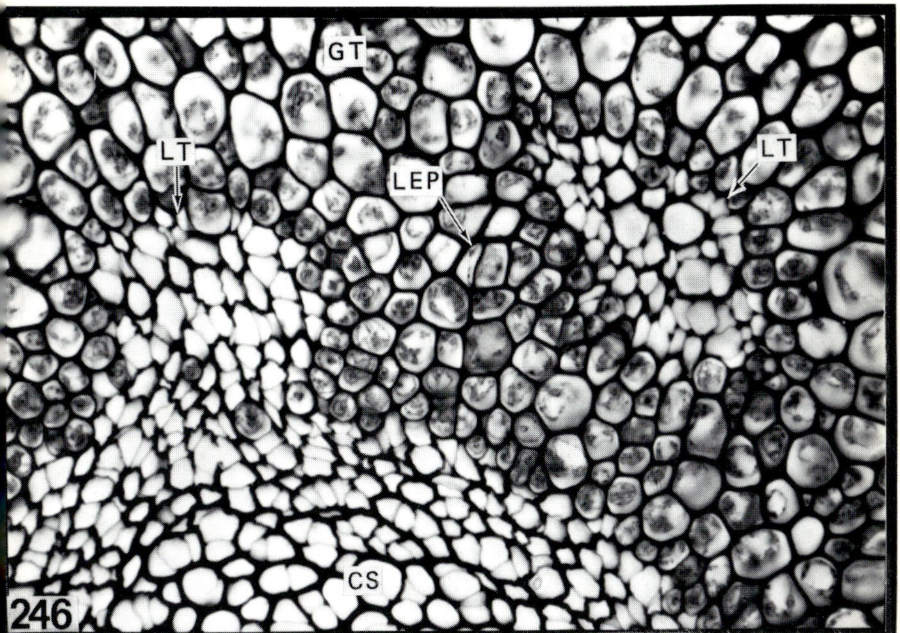
GT
LT
LEP
LT
CS
246

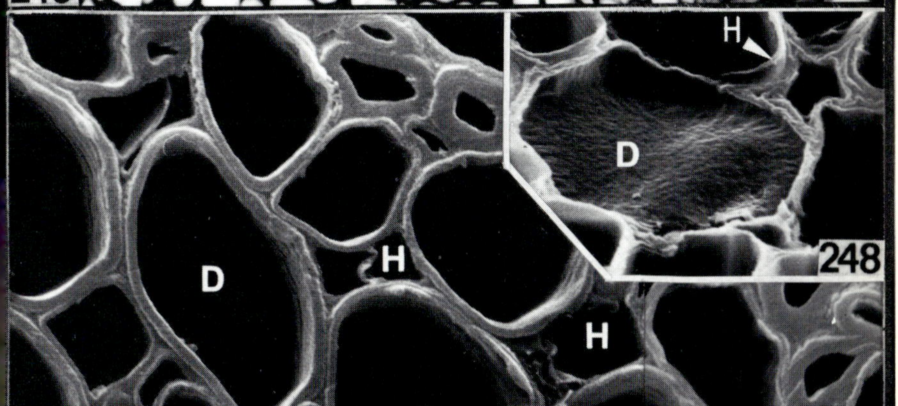
H
D
248
D
H
H
D
D
STER
247

PLATE 61

Fig. 249. *Polytrichum commune* (Polytrichales, Musci). Leafy stem. Ultrastructure of a leaf trace as seen in transection (low power electron micrograph). The trace shown here corresponds approximately to that numbered n-7 in fig. 101. Hydroids (HYD), deuters (D) and socii (S) are identifiable. Observe the contrast between the ultrastructure of deuters and that of socii. Note also multilobed (presumably degenerating) nucleus in one of the deuters (second one starting from the left). x 4,900.

D, deuter; HYD, hydroid; N, nucleus; P, plastid; PD, plasmodesmata; S, socius.
After Hébant (1972b).

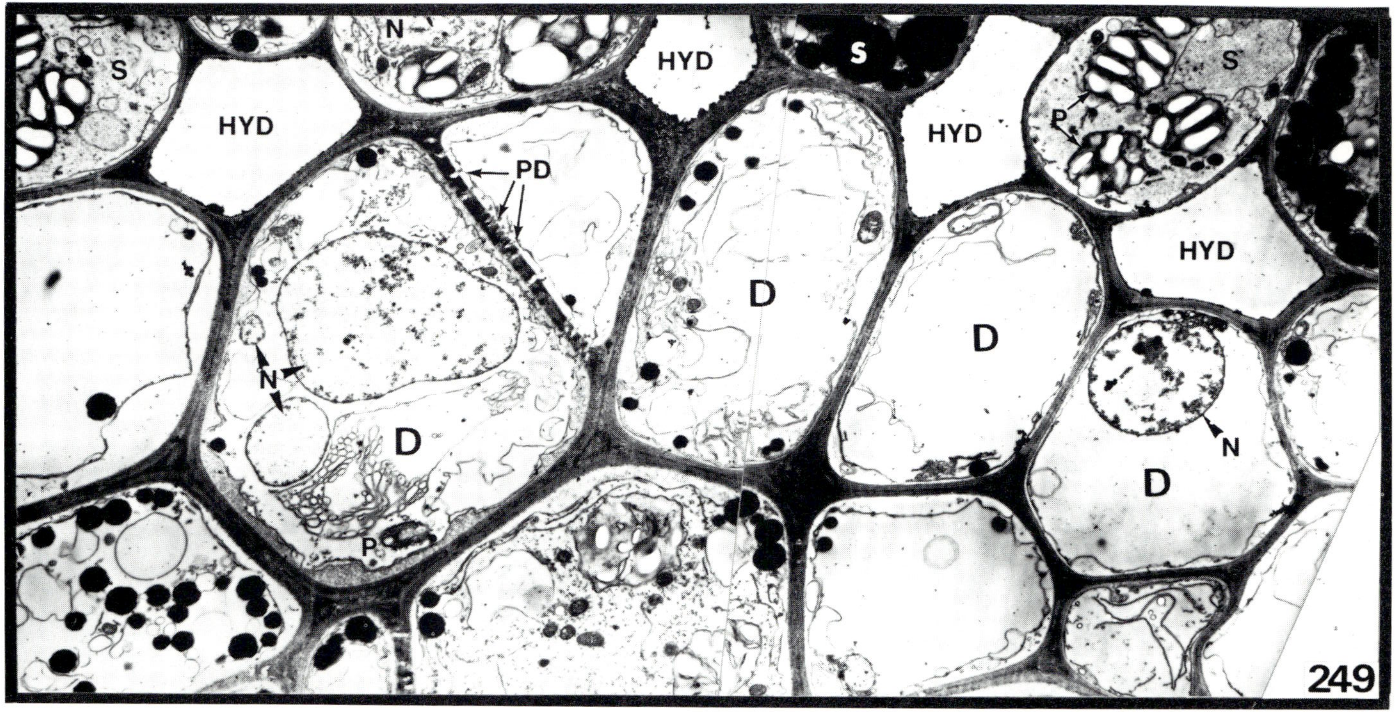
S
HYD
N
HYD
S
HYD
S
P
HYD
PD
D
D
D
N
N
D
P
D
249

PLATE 62

Fig. 250. *Polytrichum commune* (Polytrichales, Musci). Leaf. Free-hand median longisection of nerve stained with iodine-potassium iodide reagent. Specialized food-conducting cells (deuters, D) can easily be distinguished from the associated parenchyma cells (including the so-called socii, S). x 330.

Fig. 251. *Dawsonia polytrichoides* (Polytrichales). Leafy stem. Contact end wall between two deuters of a leaf trace. Many plasmodesmata are present, which show associations with the endoplasmic reticulum. x 19,000.
After Hébant (1976a).

Fig. 252-253. *Polytrichum commune* (Polytrichales). Leafy stem. Ultrastructural features of deuters : plastids with a regressed structure (fig. 252; x 16,000) and stacks of endoplasmic reticulum (fig. 253; 13,500). After Hébant (1972b).

Fig. 254. *Dawsonia superba* var. *superba* (Polytrichales). Leaf; transection. Two hydroids separated by hydrolysed contact end walls. x 10,700.
After Hébant (1976a).

Fig. 255. *Funaria hygrometrica* (Funariales, Musci). Leafy stem. Portion of transection of leaf trace showing hydroids and parenchyma cells of ground tissue. x 2,300 approx.
After Hébant (1969b).

D, deuter; ER, endoplasmic reticulum; HYD, hydroid; HW, hydrolysed wall; P, plastid; PAR, parenchyma cell; PD, plasmodesmata; S, socius.

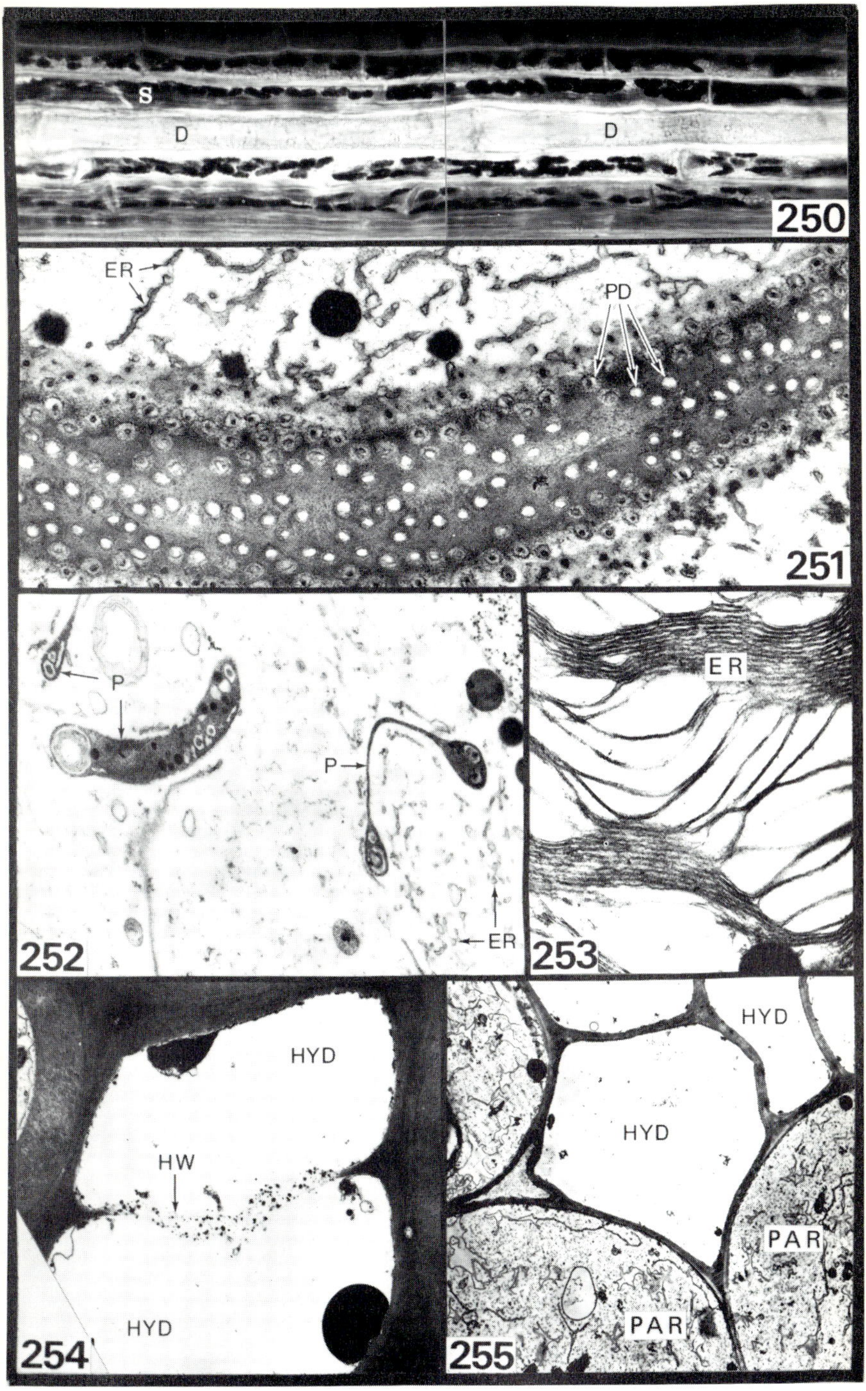
S
D
D
250
ER
PD
251
P
P
ER
252
ER
253
HYD
HW
HYD
254
HYD
HYD
PAR
PAR
255

PLATE 63

Conducting tissues in sporogonia of mosses

Figs. 256-257. *Polytrichum commune* (Polytrichales). Sporophyte.
Fig. 256. Portion of median longisection in lower region of capsule and part of seta (S). Termination of central strand (CS) of seta within capsule is seen. Note enlarged end of water-conducting strand, forming small "ampulla" (arrow). Leptoids of seta are connected to cells of conducting parenchyma. x 28.
Fig. 257. *idem.* detail of "ampulla" (arrowed). x 68.

Fig. 258. *Funaria hygrometrica* (Funariales). Median longisection of young capsule. The organization of the conducting tissues is here comparable with that in *Polytrichum* (figs. 256-257) although in a somewhat simpler form. "Ampulla" arrowed. x 28.

Figs. 259-260. *Dawsonia superba* var. *superba* (Polytrichales). Transection (fig. 259) and longisection (fig. 260) of central strand of seta. Fig. 259, x 290; fig. 260, x 400.
After Hébant (1975a).

Fig. 261. *Dendroligotrichum dendroides* (Polytrichales). General organization of central strand of seta, as it appears in transection under the scanning electron microscope. x 600.
After Hébant (1975a).

Fig. 262. *Dawsonia superba* var. *superba* (Polytrichales). Ultrastructure of hydroids in central strand of seta. x 5,300.
After Hébant (1975a).

CS, central strand; HYD, hydroid; LEP, leptoid; PAR, parenchyma cell; S, seta.

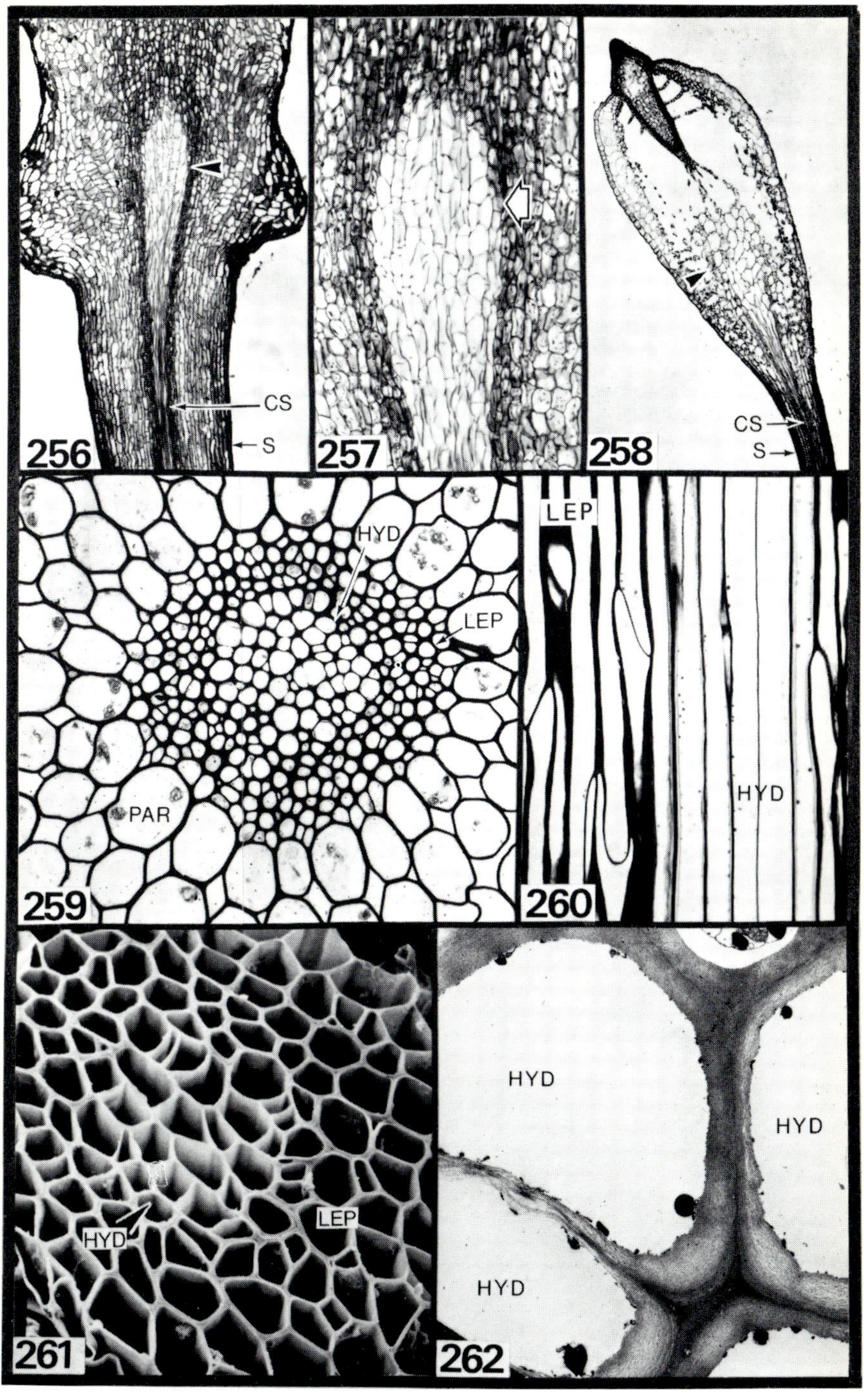
256
CS
S
257
258
CS
S
HYD
LEP
PAR
259
LEP
HYD
260
HYD
LEP
261
HYD
HYD
HYD
262

PLATE 64

Hydroids of Mosses. I.

Figs. 263-264. Similarity in appearance between a hydroid of *Mnium* (Eubryales) (fig. 263) and a tracheid of *Equisetum* (fig. 264). x 630.
After Hébant (1967a).

Fig. 265. *Polytrichum commune* (Polytrichales). Gametophyte. Transverse view of hydroids of central strand. Scanning electron micrograph; x 5,200.
After Hébant (1974).

Fig. 266. *Funaria hygrometrica* (Funariales). Portion of transection of leafy stem showing part of conducting strand of hydroids (HYD) and the parenchyma cells (PAR) of inner cortex. Note that the latter protrude within central strand. x 3,200.

263

264

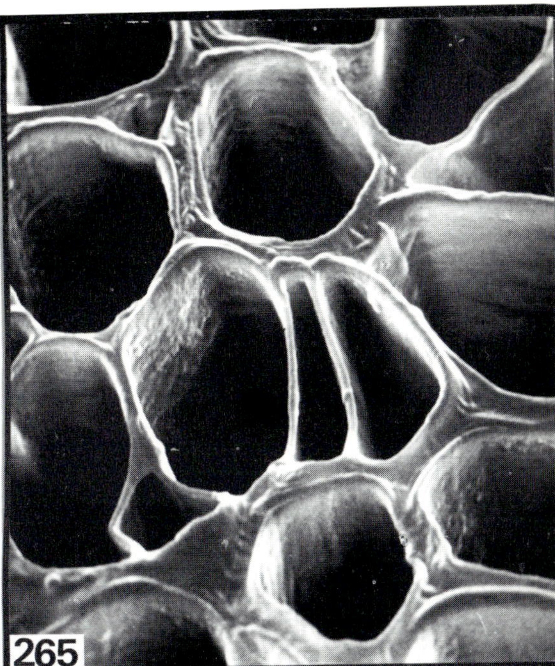
265

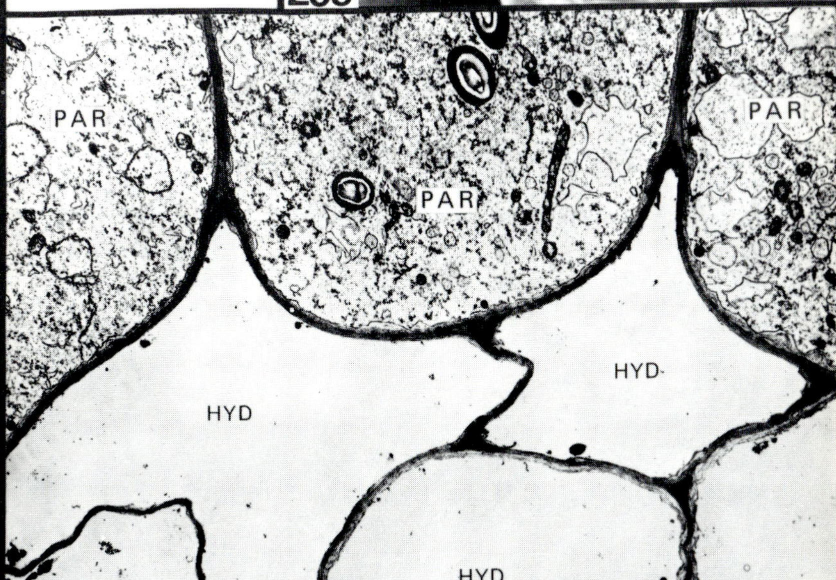

266

PLATE 65

Hydroids of Mosses. II.
Some ultrastructural features of differentiating hydroids

Fig. 267. *Dawsonia superba* var. *superba* (Polytrichales). Longisection of apical region of leafy stem. Note contrast in structure of nuclei (N) of a young developing hydroid (HYD) and a young stereid (STER). Differences in the structure of the plastids (P) are also seen (those of the stereid possess starch grains). x 3,300.
After Hébant (1976a).

Fig. 268. *Polytrichum commune* (Polytrichales). Gametophyte; longisection. Dictyosome (D) and golgi-derived vesicles (GV) in differentiating hydroid. Thin section treated by the method of Thiéry (1967) : the polysaccharidic content of many of the vesicles is demonstrated. Arrows indicate areas where such vesicles are presumably incorporated into wall (W) through plasmalemma. x 24,000.
After Hébant (1974).

Fig. 269. *Polytrichum juniperinum*. Gametophyte. Lomasome-like structure in developing hydroid. Thiéry method for the demonstration of polysaccharides. W, wall. x 32,000.

Fig. 270. *Polytrichum commune.* Gametophyte; longisection. Hydroids at an advanced stage in their development. Vacuoles (V) reach such a large volume that the cytoplasm is restricted to a thin parietal layer. A few inclusion bodies (IB) are seen in these vacuoles. T, tonoplast; x 11,000.
After Hébant (1974).

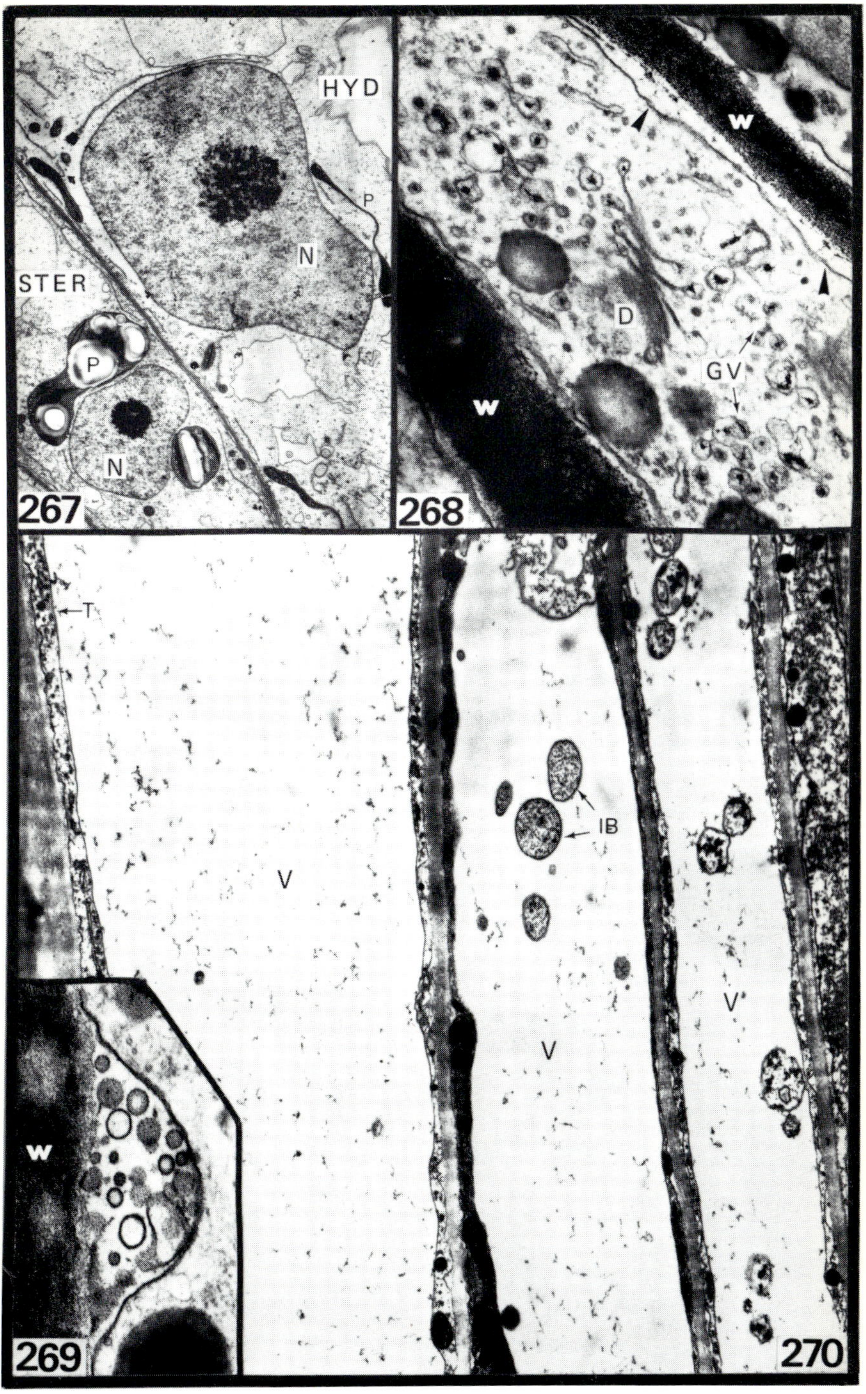
HYD
P
N
STER
P
N
267
W
D
GV
W
268
T
V
IB
V
V
W
269
270

PLATE 66

Hydroids of Mosses. III. Hydrolysed walls

Fig. 271. *Polytrichum commune* (Polytrichales). Gametophyte. Early stage in the hydrolysis of the thin end wall between two hydroids. Method of Thiéry (1967) for the demonstration of polysaccharides. x 37,000.
After Hébant (1974).

Fig. 272. *Polytrichum commune.* Gametophyte. Hydrolysed end wall of contact between two mature hydroids. A complex residual body (RB) can also be seen. x 41,000.
After Hébant (1974).

Fig. 273. *Polytrichum commune.* Gametophyte. Freeze-etched specimen. Transverse view of hydrolysed end wall between two hydroids showing loose arrangement of microfibrils. From a replica prepared by R.P.C. Johnson (Aberdeen) (see also Hébant & Johnson 1976). x 20,000.

RB, residual body; HYD, hydroid; HW, hydrolysed wall; W, wall.

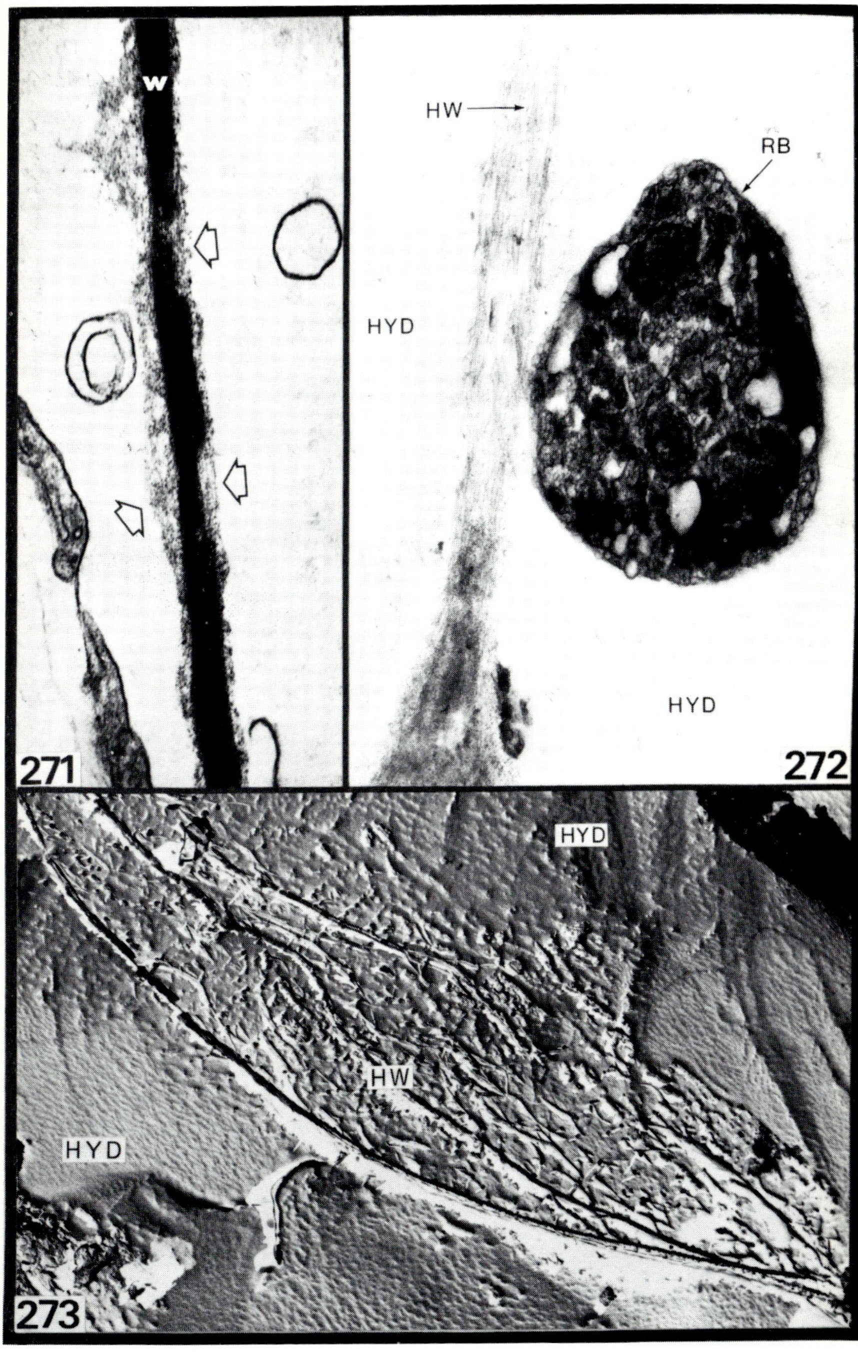

W
271
HW
RB
HYD
HYD
272
HYD
HW
HYD
273

PLATE 67

Hydroids of Mosses. IV.
Ultrastructure of hydroids in central strand of Polytrichales
(see also fig. 100)

Fig. 274. *Polytrichum commune.* Leafy stem. Portion of transverse section at periphery of central strand of hydroids. Parenchyma cells of the hydrom-sheath can also be seen. x 6,350.

Fig. 275. *Pogonatum urnigerum.* Leafy stem. Transverse view of the hydroids from the median portion of the central strand. The "secondary" walls of these hydroids show a clear stratification. x 5,400.
After Hébant (1974).

HS, hydrom-sheath; HYD, hydroid; HW, hydrolysed wall; ML, middle lamella; SW, "secondary" wall.

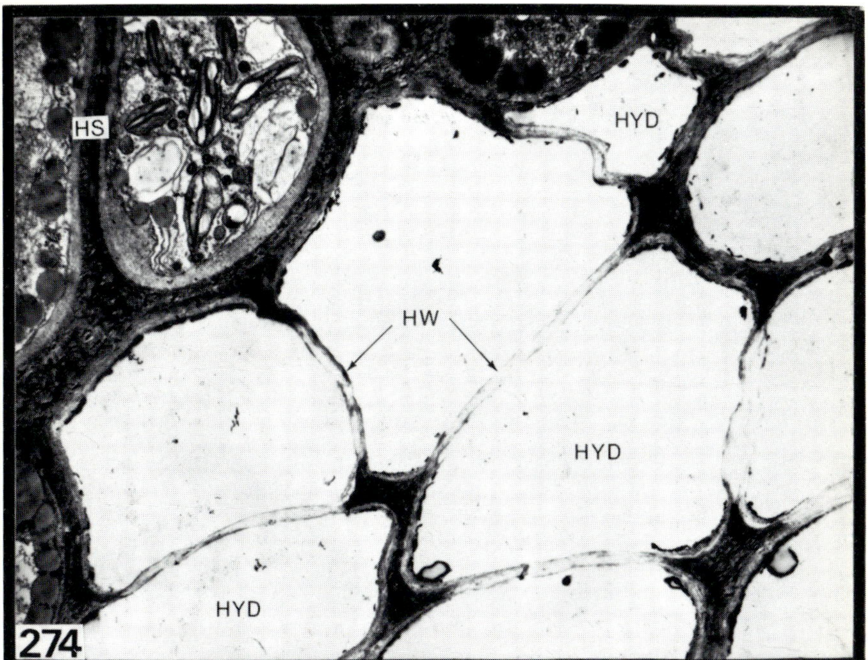
HS
HYD
HW
HYD
HYD
274

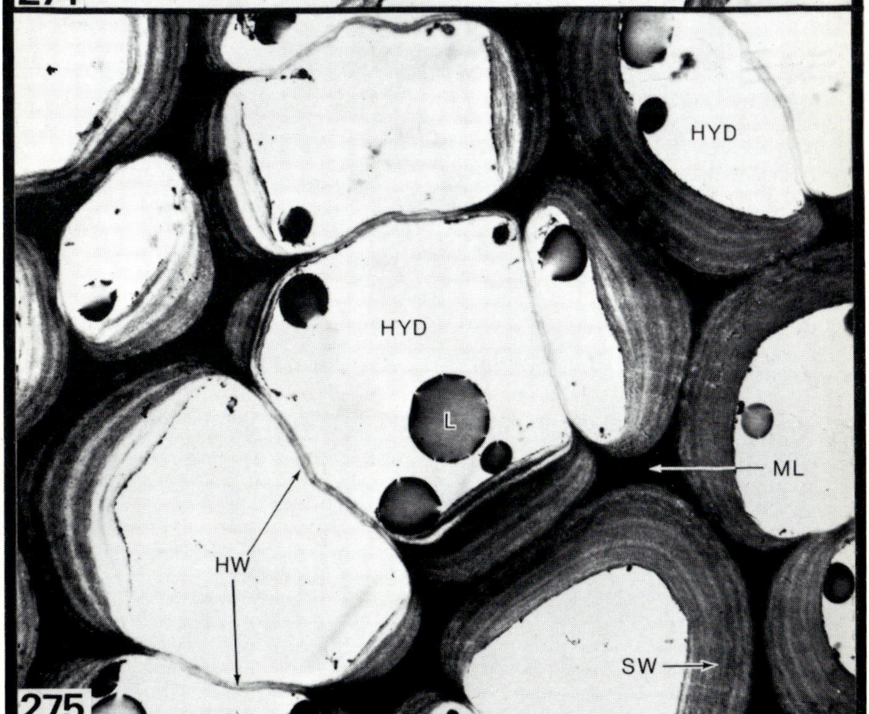
HYD
HYD
L
ML
HW
SW
275

PLATE 68

Hydroids of Mosses. V.
Lysosomal degeneration of the protoplast of the hydroids

Fig. 276. *Polytrichum commune* (Polytrichales). Longisection of aerial leafy stem. A peak of strong acid phosphatase activity is seen at the level (arrowed) where the protoplast of the hydroids degenerates. Burstone method. CS, central strand; x 35.
After Hébant (1970b).

Fig. 277. *P. commune.* Transverse section of leafy stem at a short distance below apex. Strong lysosomal acid phosphatase activity in differentiating hydroids of central strand (CS) and of "true" leaf traces (LT). Burstone method. x 175.

Fig. 278. *Polytrichum formosum.* Strong acid phosphatase activity in differentiating hydroids of central strand (CS) of seta, as seen in longitudinal section. Burstone method. x 115.

Fig. 279. *Mnium affine* (Eubryales). Transverse section of leafy stem at a short distance below apex. Strong lysosomal acid phosphatase activity in differentiating hydroids of central strand (CS) and of "false" leaf traces (LT). Gomori method. x 92.

Figs. 277 to 279 after Hébant (1973b).

Fig. 280. *Thuidium tamariscinum* (Hypnobryales). Transverse section in gametophyte. Acid phosphatase activity present in differentiating vestigial central strand (CS). Burstone technique. x 66

Fig. 281. *Polytrichum juniperinum* (Polytrichales). Gametophyte. Acid phosphatase activity in dictyosome (D) and golgi-derived vesicle (GV) in differentiating hydroid. x 45,000.

Fig. 282. *P. juniperinum.* Gametophyte; differentiating hydroid. Strong acid phosphatase acitivity in small vesicles (VS) (=primary lysosomes ?). x 31,500.

Fig. 281-282 after Hébant (1974).

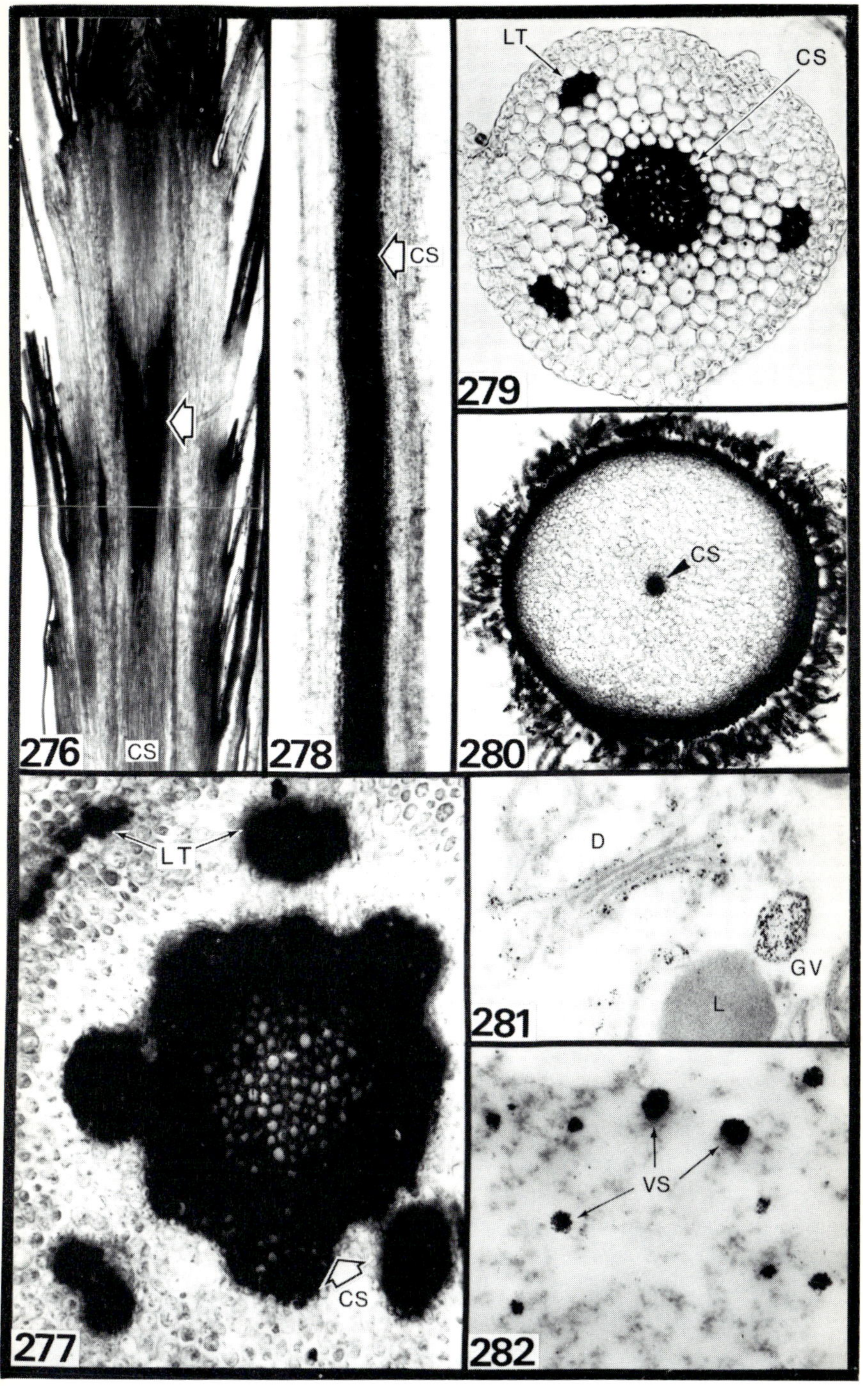
276
CS
278
CS
LT
CS
279
CS
280
LT
CS
277
D
GV
L
281
VS
282

PLATE 69

The leptom of gametophytes of polytrichaceous mosses. I.

Fig. 283. General appearance of leptom at periphery of central strand of hydroids (CS) in aerial leafy stem of *Polytrichum commune* (transverse section). A large "true" leaf trace (LT) merges with the central strand. Two groups of leptoids (LEP), one each side of the leaf trace are identifiable. x 330.

Fig. 284. *P. commune.* Portion of longisection in apical region of leafy stem showing young developing leptoid and associated parenchyma cells. Note contrast in structure of nuclei in both cell types (see also plate 39). Phase contrast. x 800.

Fig. 285. *Dendroligotrichum dendroides.* Portion of longisection of upright leafy stem showing hydroids (HYD) of central strand, leptoids (LEP) and associated parenchyma cells. x 450.
After Hébant (1973a).

Fig. 286. *P. commune.* Part of radial longisection of aerial leafy stem. Hydroids of central strand, leptoids, and associated parenchyma elements are shown. Nomarsky phase contrast (compare with fig. 21, plate 2, in Tansley & Chick 1901 !). x 420.

Fig. 287 to 289. *P. commune.* Leafy stems; longisections stained with the PAS reaction to show plasmodesmata in end walls (figs. 288-289) and lateral wall (fig. 287) of leptoids. Note that these plasmodesmata are quite scarce on lateral walls (fig. 287), whereas their abundance varies greatly in end walls (compare figs. 288 and 289). Figs. 287, x 1,750; fig. 288, x 1,250; fig; 289, x 1,640.
Fig. 289 after Hébant (1967ab).

Fig. 290. *P. commune.* Longisection of leafy stem; demonstration of callose by its aniline blue fluorochrome reaction. The steeply inclined end walls of contact between leptoids, and the terminal walls of neighbouring parenchyma cells appear to be covered with callose. x 380.
After Hébant (1968).

CS, central strand of hydroids; HYD, hydroids; LEP, leptoid; LT, leaf trace; PAR, parenchyma cell.

Note : Apparent discrepancies in the magnifications given in this plate are due to the fact that specimens from various origins are shown. Cell sizes in them may vary widely.

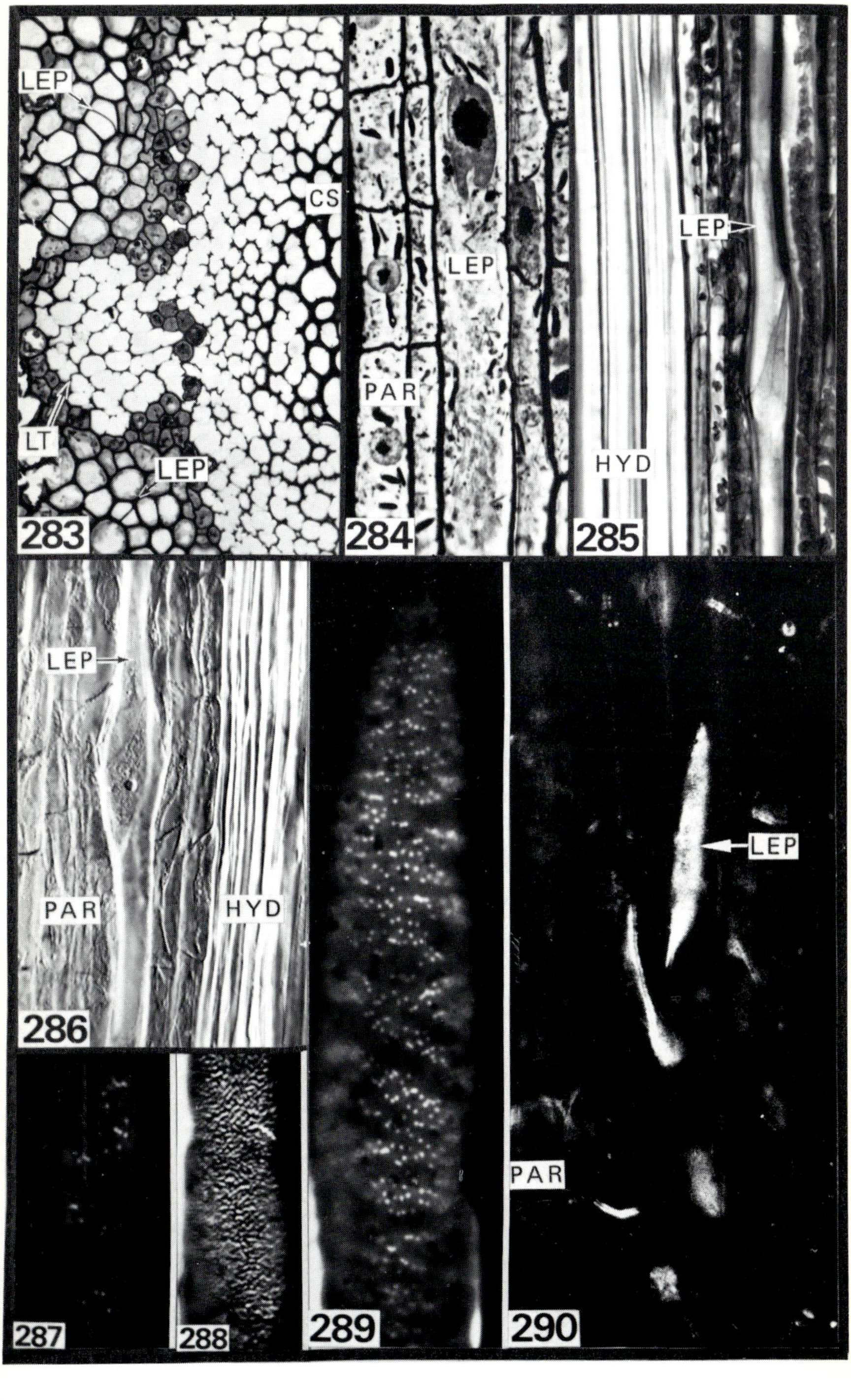
LEP
CS
LT
LEP
283
LEP
PAR
284
LEP
HYD
285
LEP
PAR
HYD
286
LEP
PAR
287
288
289
290

PLATE 70

The leptom of gametophytes of polytrichaceous mosses. II.

Fig. 291. *Polytrichum commune.* Leafy stem; transverse section. Low power electron micrograph showing contrast between organization of protoplasm of leptoids *sensu-stricto* and of adjacent parenchyma cell (lower left). The leptoid at lower right appears somewhat disturbed : the accumulation of endoplasmic reticulum, vesicles and lipid droplets against the end wall could thus possibly be an artifact (compare with that in fig. 294). x 13,000.

Fig. 292. *P. commune.* Enlarged plasmodesmata in end wall of contact between two leptoids. Transverse section; x 19,300.

Fig. 293. *Dendroligotrichum dendroides.* Small "pore" in end contact wall between two leptoids. x 36,000.

Fig. 294. *D. dendroides.* Highly organized system of endoplasmic reticulum in "accumulation" against the end wall of a little disturbed leptoid. Transverse section; x 27,500.

Figs. 293-294 after Hébant (1976b).
ER, endoplasmic reticulum; L, lipid droplet; LEP, leptoid; N, nucleus; P, plastid; PAR, parenchyma cell; PD, plasmodesma; TU, tubule; W, wall.

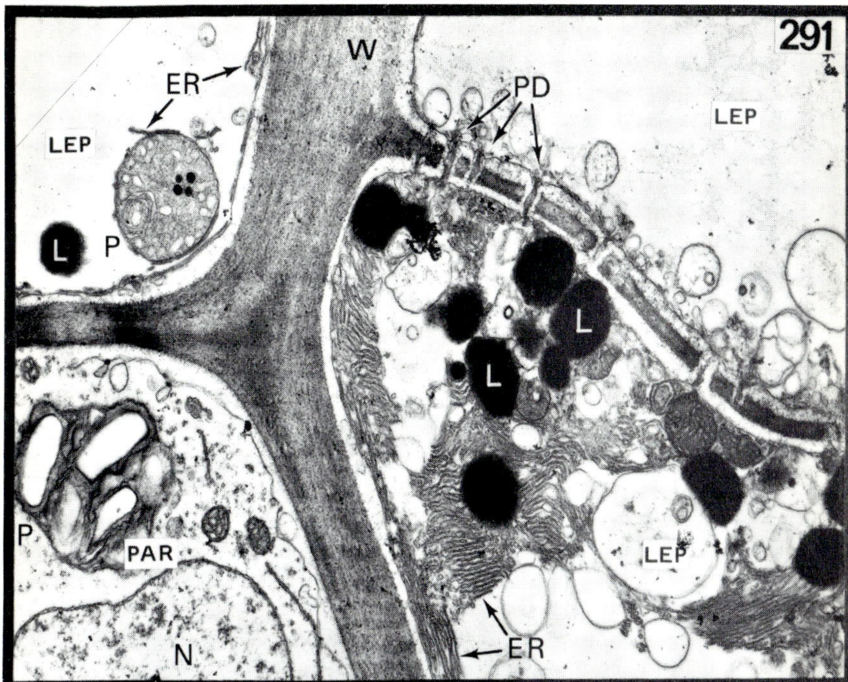
291
W
ER
ER
LEP
PD
LEP
L
P
L
L
P
PAR
N
LEP
ER
ER

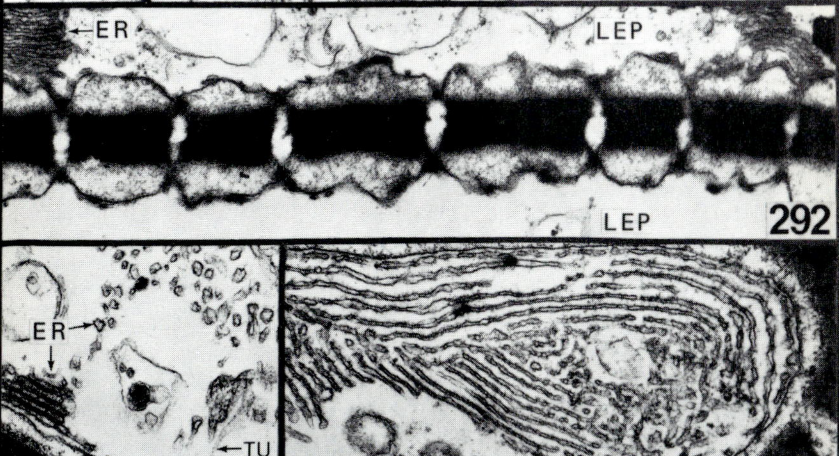
ER
LEP
LEP
292

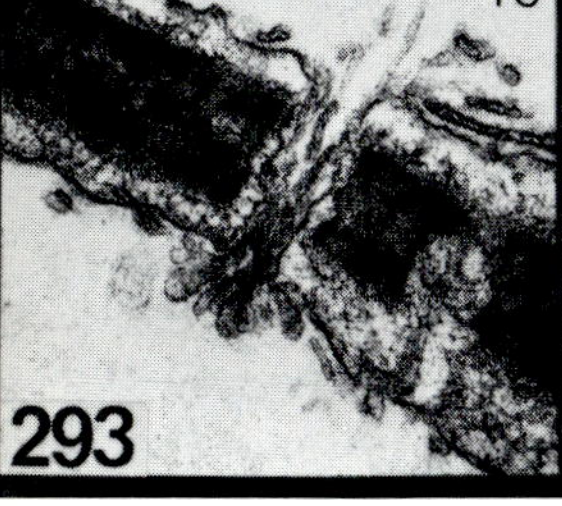
ER
ER
TU
293

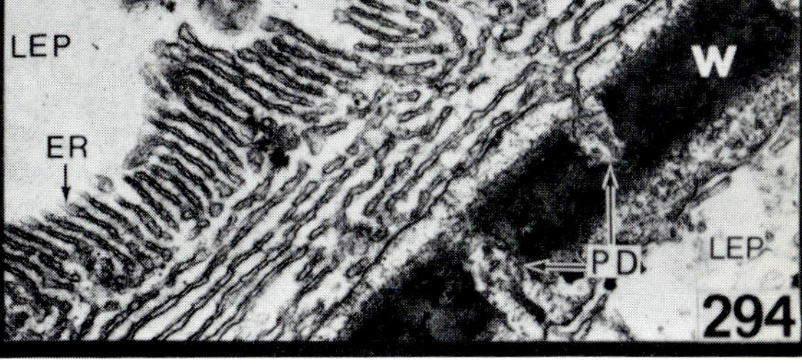
LEP
ER
W
PD
LEP
294

PLATE 71

The leptom of gametophytes of polytrichaceous mosses. III.

Fig. 295. *Polytrichum commune.* Plasmodesma in lateral wall between a nearly mature leptoid and adjacent parenchyma cell. x 21,500.

Fig. 296. *Atrichum ligulatum.* Plamodesma in lateral wall with "nacreous" thickenings between two leptom cells. Transection of leafy stem; x 21,000.

Fig. 297. *Polytrichadelphus magellanicus.* Plastid of mature leptoid *s.str.* Note absence of starch, regression of thylakoids and presence of dense globules in stroma. This plastid also shows looped invaginations of its inner membrane. x 37,500.

Fig. 298. *Polytrichum commune.* Mature leptoid. The plastid shown here has retained small grana in its stroma – an uncommon feature in plastids of aged leptoids. x 37,000.

Fig. 299. *Dendroligotrichum dendroides.* Plastids of highly differentiated leptoid. x 47,000.

Fig. 300. *Atrichum ligulatum.* Plastid with regressed structure in a leptom cell (= a leptoid *sensu stricto* ?). x 28,000.

ER, endoplasmic reticulum; LEP, leptoid; M, mitochondrion; N, nucleus; NW, nacreous wall; P, plastid; PAR, parenchyma cell; PD, plasmodesma; W, wall.

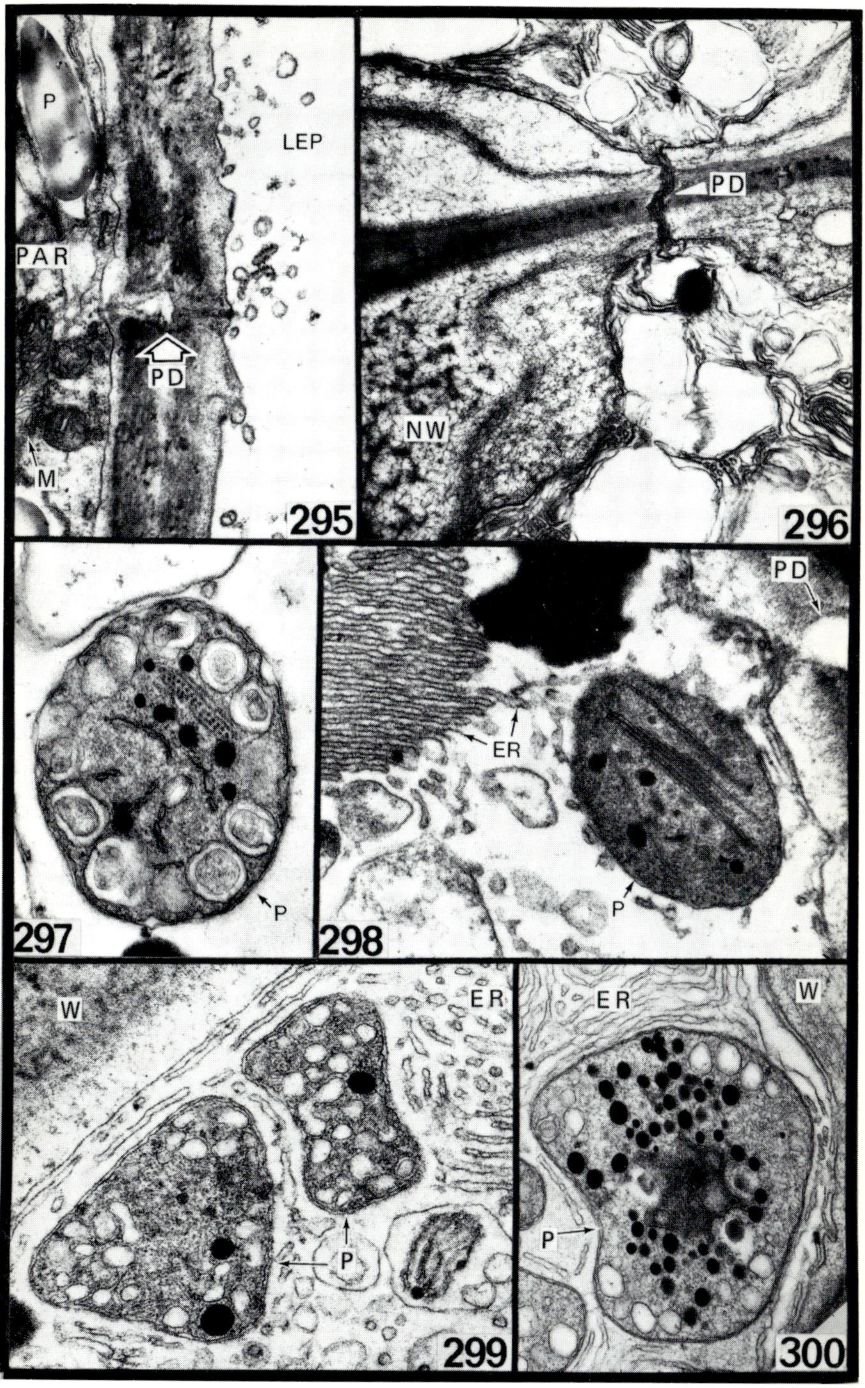
P
LEP
PAR
PD
M
295
PD
NW
296
P
297
PD
ER
P
298
W
ER
P
299
ER
W
P
300

PLATE 72

The leptom of gametophytes of polytrichaceous mosses. IV.

Fig. 301. *Dendroligotrichum dendroides.* Longisection in *aged* portion of upright stem showing leptoid (LEP) and associated parenchyma cell (PAR). Feulgen reaction. Note difference in chromaticity between nucleus of leptoid (NL) and of parenchyma cell (NP). x 800.
After Hébant (1973a).

Fig. 302. *Polytrichadelphus magellanicus.* Autoradiograph of longisection of specimen previously incubated in tritiated uridine (10μCie/ml; 6 hours incubation of stem fragment). The nucleus (NL) of a fully mature leptoid has not incorporated the precursor, whereas the nuclei (NP) of three neighbouring parenchyma cells are heavily labelled. x 650.

Fig. 303. *Polytrichum alpinum.* Leptoid in final stage of differentiation. The nucleus seen here is presumably degenerating. Specialized plastids are also seen. Longisection, x 15,000.

Fig. 304. *Polytrichum commune.* Organelle similar to "refractive spherules" of pteridophyte phloem apparently originating from the endoplasmic reticulum in a young leptoid. x 57,500.
After Hébant (1976b).

Fig. 305. *Polytrichum commune.* Organelles (G) similar in appearance to "refractive spherules" of pteridophyte phloem persisting in "mature" leptoid *s.str.* x 23,000.
After Hébant (1970a).

ER, endoplasmic reticulum; G, organelle similar in appearance to "refractive spherules" of pteridophytes; LEP, leptoid; M, mitochondrion; NL, nucleus of a leptoid; NP, nucleus of a parenchyma cell; P, plastid; PAR, parenchyma cell.

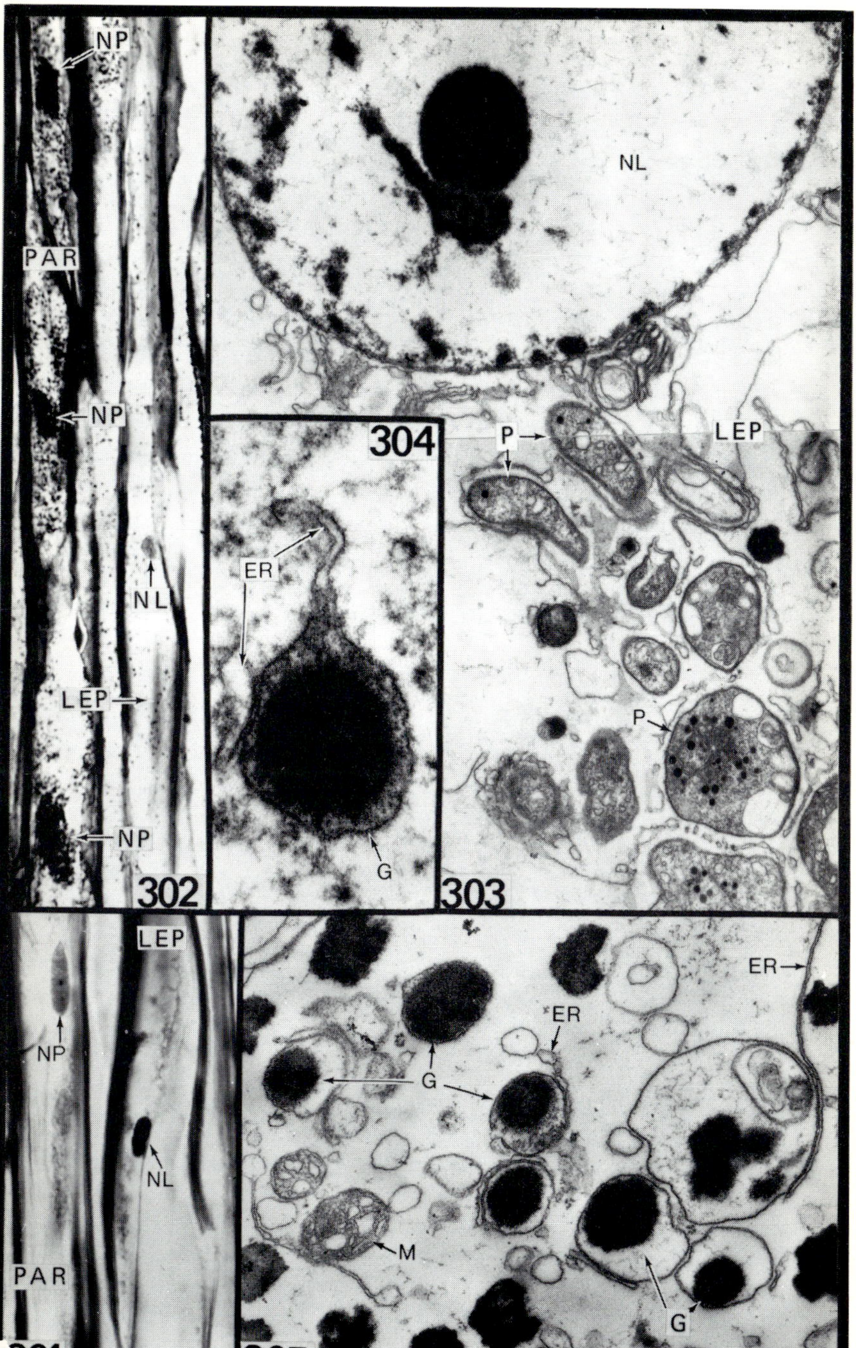
NP
PAR
NP
NL
LEP
NP
302
NL
304
ER
ER
G
P
P
LEP
P
303
LEP
NP
NL
PAR
301
ER
ER
G
G
M
G
305

PLATE 73

The leptom of gametophytes of polytrichaceous mosses. V.

Fig. 306. *Dendroligotrichum dendroides.* Complex network and stacks of endoplasmic reticulum in the lumen of a mature leptoid *s.str.* of main upright stem. x 27,000.

Fig. 307. *Polytrichum commune.* Extensive system of stacked endoplasmic reticulum applied to plasmalemma in peripheral region of mature leptoid *s.str.* It is not known with certainty whether the lack of a tonoplast delimiting a central vacuole is a natural feature or an artifact. Transverse section; x 18,500.

Fig. 308. *Polytrichum commune.* Portion of longisection of young leptoid. Autophagic vacuoles (AV) are here identifiable, and an early stage in stacking of endoplasmic reticulum is also seen (arrow). x 28,000.
After Hébant (1969a).

Fig. 309. *Atrichum undulatum.* "Nacreous" thickenings (NW) in the lateral walls of an obviously parenchymatous element (note large chloroplastid with starch grains) of the leptom *sensu lato.* Transverse section; Thiéry method for the demonstration of polysaccharides. x 10,800.

AV, autophagic vacuoles; ER, endoplasmic reticulum; LEP, leptoid; NW, "nacreous" wall; P, plastid; PAR, parenchyma cell; W, wall.

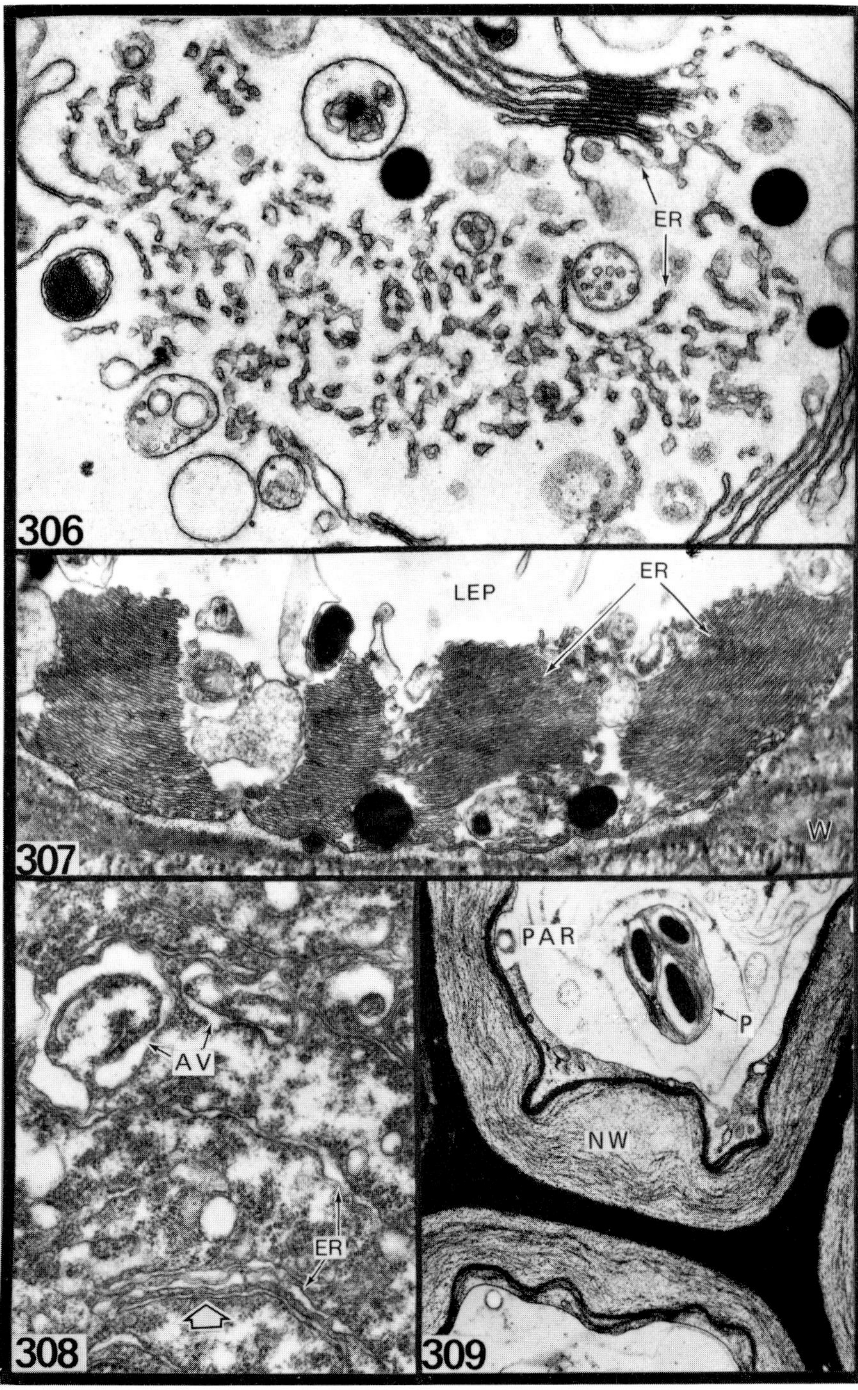
306
ER
ER
307
LEP
ER
W
308
AV
ER
309
PAR
P
NW

PLATE 74

Interphase between gametophyte and sporophyte. I.

Fig. 310. *Polytrichum formosum* (Polytrichales, Musci). Longisection of female fertile plant showing developing embryo (2n) within accrescent archegonium (AR). The lower region of the embryo apparently plays an important role during the early stages of sporogonium development, especially during its implantation within gametophytic tissues; it persists in the form of a more or less necrotic "appendage" at the base of the foot of the mature sporophyte. x 85.

Fig. 311. *Dendroligotrichum dendroides* (Polytrichales, Musci). Median longisection of lower region of vaginula (gametophytic tissue, (n)) and foot of sporophyte (2n). Note that extremity of sporogonium is deeply inserted into central strand (CS) of gametophyte. x 110 approx.
After Hébant (1975a).

Figs. 312 to 316. General appearance of the interphase between gametophyte and sporophyte in some selected bryophytes :

Fig. 312. *Diphyscium foliosum* (Buxbaumiales, Musci). Longisection; x 54.

Fig. 313. *Andreaea sp.* (Andreaeales). Longisection; interphase arrowed; x 82.

Fig. 314. *Riella sp.* (Sphaerocarpales, Hepaticae). Longisection; interphase arrowed; x 82.

Fig. 315. *Funaria hygrometrica* (Funariales, Musci). Transverse section; x 330.

Fig. 316. *Symphyogyna undulata* (Metzgeriales, Hepaticae). Vertical longitudinal section; x 33.

APG, "appendage" of foot; AR, archegonium; CS, central strand; TC, transfer cells; (n), gametophytic phase; (2n), sporophytic phase.

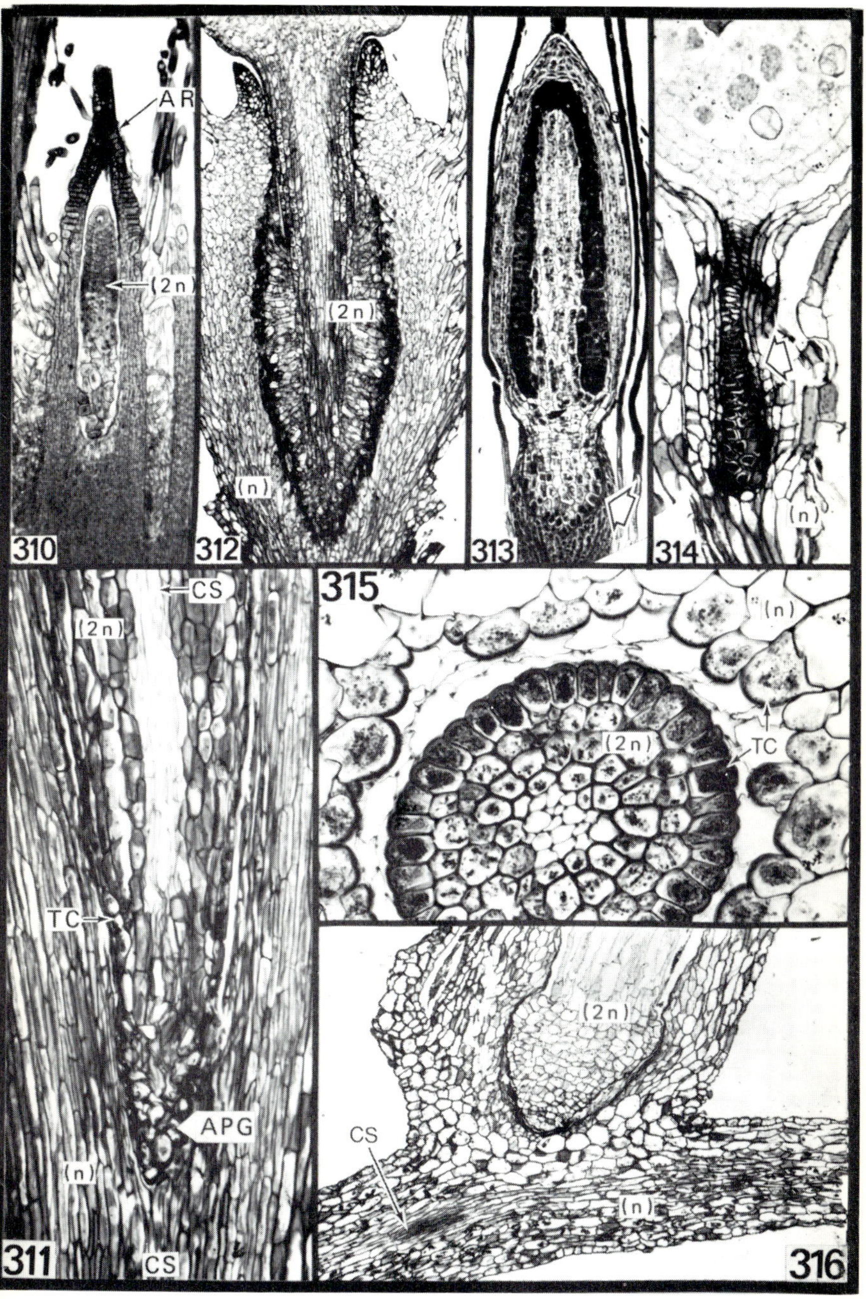
310
AR
(2n)
312
(2n)
(n)
313
314
(n)
311
CS
(2n)
TC
APG
(n)
CS
315
(n)
(2n)
TC
316
(2n)
CS
(n)

PLATE 75

Interphase between gametophyte and sporophyte. II.

Fig. 317. *Mnium cuspidatum* (Eubryales, Musci). An electron micrograph of the cells located at the juncture of sporophyte (2n) and gametophyte (n) (longisection). Wall ingrowths can be seen on both gametophytic and sporophytic sides. Note also abundance of mitochondria, especially in sporophytic cell. x 9,000.
After Eymé & Suire (1967).

Fig. 318. *Dawsonia superba* var. *superba* (Polytrichales, Musci). Transection of outer region of foot of sporophyte. Detail of ultrastructure of transfer cell. Many organelles, especially mitochondria and small plastids are seen in the vicinity of well developed wall ingrowths. Method of Thiéry (1967) for the demonstration of polysaccharides. x 11,700.

Fig. 319. *Dawsonia superba* var. *superba.* Portion of transfer cell : numerous mitochondria and small plastids are seen. x 18,000.

Figs. 318-319 after Hébant (1975a).

ER, endoplasmic reticulum; M, mitochondrion; P, plastid; W, wall; (n), gametophytic phase; (2n), sporophytic phase.

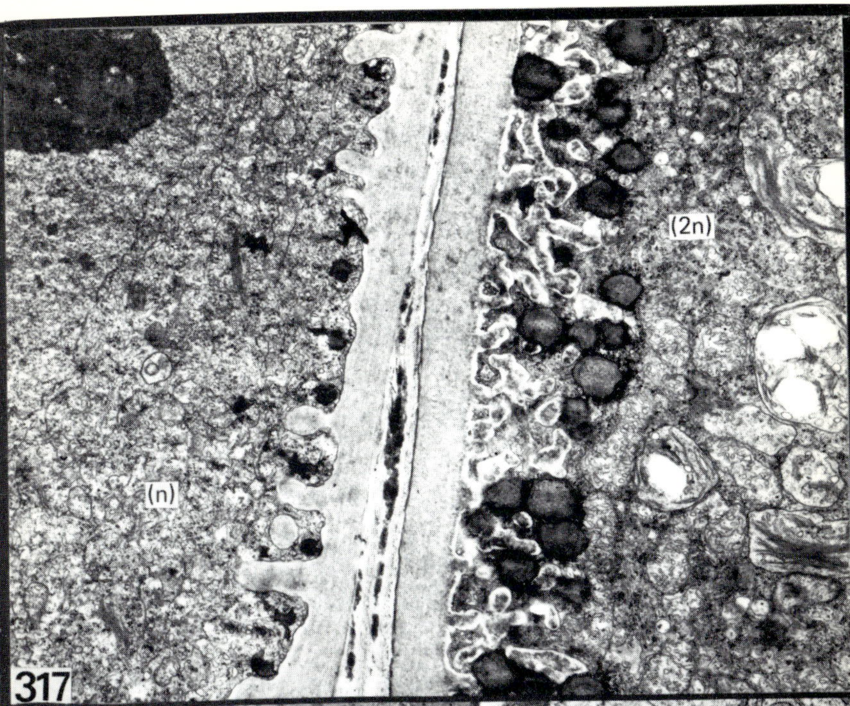
(2n)
(n)
317

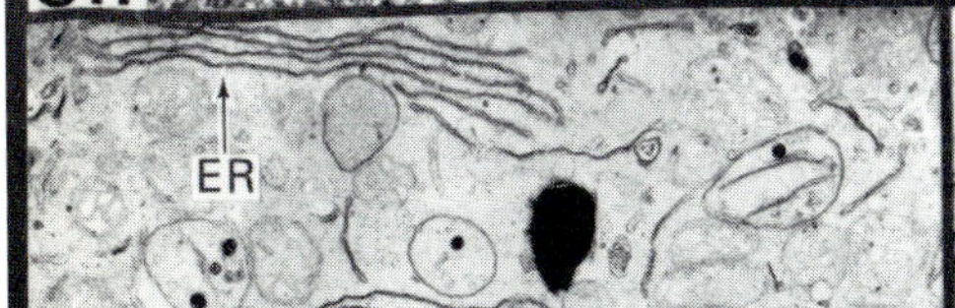
ER
M
P
W
318

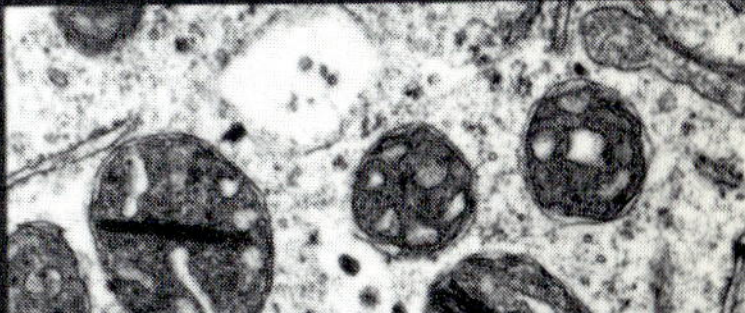
ER
M
P
319

PLATE 76

Interphase between gametophyte and sporophyte. III.

Fig. 320. *Dawsonia superba* var. *superba* (Polytrichales, Musci). Transverse section through vaginula (gametophytic tissue, (n)) and foot of sporophyte (2n). Note lacunate cortex of the foot. The central strand (CS) of the sporophyte is also visible. x 250. After Hébant (1975a).

Fig. 321. *Polytrichum commune* (Polytrichales). Median longisection through lower region of foot and vaginula. Method of Nachlas et al. (1957) for the demonstration of succinic dehydrogenase activity. Many cell types in the sporophyte react strongly, i.e. have intense respiratory activities; this concerns more particularly the transfer cells (TC) at the periphery of the foot, and the cells which surround the central strand in its median region. The foot of the sporophyte thus appears as a very active organ. x 62.

Fig. 322. *Scapania undulata* (Jungermanniales, Hepaticae). Longisection of the foot of the sporophyte (2n) surrounded by gametophytic tissue (n). Intense acid phosphatase activity on both sides of the interphase are revealed. Burstone method. x 96.

Figs. 321-322 after Hébant & Suire (1974).

CS, central strand; TC, transfer cell; (n), gametophytic phase; (2n), sporophytic phase.

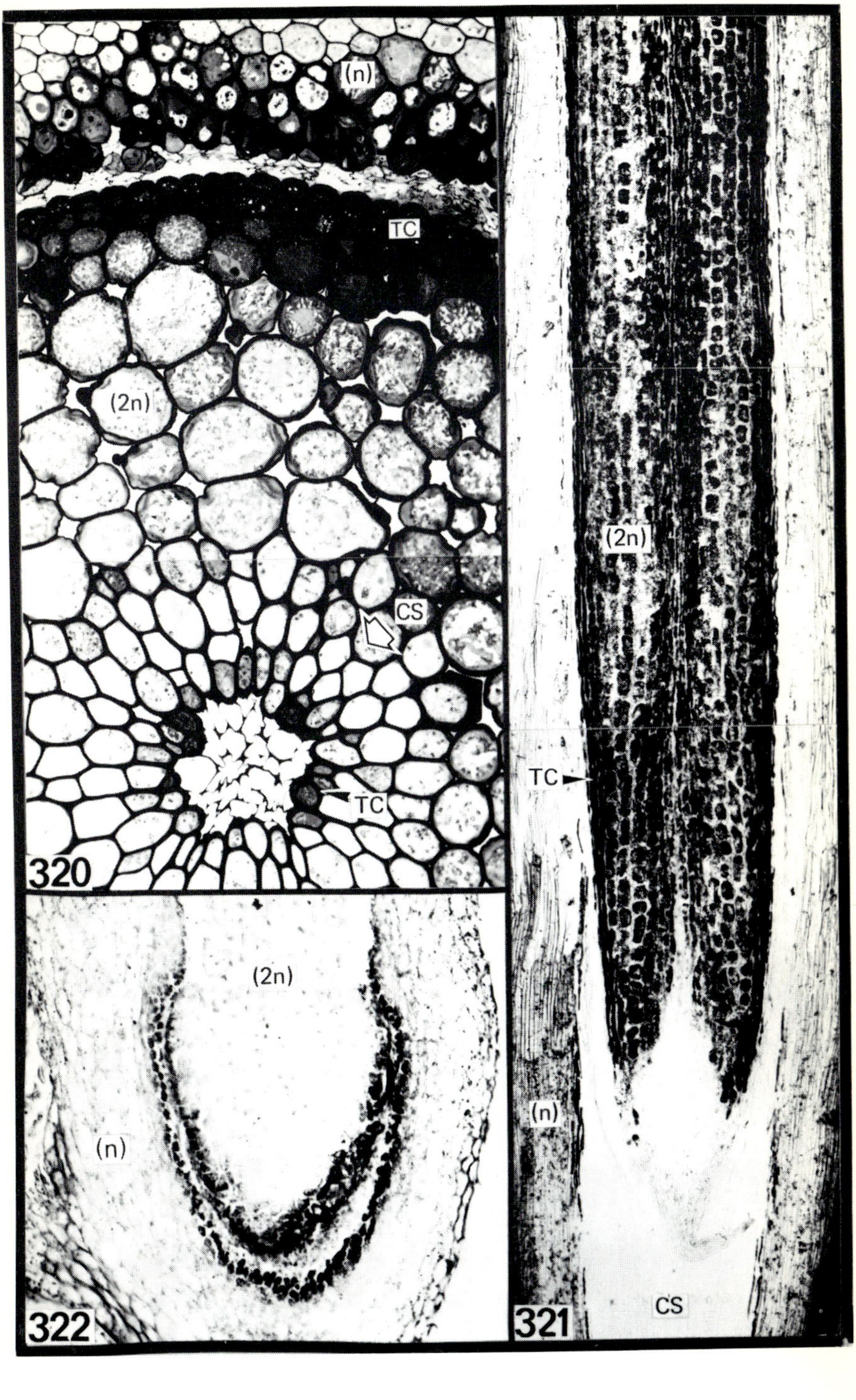

(n)
TC
(2n)
CS
TC
320
(2n)
(n)
322
(2n)
TC
(n)
CS
321

PLATE 77
Internal conduction of water

Fig. 323. A repeat of an ancient experiment by Haberlandt (1886) to demonstrate the conducting role of the central strand in the leafy stem of *Mnium undulatum* (Eubryales, Musci) : when the base of a cut stem is fed with an eosin solution, the stain is found to occur at higher levels in the central strand of the upright gametophyte. The section shown here was photographed with an incident-light microscope, using a green filter to enhance contrast. x 80.

Fig. 324. *Polytrichum commune* (Polytrichales, Musci). Leafy stem. Demonstration of the conducting role of the hydroids of the central strand (CS) and of the leaf traces (LT) with calcofluor white. (Freehand transection mounted in non-fluorescent oil; UV fluorescence microscopy). x 55 approx.

Fig. 325. *Polytrichum commune.* Leafy stem. Autoradiography of transection of freeze-dried specimen; detail. The passage of a radioactive solution from the central strand (CS) to a leaf trace (LT) is shown. In this experiment uranyl nitrate was used as a tracer for water movement, as it is not metabolized. x 180 approx. Original by courtesy of Marthe and Michel Penot (Brest).

Fig. 326. *Dawsonia superba* var. *superba* (Polytrichales). Leafy stem. Demonstration of the conducting role of the hydroids of the central strand (CS) with calcofluor white. Same method as in fig. 324; transection; x 450.
After Hébant (1974).

Fig. 327. *Dendroligotrichum dendroides* (Polytrichales). Gametophyte. Demonstration of the conducting role of the hydroids of the main strand and of secondary strands with calcofluor white. Same method as in fig. 324. Transection; x 110 approx.

Fig. 328-330. *Takakia lepidozioides* (Takakiales, Hepaticae).
Fig. 328. Progression of a solution of fluorescein within central strand of a gametophytic axis. The arrow indicates the direction of the apex of the stem. Blue light fluorescence microscopy.
Figs. 329 and 330. Two views of the same specimen taken after ca. 25 minutes interval. In fig. 329, the fluorochrom (fluorescein) is present in the central strand only. In fig. 330, it has spread through the whole cortex of the stem by diffusion.
Figs. 328-330 after Hébant (1972a); x 48.

C, cortex; CS, central strand; LT, leaf trace.

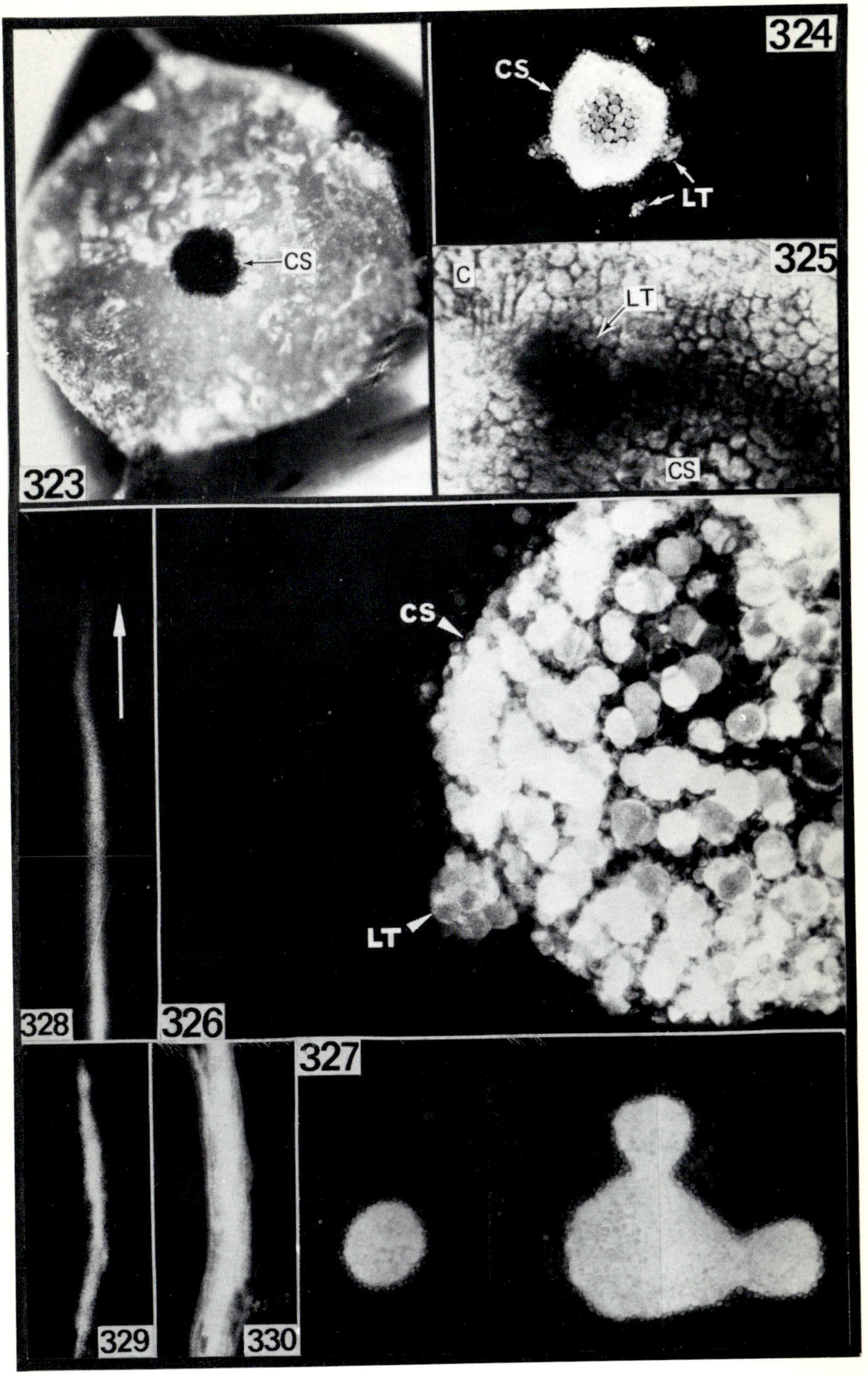
323
CS
324
CS
LT
LT
325
C
LT
CS
326
CS
LT
327
328
329
330

PLATE 78

Figs. 331-332. Exudation in gametophytes of *Polytrichum commune* (Polytrichales, Musci). In fig. 331, a clear exudation is seen in the leptom zone around the central strand of water-conducting elements (hydroids). In fig. 332, exudation is rather weak, but its precise origin can be seen : exudate originates not only from the leptoids *sensu stricto*, but also from the many cells of specialized conducting parenchyma which occur in association with them, and are related to the leaf traces (compare with fig. 101). x 50 approx. After Hébant (1975c).

Figs. 333 to 335. *Polytrichum commune.* Leafy stems; longisections. Enzyme activities in parenchyma cells associated with central strands (viz. cells of the hydrom-sheath; elements of the leptom *sensu lato*) :

Fig. 333, peroxidases (DAB method), x 35

Fig. 334, acid phosphatase (Gomori test), x 34

Fig. 335, succinic dehydrogenase (method of Nachlas et al. 1957), x 200

In fig. 333 (peroxidase activities) certain cells of a leaf trace (LT) are also stained.

Figs. 333-334, after Hébant (1970b).

CS, central strand; HYD, hydroid; LT, leaf trace.

Fig. 336 A-C. *Hedwigia ciliata* (Isobryales, Musci). Rise of a solution of fluorochrome on the surface of the leafy stem (from A to C). The base of a dry gametophyte was soaked in 0.05% K-fluorescein, the specimen being laterally illuminated with blue light; photographs were taken at intervals through an orange-yellow barrier-filter. It takes approximately 30-40 mn for the solution to reach the top of the stem (actual size : 5 mm). [This in fact is a repeat, using modern methods, of an observation made a long time ago on the same species by J. Hedwig (1782) !].

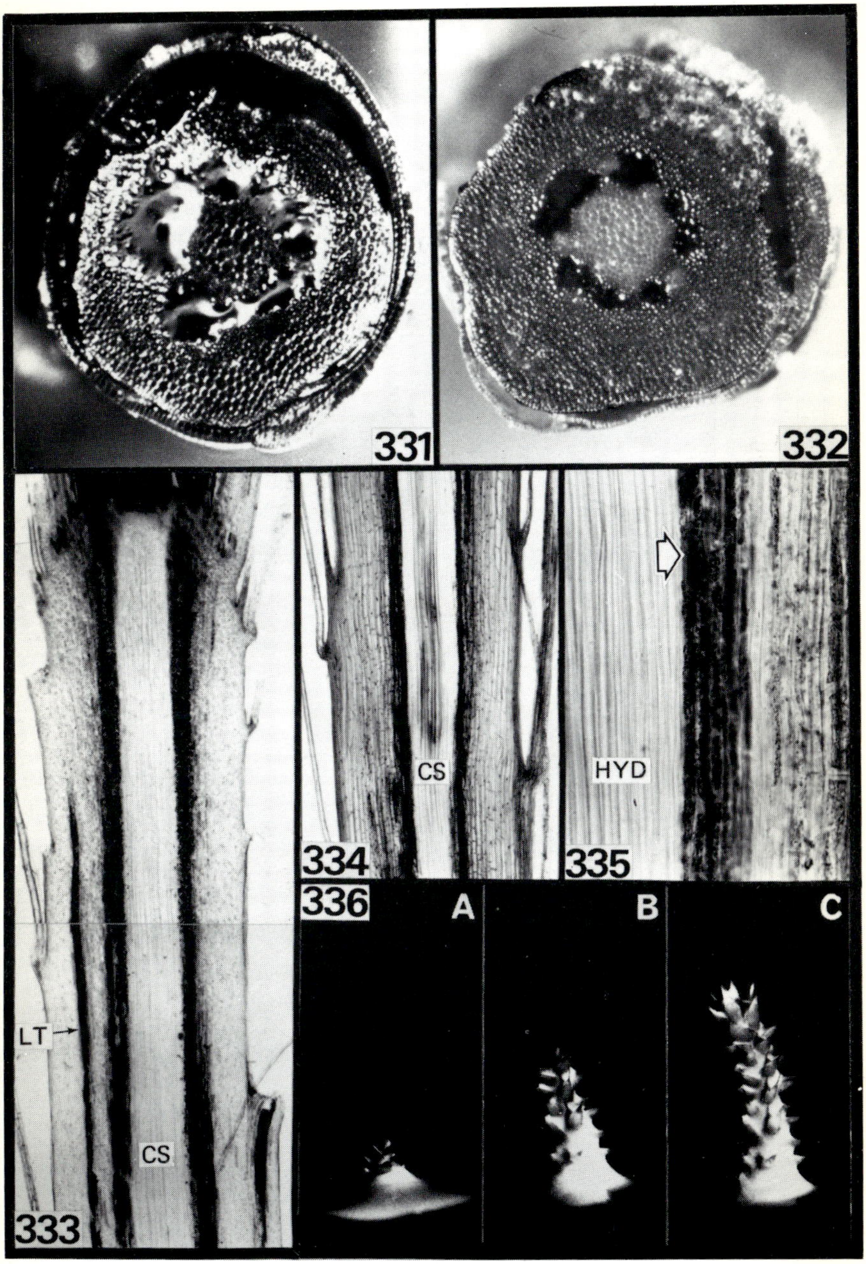
331
332
333
LT
CS
334
CS
335
HYD
336
A
B
C

PLATE 79

Physiology of conduction. III.
Translocation of organic compounds

Fig. 337. *Polytrichum alpinum* (Polytrichales, Musci). Dried gametophyte (left) with the associated autoradiograph (right). The green above-ground shoot has been exposed to $^{14}CO_2$ in the field. The subsequent autoradiograph reveals preferential translocation of the labelled compounds into the new shoot (arrow) arising from the underground stem system. x 1.5.
After Collins & Oechel (1974).

Fig. 338. *Polytrichum commune.* Histoautoradiograph of transection through leafy stem fed with ^{14}C. The section was cut 5 cm below the apex and 2 cm above the activated leaf. A strong labelling of the leptom zone is evident (arrow). (Duration of the incubation : 15 mn). x 155.
After Eschrich & Steiner (1967).

337

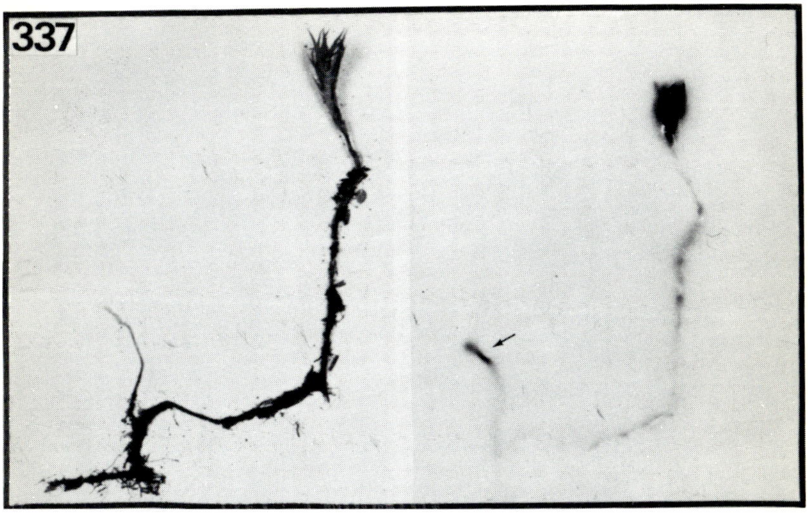

338

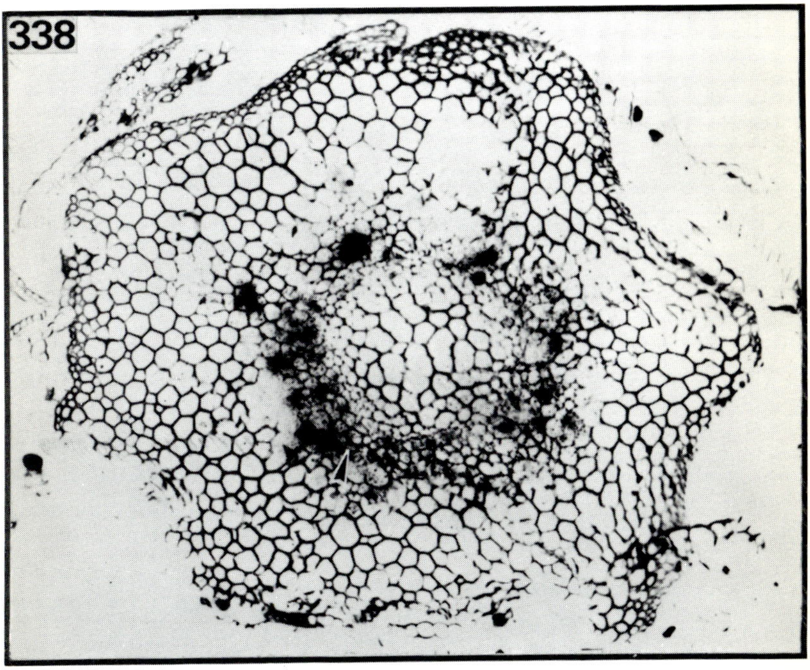

PLATE 80

Evolutionary problems

Figs. 339 to 341. Similarity of appearance between steles of small axes of *Rhynia gwynne-vaughanii* (figs. 339 and 340) and the water-conducting strand of *Funaria hygrometrica* (Funariales, Musci) (fig. 341). Transections; figs. 339 and 340, x 100 approx.; fig. 341, x 240 approx.

Fig. 342. *Rhynia gwynne-vaughanii.* "Dichotomous" branching of the stele. x 33.

Figs. 343-344. Similarity of appearance between the more differentiated elements of *Rhynia* phloem (SE ? = putative sieve element) (fig. 343), and the leptoids (LEP) of a polytrichaceous moss (gametophyte of *Polytrichum juniperinum*) (fig. 344). PAR, parenchyma cell; XYL, xylem; fig. 343, x 265; fig. 344, x 530.
Fig. 343, after Hébant (1976c); fig. 344, after Hébant (1964).

Fig. 345. *Sphagnum palustre* (Sphagnales). Stomates in the epidermis of the sporogonium wall. x 230.

Fig. 346. *Sphagnum palustre.* Trilete spore. x 400.

Fig. 347. Transverse section of a smaller axis from the Rhynie Chert with no conducting strand visible at this level. x 57.

Fig. 348. *Horneophyton lignieri.* Transverse section of sporangium. A columella (COL) comparable to that in the sporogonium of diverse contemporary bryophytes can be seen in its median region. SP, spores; x 25.
(Slide courtesy J. Guiraud, Montpellier).

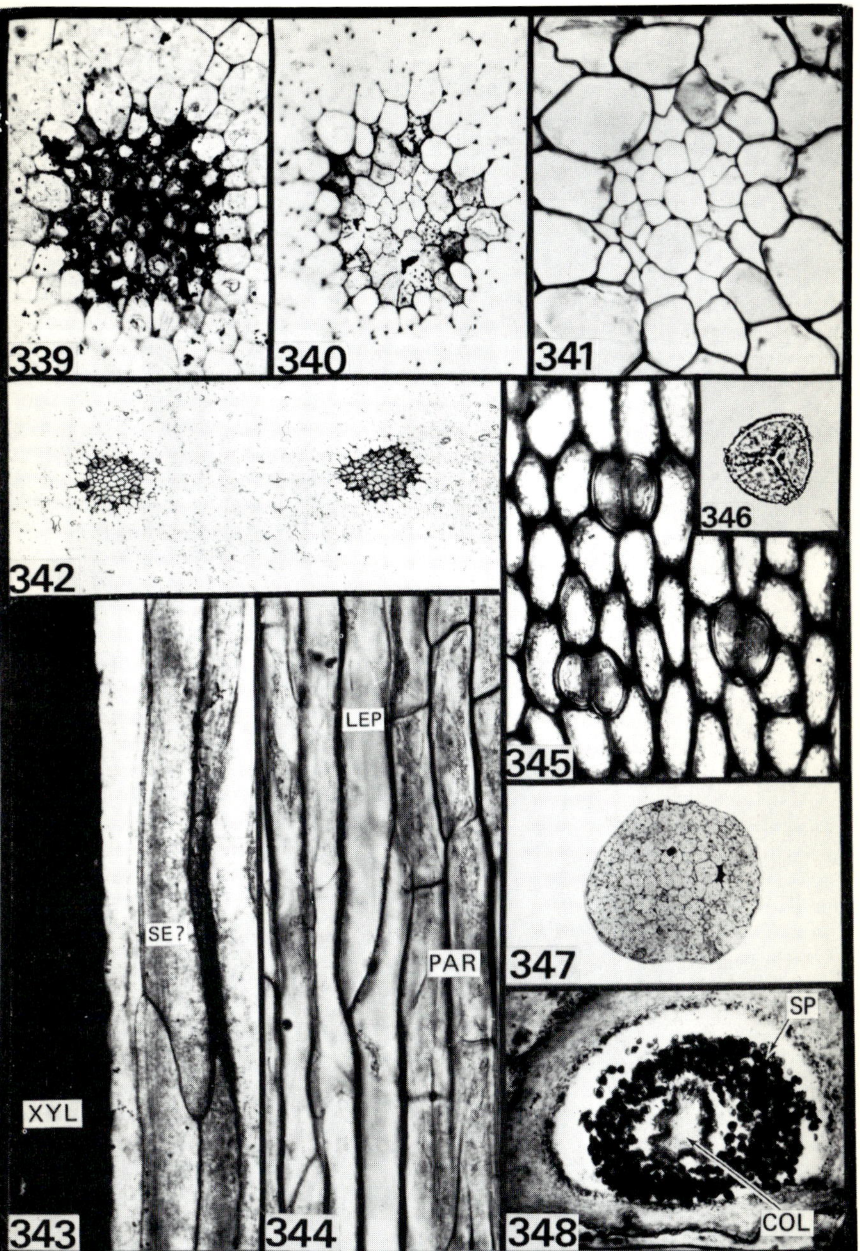
339
340
341
342
345
346
347
LEP
SE?
PAR
SP
XYL
COL
343
344
348